T0235828

Hardware Security Primitives

Mark Tehranipoor • Nitin Pundir •
Nidish Vashistha • Farimah Farahmandi

Hardware Security Primitives

 Springer

Mark Tehranipoor
University of Florida
Gainesville, FL, USA

Nidish Vashistha
University of Florida
Gainesville, FL, USA

Nitin Pundir
University of Florida
Gainesville, FL, USA

Farimah Farahmandi
University of Florida
Gainesville, FL, USA

ISBN 978-3-031-19187-9   ISBN 978-3-031-19185-5   (eBook)
https://doi.org/10.1007/978-3-031-19185-5

This Springer imprint is published by the registered company Springer Nature Switzerland AG
The registered company address is: Gewerbestrasse 11, 6330 Cham, Switzerland

Dedicated to:
 Mark Tehranipoor: My students
 Nitin Pundir: The opportunities I got
 Nidish Vashistha: My family and friends
 Farimah Farahmandi: My mentor
Dr. Tehranipoor, students, and family

Preface

In this modern world of highly connected electronic devices and online cloud data storage, a user can access any sort of information anytime and anywhere with the help of smart devices. Besides storing users' personal information, these devices also store proprietary data such as encryption keys, soft intellectual property, and confidential information to perform artificial intelligence specialized tasks such as an autopilot for self-driving cars. The presence of confidential and intellectual information on hardware devices makes them a lucrative target for hackers' attacks. An adversary can compromise the security of these hardware devices, hijack information to achieve financial gains, and steal intellectual property to perform reverse engineering for manufacturing counterfeit cloned devices. Besides these cloned counterfeits, recycled and refurbished devices can be sold as new ones. These recycled devices cause a revenue loss to manufacturers and pose safety issues due to their reduced life span and reliability. Hence, these devices need to be secured from these attacks. One of the possible solutions to ensure hardware security is to physically embed secure circuits for device authentication, random passkey generation, and anti-counterfeiting detection. These circuits have a unique signature as analogous to human retinal/finger imprints and DNA. These signatures are random, hard to predict, and nearly impossible to clone. Hence, it prevents unauthorized access to data and ensures reliable hardware platforms for secure communications, device authentication, and defense against many software and hardware risks and attacks. Physically unclonable functions (PUFs) and True Random Number Generators (TRNGs) are widely used as hardware security primitives to secure hardware devices and counterfeit detection. Therefore, it is necessary to understand their types, applications, and functions for securing hardware devices.

This book will be a comprehensive reference for circuits and systems designers, graduate students, academics, and industrial researchers interested in hardware security and trust. It will include contributions from experts and researchers in the field of secure hardware design and assurance. In addition, this book will cover various security primitives, design considerations for a secure SoC design, and their applications in counterfeit hardware detection.

This volume will provide the most comprehensive coverage of various hardware security primitives, their roles in hardware assurance and supply chain from the integrated circuit to the package level. Chapters 1–7 cover different types of physical unclonable functions (PUFs), which are fundamental components to hardware security. Chapter 8 discusses true random number generators (TRNSs) developed by exploiting the entropy in hardware manufacturing. Chapter 9 discusses hardware security primitives developed using emerging technologies other than CMOS, such as carbon nanotubes. Chapters 10 and 11 present various techniques for hardware camouflaging and watermarking, respectively. Chapter 12 covers various lightweight cryptographic algorithms that can be alternatives to PUFs on resource-constraint devices. Chapter 13 discusses growing virtual proof of reality to provide security based on blockchain and smart contracts. Chapter 14 covers analog security, usually neglected during hardware security discussions. Chapters 15, 16, and 17 cover various IC and package level methods for tempering, counterfeit, and recycled detection. Finally, Chaps. 18 and 19 cover various side-channel and fault-injection resistant primitives for security in cryptographic hardware.

Gainesville, FL, USA   Mark Tehranipoor
Gainesville, FL, USA   Nitin Pundir
Gainesville, FL, USA   Nidish Vashistha
Gainesville, FL, USA   Farimah Farhmandi
July 2022

Acknowledgments

This work would not have been possible without the assistance of individuals affiliated with the Florida Institute for Cyber Security (FICS) Research at the University of Florida. We appreciate their persistent work, review, and scientific contributions in several chapters:

Upoma Das, Chap. 1
Rasheed Kibria, Chap. 2
Nurun N. Mondol, Chap. 3
Md. Latifur Rahman and Md. Saad Ul Haque, Chap. 4
Pantha Protim Sarker, Chap. 5
Shams Tarek and Azim Uddin, Chap. 6
Sourav Roy, Chap. 7
Upoma Das, Chap. 8
Liton Kumar Biswas, Chap. 9
Nasmin Alam, Chap. 10
Shuvagata Saha and Hasan Al-Shaikh, Chap. 11
Amit Mazumder Shuvo, Chap. 12
Rasheed Kibria, Chap. 13
Rui Guo, Chap. 14
Rui Guo, Chap. 15
Md Kawser Bepary and Dipayan Saha, Chap. 16
Sajib Ghosh and Mridha Md Mashahedur Rahman, Chap. 17
Mohammad Shafkat M. Khan, Chap. 18
Tanvir Rahman, Chap. 19

In addition, we would like to thank Erika Clesi and Dr. Dhwani Mehta for their help in reviewing the book.

Contents

Chapter 1
Intrinsic Racetrack PUF

1.1 Introduction

The ever-growing electronic industry has created a high demand for systems security and authenticity of electronic products. However, there are several risks and vulnerabilities associated with electronic devices. Intellectual Property (IP) piracy, Integrated Circuit (IC) overproduction, and IC counterfeiting are some of the threats to commercial products [7, 22, 23]. Traditionally, key-based cryptography systems were used for IP protection and licensing applications. However, key storage and management make an IC susceptible to various invasive and non-invasive physical attacks and tampering [1, 19]. Meanwhile, the semiconductor industry focused on developing tamper-resistant ICs for the time being [5]. However, due to the diversity of attacks, designing full tamper-proof ICs is challenging and suffers substantial area, power, and performance overhead. Additionally, temper-proof devices are costly, which makes them commercially unattractive. So, researchers started looking for viable alternative solutions.

Secret keys protection motivated academia and industry to implement one-way functions that are easy to implement in one direction but hard to reverse [2, 17, 18, 20]. Furthermore, these one-way functions must be inexpensive to fabricate, inherently random, difficult to duplicate, without a definite mathematical representation, and be intrinsically tamper-resistant. Other desired characteristics include cost-effectiveness, inexpensiveness, and ease of implementation on resource-constrained platforms. The solution to all of these was exquisite, the process variation of the circuits. Due to process variation, each IC is unique. Thus, the idea was to devise systems capable of extracting keys from these unique devices and make the responses of the one-way function dependent on the process variation. Further, this idea is how Physical Unclonable Function (PUF) was conceptualized. PUF is based on a physical system and is easy to evaluate. Moreover, its output is a random function of process variation and unpredictable even for an attacker with unrestrained physical access. Because of these factors, researchers have been motivated to develop different PUF systems, increase their reliability and robustness, eliminate problems associated with practical implementations, and make PUFs tamper-proof against different kinds of attacks.

Gassend et al. [3] and Lee et al. [9] first attempted to bring together the idea of PUF. The idea was to bind the response to a random physical system in a way that was unique to the specific hardware and not replicable. PUFs can be implemented using various techniques, and Racetrack PUF is one of them. Racetrack PUF is a family of PUFs that count on comparing time delays of the signal paths. This chapter reviews the basic concepts, implementations, architecture, limitations, and applications of various Racetrack PUFs. Racetrack PUFs are a family of PUFs that exploit time delays across similar signal paths due to process variation. Racetrack PUFs consist of Arbiter PUF (APUF), ClockPUF, and Ring Oscillator (RO) PUF. First, this chapter introduces the idea of a silicon PUF that exploits the manufacturing process variation of complex ICs to examine its response over time to a specific challenge, thereby authenticating individual ICs. Then, the methodology of building a unique secret key is reviewed (statistical delay variations of transistors and wires to build a unique secret key are reviewed). Then, different basic Racetrack PUFs are reviewed (i.e., APUF, ROPUF, and ClockPUF) and improvements for these PUFS are evaluated [12, 14, 24].

The rest of the chapter is organized as follows: Sect. 1.2 provides all the necessary basic information on Racetrack PUFs. Section 1.3 aims to provide a detailed overview and survey of different Racetrack PUFs. Moreover, the overview especially focuses on the challenge–response pair (CRP) generation mechanisms, underlying physical process variations, degrees of security, reliability with the environmental process variations, and resiliency against different attack models. It also discusses the practical implementation issues and their potential solutions. Applications of various Racetrack PUFs are also briefly discussed. Finally, Sect. 18.8 concludes this chapter.

1.2 Background

In this section, some of the basic concepts related to PUF are briefly described. First, we introduce the basic challenge–response pair generation. As previously mentioned, PUF refers to a physical object that, for a given input (known as a challenge), provides a physically defined signature output (known as a response), as shown in Fig. 1.1. During device fabrication, the process variation is introduced in the manufacturing step and is unavoidable. Each chip is slightly different due to process variation and thus helps the unclonable functionality. PUF exploits this manufacturing variation to identify and authenticate a device uniquely. As the challenge and response of a PUF come in a pair, it is called the Challenge–Response Pair (CRP). Since the PUF is hard to characterize and easy to identify, it is used in hardware security for the applications like device identification, authentication, authorization, etc. An overview of the authentication process is shown in Fig. 1.2.

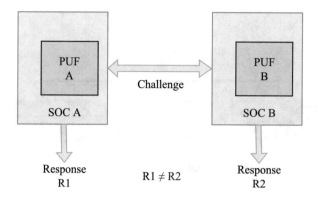

Fig. 1.1 Illustration of challenge–response Pair (CRP) generation by exploiting the chip's process variation

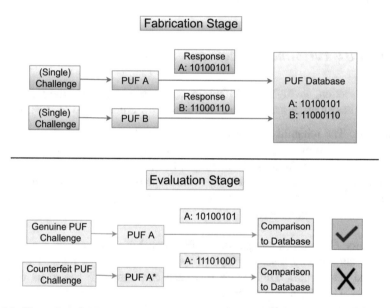

Fig. 1.2 Illustration of chip authentication flow using PUF's CRPs

There are several families of PUF-based on their working principle, architecture, and implementation. Racetrack PUF counts on comparing time delays of the different signal paths. As shown in Fig. 1.3, there are three types of Racetrack PUFs—Arbiter PUF, Ring-Oscillator PUF (RO-PUF), and Clock-PUF. An Arbiter PUF generates a response based on the propagation time difference between two electrical signals transmitted through theoretically symmetrical courses. Ring-Oscillator PUFs produce CRPs by manipulating the frequency differences of on-chip oscillators [21]. Finally, Clock-PUFs assess the difference in clock signal

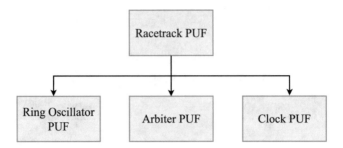

Fig. 1.3 Different kinds of Racetrack PUFs

transmission speed through different branches of a signal line, derived from the manufacturing variation of these lines [24].

1.2.1   PUF Performance Metrics

In this subsection, we highlight different metrics used to measure the performance of PUFs. These metrics are used to characterize how efficient and secure a PUF is.

Reliability The reliability or steadiness of a PUF describes how efficient it is in reproducing the same response if the same challenge bits are passed through the PUF multiple times. To uniquely identify a PUF, or for a PUF to be reliable, we expect that a PUF will always give the same response for the same challenge bits. However, due to environmental variation, this unique response can change, and the reliability of a PUF can decrease.

Uniqueness Uniqueness represents a PUF's probability of uniquely identifying a particular chip among a set of chips of the same type (same mask, technology, wafer, and same chip). This means that each chip response has to be uniquely identifiable. Usually, a single challenge passed through various PUFs will result in unique responses in each PUF. However, if two PUFs produce the same response for the same challenge, they are not unique, and uniqueness decreases.

Randomness The randomness of a PUF represents its ability to produce a random response for any given challenge. The response must be truly random and cannot be biased toward any particular output. The true randomness of a PUF makes it unclonable.

1.3 Racetrack PUF

Racetrack PUF is a family of PUFs that count on comparing time delays of the different signal paths. As already mentioned in Sect. 1.2, there are mainly three types of Racetrack PUFs—Arbiter PUF, Ring-Oscillator PUF (RO-PUF), and Clock-PUF. Here we will discuss these PUFs, in detail, with their architecture and working functions.

1.3.1 Arbiter PUF

Figure 1.4 illustrates the block diagram of the candidate APUF implementation. Here, two delay paths are excited simultaneously with a rising edge signal, where the transistors on each delay path race against each other to arrive at the arbiter. The arbiter (transparent latch) gives an output of 0 or 1 depending on which transistor path reaches the arbiter with the rising edge first. Figure 1.5 shows the switching component in detail. Each Switch component has a pair of 2-to-1 Mux and buffers. The n-bit challenge is fed into the PUF. Each bit (b_i) of the N-bit challenge is the control bit for a single switch component; this b_i dictates the data path configuration of the MUX (in Fig. 1.4, if $b_i = 0$, data path goes straight; if $b_i = 1$, data path is crossed). Therefore, the upper and lower racing paths will have different path configurations depending on the challenge bits. Due to manufacturing process variation, the path delay in these racing paths will be different for different chips for the same challenge. As a result, the same challenge will result in different arbiter outputs for different PUFs, which can be used for device identification and authentication.

The PUF assumes that an adversary has physical access to the device but cannot access challenges and guess the corresponding responses from the PUF circuit [9]. Therefore, the challenges are stored in secret storage and sent to the physical chips only when needed to generate the response. To ensure that APUF response is a function of process variation, the APUF's path delay should be significant enough that chips made from the same mask, technology node, and even wafer can be uniquely identified. Due to this reason, it is better to consider the relative path-delay comparison of each path rather than taking the absolute delay of the whole circuit. Since probing the PUF for eavesdropping will change the coupling capacitance, it will thus change the path delay, where the adversary cannot measure the path delay through direct probing. Furthermore, as manufacturing process variation is random and cannot be exactly replicated, an adversary cannot make a clone of the PUF.

However, it was shown that conventional APUFs could be mathematically modeled using additive delay timing models. Therefore, further research [9] suggested a feed-forward arbiter model that introduces non-linearity in the data paths, so that delay time modeling is not possible. As a result, they could virtually model the basic PUF with 97% accuracy. However, they could not model the feed-forward path for

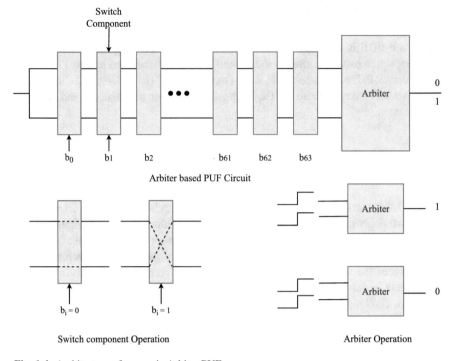

Fig. 1.4 Architecture of a generic Arbiter PUF

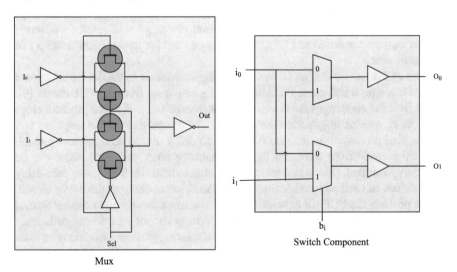

Fig. 1.5 Example circuit showing the switching component in Arbiter PUF

its non-linearity in delay. Though this technique has overcome the above difficulties, the path delays can change due to changes in power supply voltage, temperature, aging, or measurement noise, affecting the PUF's reliability.

1.3.2 Ring-Oscillator PUF

Figure 1.6 illustrates the diagram of a generic ROPUF implementation. Here, a delay circuit is placed in a parameterized self-oscillating circuit. The frequency of the self-oscillating loop is a function of the delay circuit itself. A counter uses this frequency as output after a predefined clock cycle. Multiple self-oscillating loops are placed in a commercial chip, and the delay is calculated from the frequency of those loops.

Figure 1.7 shows the delay circuit in detail. Each switch block has a pair of 2-to-1 Mux and buffers and two data paths: upper and lower. The n-bit challenge is fed into the PUF. Each bit of the challenge then controls the path configuration of the MUX. Therefore, the upper and lower racing paths will have different path configurations depending on the challenge bits. To prevent time-accurate modeling of the PUF, buffers are added. One buffer will always be on, whereas the other buffer will be conditionally on. This means buffers will introduce non-monotonic behavior, i.e., the total delay is not a function of device and wire delay. Due to manufacturing variation, when the same input will be provided to the upper and lower path, path delay in these racing paths will be different for different chips with

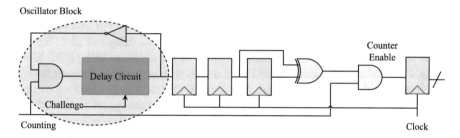

Fig. 1.6 Illustration of self-oscillating loop circuit for ROPUFs

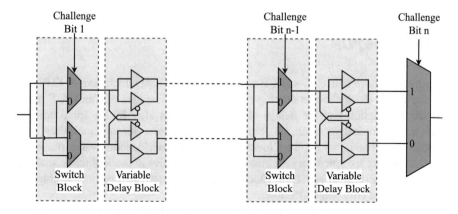

Fig. 1.7 Illustration of non-monotonic delay circuit

the same challenge. As a result, the same challenge will result in different outputs depending on the individual path-delay characteristic, which can be used for device identification and authentication.

In an ideal scenario, the same challenge to a ROPUF should always result in the same output. However, in a practical case, since an analog physical system is used for a PUF's calculation and its analog signal is converted into a digital signal, the response might vary. Therefore, Error Correction Code (ECC) can be used to correct the bit errors. Furthermore, though process variation causes the path delays to be unique, to definitely ensure that no two chips are alike, the Electronic Chip ID (ECID) of the chips can be combined with the PUF challenges [3]. Moreover, multiple PUFs can be used together in multiple rounds to strengthen the PUF, i.e., the output of round 1 PUF is used as input for round 2 PUF. However, if an adversary gains access to CRPs of the PUF, it can model the PUF through linear or non-linear additive delay models. Though a single CRP can lead to an unsolvable equation, a small subset of CRPs could give the solvable mathematical model to break the PUF. Therefore, original CRPs outside the PUF system should be stored in a hashed state. That way, even if an adversary gains access to hashed challenges, it cannot make a query to the PUF to obtain the original responses. Similarly, the PUF system can be a target of side-channel attacks such as fault injection, glitch injection, and power side-channel.

The ROPUF logic circuit can be easily implemented on an FPGA to evaluate its performance. The consecutive measurement of the same delay can give slightly different results due to measurement error, which can be corrected using ECC. Similarly, the standard deviation in inter-FPGA delays with compensated measurement can depend on the pair of loops used. Additionally, the frequency of the loop can be influenced by the nearby circuitry, [3]. Frequency can also change with temperature but could be compensated by taking ratios of the delays of particular loops instead of absolute delay. Similarly, voltage regulators can compensate for delay variation due to voltage fluctuation.

1.3.3   ClockPUF

A novel approach called ClockPUF was introduced in [24] which derives PUF bits by exploiting clock skew in on-chip clock networks. The idea of ClockPUF relies on the assumption that the delays of all clock routes alter by the same amount with environmental variation, i.e., change in temperature. So, the clock skew will remain unchanged, making the ClockPUF robust to environmental variation. Also, clock networks are tuned very accurately and are sensitive to clock glitches, making the ClockPUF tamper-resistant. ClockPUFs can be used in unclonable chip ID generation and chip authentication.

Figure 1.8 shows the architecture of a ClockPUF. The clock signals are tapped out from the sinks and then brought together by the buffered return path to produce PUF bits comparing the arrival times of certain clock signals. A small AND gate is

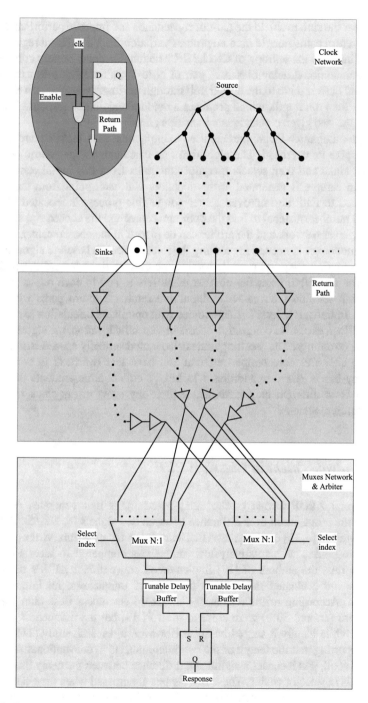

Fig. 1.8 Illustration of sample ClockPUF architecture

placed near the sink to shield the tap-out capacitance and improve performance. The buffered return paths sample random process variation from a different region of the chip and increase the entropy of ClockPUF. The tunable delay buffers compensate for the unintentional delay biases. A pair of N-to-1 multiplexors select two clock signals and deliver them to the arbiter unit through the tunable buffers. An S-R latch is used as the arbiter unit, which produces a random and unique response based on which of the two given clock signals transitions first.

Design-automation techniques are used to implement the ClockPUF architecture and minimize its overhead. The algorithm first calculates the geometric center of the given sinks and then selects n equidistant sinks from this initial center. Then a gaussian sample is generated at this location, and each point from the sample is evaluated to find the improved center point. This process is repeated M times (maximal number of steps) to find the optimum center and the chosen sinks. Finally, the multiplexer network and the arbiter can be placed in the chosen center.

A methodology is needed for routing the return paths. A baseline algorithm can minimize routing congestion while routing the chosen sinks toward the optimum center. The algorithm takes the number of buffers to add to each return path, the chosen sink set, and their center as input and returns n return paths with buffer locations. Initially, it selects L-shape routes with optimized orientation to minimize overlap. When excessive congestion occurs in a specific location, the algorithm opts for Z-shape routings. The resulting routes also include equally spaced buffers.

ClockPUFs are more tamper-resistant and have less overhead in comparison with delay-based PUFs. In addition, ClockPUFs collect large amounts of process variation from different chip areas and exploit any skew in the clock that helps thwart different attacks.

1.3.4 Advancement in Classical PUF

Conventional APUF is one of the delay-based PUFs that generates responses based on the propagation time difference between two signals [6, 13, 15]. Previous research studies [6, 15] confirm that APUFs implemented on Xilinx Virtex-5 FPGAs produce responses with relatively low uniqueness compared to ideal PUFs. To reconfirm this, the authors of [12] implemented conventional APUFs on Virtex-5 FPGAs and evaluated these PUFs in terms of uniqueness, randomness, and steadiness. According to the results, the measured steadiness (less than 1%) and randomness (around 50%) were close to ideal PUFs, but the uniqueness was less than 5%, while ideally it should be 50%. However, a research study [11] clarifies this by reporting that the length of the two interconnects in conventional APUF are not equal at all, and it causes a significant difference between the delay times of the two signals in selector chains. These two signals are crossed when the challenge bit (c_i) is 1. Therefore, the proportion of 1s or 0s in the response becomes deterministic, depending on the hamming weight of the challenge. If the hamming weight is odd, then there is a high probability that the proportion of 1s in responses will be higher and vice versa. So, under the condition of a randomly chosen challenge, the high randomness value demonstrated by conventional APUFs is just superficial.

So, it is evident that the explanation behind the low uniqueness of conventional APUFs is the imbalanced length of the two interconnects. To overcome this short-coming, a technique called Double Arbiter PUF (2–2 DAPUF) is proposed in [13], which was designed for balancing the length of the two interconnects. However, among the DAPUF pairs, one pair has comparatively low-unique responses. To resolve this, a novel concept was introduced in a research study [12]: *mode of operation for APUF* governed by the connection method of the interconnect to an arbiter. Another type of DAPUF, named 3–1 DAPUF, was also proposed in this same study, which is an improved version of 2–2 DAPUF. Here, m-n APUF, or DAPUF, means it has m selector chains and generates an n-bit response.

Figure 1.9a shows the mode of operation for APUF of 2–2 DAPUF. $S_{1,L}$ and $S_{1,R}$ are the inputs of the first selector chain. So, no matter what is the challenge, the signals on $S_{1,L}$ will either reach $W_{1,L}$ or $W_{1,R}$ and similarly, the signals on $S_{2,L}$ will reach $W_{2,L}$ or $W_{2,R}$. Therefore, there is no chance of crossing the signals on $S_{1,L}$ and $S_{2,L}$ (or $S_{1,R}$ and $S_{2,R}$) even if any challenge bit $c_i = 1$ ($0 < i < 64$). On the other hand, 2–2 APUF is basically two APUFs generating 2-bit responses as shown in Fig. 1.9b. It has the same circuit costs as DAPUFs of m = 2. But, the signals on $S_{2,L}$ and $S_{2,R}$ are crossed when $c_i = 1$ which is different from 2–2 DAPUF. Due to the difference of these wire connections, 2–2 DAPUF response does not depend on the hamming weight of the challenge. It may cause low randomness if a deterministic difference between the delay times of the two signals is produced. As a consequence of these biased responses, one pair of the 2–2 DAPUFs have lower uniqueness.

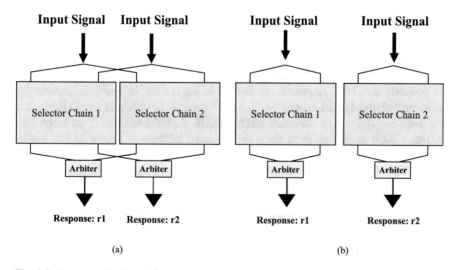

Fig. 1.9 Example circuit highlighting the 2–2 PUF's structure. (**a**) 2-2 DAPUF. (**b**) 2-2 APUF

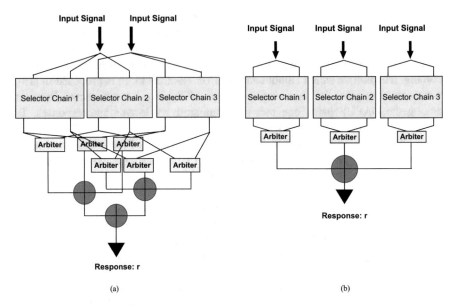

Fig. 1.10 Example circuit highlighting the 3–1 PUF's structure. (**a**) 3-1 DAPUF. (**b**) 3-1 APUF

For further improvements, 2–1 DAPUF is proposed in [12], which generates a 1-bit response by XORing the 2-bit responses generated by a 2–2 DAPUF. XORing responses help achieve a less-biased response. In the same way, the 2-bit responses of 2–2 APUF can also be XORed into 1-bit responses to produce 2–1 APUF (2-XOR APUF). The mode of operation of 2-XOR APUF and 2–1 DAPUF are different, and they were compared in terms of randomness, uniqueness, and steadiness. The 2–1 DAPUFs demonstrate higher uniqueness but lower steadiness than the 2–1 APUFs. It is considered that the large difference in delay times arising from the unequal wire length of the APUFs results in high steadiness but low uniqueness.

To improve uniqueness, DAPUFs can be designed with three selector chains to generate six 1-bit responses as shown in Fig. 1.10a. In order to reduce the influence of the biased responses, a 1-bit response is obtained by XORing the six 1-bit responses. On the other hand, 3-XOR APUF generates 1-bit responses XORing 3-bit responses from three conventional APUFs as shown in Fig. 1.10b. The randomness, uniqueness, and steadiness of 3–1 DAPUF were compared to that of 3-XOR APUF. Both demonstrate higher randomness, but the uniqueness of 3–1 DAPUF outmatches 3-XOR PUF.

Using the new mode of operation, 3–1 DAPUFs generate responses with high uniqueness among all pairs of FPGAs, which overcomes the shortcoming of [13]. The uniqueness of APUF and DAPUF also improves with additional duplicated selector chains. However, there is a trade-off between steadiness and uniqueness. The steadiness of responses from DAPUF degrades with an increasing number of selector chains. Furthermore, added selector chains also increase the power and area overhead of design. Future research might involve implementing 4–1 or 5–1 DAPUF and evaluating the obtained responses according to uniqueness, randomness, and steadiness. The higher uniqueness of DAPUF makes it suitable for chip authentication.

Now, how RO-PUFs can be improved using FPGAs will be discussed. Since FPGAs are reconfigurable, they serve as perfect candidates for faster hardware development. However, there are several challenges to implementing PUFs on FPGAs. Since only the top-level design blocks (e.g., memory blocks, Look Up Tables (LUTs), etc.) can be configured on FPGAs, the layout level design techniques cannot be exploited using this platform. Moreover, a good deal of PUF topologies [16] demand a prudent routing that is hard to synthesize on FPGAs.

RO-PUFs can be used as FPGA-friendly secure primitives since RO-PUFs have ring oscillators that can be implemented on FPGAs leveraging hard macro design strategies. However, systematic or correlated process variation (PV) can negatively impact a PUF's uniqueness, and environmental noise due to temperature and voltage variations can downgrade a PUF's response reliability. Therefore, a compensation method [14] can be used to alleviate the effect of a systematic PV, which can boost the PUF response uniqueness by almost 18%. Moreover, compensation methods feature innovative configurable ring-oscillator (CRO) design techniques that are area efficient and immune to noise on PUF responses, thus solving PUF reliability issues on FPGA.

This PUF implementation on FPGA addresses the effect of systematic PV in PUF responses as compared to previous PUF designs (i.e., SRAM PUF [4], Butterfly-PUF [8], Arbiter PUF (APUF) [10], etc.) Moreover, the previous PUF response improvement techniques involved post-processing, but the novel CRO technique [14] can be implemented in the pre-quantization phase. Moreover, the CRO technique is based on circuit-level implementation, where it does not add any overheads, unlike existing RO-PUF designs on FPGAs.

For an ideal PUF, uniqueness should be 50%; if the response is truly random, the probability of having 0 or 1 in each response bit would be equal. Therefore, random bits are generated if there is only random PV. However, due to the continuous scale down of device feature size, systematic PV exists with the random PV. The systematic PV produces bit aliasing among different chips, which reduces the inter-die Hamming Distance (HD), causing an ID collision or false positives to authenticate the device [14].

Since the systematic PV has a spatial characteristic, the interconnects and process variations will influence logic blocks close to each other similarly. Thus, for two ROs located nearby, the difference in systematic delay variation will be very low. Thus, their difference in random PV will be more likely to offset the difference in their systematic PV, improving the PUF uniqueness. This difference can be compensated in two steps: the first step is to place the group of ROs as close as possible to each other, and the second step is to pick the physically adjacent pair of ROs to evaluate a response bit [14]. These two steps can mitigate the systematic PV by preventing bit aliasing and improving PUF uniqueness. However, if an adversary knows the frequency distribution of the frequency of all the ring oscillators in a PUF, they can avoid systematic variation by analyzing the distribution, which is a very

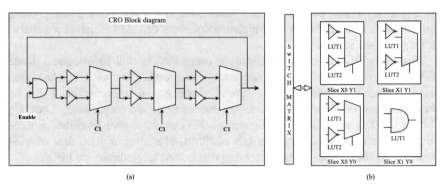

Fig. 1.11 (**a**) Block diagram of a configurable ring oscillator (CRO). (**b**) Illustrating the implementation of CRO in logic blocks of a Xilinx FPGA

time-consuming, costly, and impractical process in a larger RO-PUF. The advantage of the compensated method is that it is easy to implement and does not require any previous information on the nature of correlation variation. However, the limitation of the above method is that it limits the highest number of independent response bits from n to (n-1).

The configurable RO (CRO) technique can improve PUF reliability [14]. In this RO structure, a 2to1 MUX is placed at each delay level to choose one inverter between the two. 2^m different oscillator rings can be set up in such an RO with m stages.

For example, in Fig. 1.11a, eight different ROs can be configured utilizing only three control inputs and a 2to1 MUX. For maximum reliability, RO pairs with a maximum frequency difference are selected. This is shown in Fig. 1.12. The advantage of this method is that it is area efficient, possible to implement in existing FPGA design techniques, and while during response evaluation, the method only needs a one-time characterization of PUF.

Now, to validate the compensation method, PUF response has been evaluated under three methods. In Method 1 (M1), ring oscillators are spread over the FPGA randomly. Then, PUF responses are extracted by intentionally choosing RO pairs in spatially distant places from each other in an FPGA. In Method 2 (M2), ROs are aligned closely on FPGA in a 2D array with a placement constraint, and then responses are extracted from selected RO pairs, which are kept as far apart as possible within the array. Finally, in Method 3 (M3), physically close ROs are selected within the array. Method M3 has better uniqueness than method M2 and M1 [14].

Experimental data shows that random route-n-place has the minimum uniqueness, and the proposed methods yield higher uniqueness for all three updated methods. Now, to validate CRO technique, it is compared against all eight individual RO configurations under different temperature and voltage variations. It can be seen that the CRO technique provides a consistently lower number of unstable bits [14].

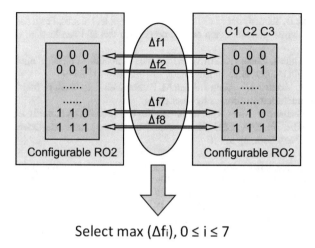

$$\text{Select max } (\Delta f_i),\ 0 \leq i \leq 7$$

Fig. 1.12 Illustration to show ring oscillator pair selection process to ensure maximum difference of frequency between two CROs

1.4   Conclusions

In this chapter, we have covered the basic working principles behind different Racetrack PUFs, operations of the devices, limitations of practical implementation, and their various applications. The experimental results in the literature have shown that the PUF devices can generate unique and reliable responses from different challenges and circuits. From the experimental results [3], it can be observed that reliable authentication can be carried out despite substantial environmental variations by using the process variation of the circuits. Another novel PUF, ClockPUF, was discussed in detail [24] and experimental data in literature proved their tamper-resistance and stability. The only limitation it suffered was a power overhead of almost 20%. Lastly, some solutions to circumvent reliability and aging issues, and their impact on other environmental conditions were overlooked.

References

1. Anderson R (2020) Security engineering: a guide to building dependable distributed systems. Wiley, New York
2. Choudhury M, Pundir N, Niamat M, Mustapa M (2017) Analysis of a novel stage configurable ROPUF design. In: 2017 IEEE 60th International Midwest Symposium on Circuits and Systems (MWSCAS). IEEE, New York, pp 942–945
3. Gassend B, Clarke D, Van Dijk M, Devadas S (2002) Silicon physical random functions. In: Proceedings of the 9th ACM Conference on Computer and Communications Security, pp 148–160
4. Guajardo J, Kumar SS, Schrijen GJ, Tuyls P (2007) Fpga intrinsic PUFs and their use for IP protection. In: International workshop on cryptographic hardware and embedded systems. Springer, Berlin, pp 63–80

5. Guin U, Forte D, Tehranipoor M (2013) Anti-counterfeit techniques: From design to resign. In: 2013 14th International workshop on microprocessor test and verification. IEEE, New York, pp 89–94
6. Hori Y, Katashita T, Kobara K (2013) Performance evaluation of physical unclonable functions on kintex-7 FPGA. IEICE Technical Report of RECONF 113:91–96
7. Hossain MM, Vashistha N, Allen J, Allen M, Farahmandi F, Rahman F, Tehranipoor M (2022) Thwarting counterfeit electronics by blockchain
8. Kumar SS, Guajardo J, Maes R, Schrijen GJ, Tuyls P (2008) The butterfly PUF protecting IP on every FPGA. In: 2008 IEEE International Workshop on Hardware-Oriented Security and Trust. IEEE, New York, pp 67–70
9. Lee J, Lim D, Gassend B, Suh G, van Dijk M, Devadas S (2004) A technique to build a secret key in integrated circuits for identification and authentication applications. In: 2004 Symposium on VLSI Circuits. Digest of Technical Papers (IEEE Cat. No.04CH37525), pp 176–179. DOI 10.1109/VLSIC.2004.1346548
10. Lim D, Lee JW, Gassend B, Suh GE, Van Dijk M, Devadas S (2005) Extracting secret keys from integrated circuits. IEEE Trans Very Large Scale Integr VLSI Syst 13(10):1200–1205
11. Machida T, Nakasone T, Sakiyama K (2013) Evaluation method for arbiter PUF on FPGA and its vulnerability. Tech. Rep., IEICE Technical report of ISEC
12. Machida T, Yamamoto D, Iwamoto M, Sakiyama K (2014a) A new mode of operation for arbiter PUF to improve uniqueness on FPGA. In: 2014 Federated Conference on Computer Science and Information Systems. IEEE, New York, pp 871–878
13. Machida T, Yamamoto D, Iwamoto M, Sakiyama K (2014b) A study on uniqueness of arbiter PUF implemented on FPGA. In: Proceedings of the 31st Symposium on Cryptography and Information Security (SCIS'14)
14. Maiti A, Schaumont P (2011) Improved ring oscillator PUF: An FPGA-friendly secure primitive. J Cryptol 24(2):375–397
15. Maiti A, Gunreddy V, Schaumont P (2013) A systematic method to evaluate and compare the performance of physical unclonable functions. In: Embedded systems design with FPGAs. Springer, New York, pp 245–267
16. Morozov S, Maiti A, Schaumont P (2010) An analysis of delay based PUF implementations on FPGA. In: International Symposium on Applied Reconfigurable Computing. Springer, Berlin, pp 382–387
17. Pundir N, Amsaad F, Choudhury M, Niamat M (2017) Novel technique to improve strength of weak arbiter PUF. In: 2017 IEEE 60th International Midwest Symposium on Circuits and Systems (MWSCAS). IEEE, New York, pp 1532–1535
18. Rahman MT, Forte D, Fahrny J, Tehranipoor M (2014) ARO-PUF: An aging-resistant ring oscillator PUF design. In: 2014 Design, Automation and Test in Europe Conference and Exhibition (DATE). IEEE, New York, pp 1–6
19. Rahman MT, Shi Q, Tajik S, Shen H, Woodard DL, Tehranipoor M, Asadizanjani N (2018) Physical inspection and attacks: New frontier in hardware security. In: 2018 IEEE 3rd International Verification and Security Workshop (IVSW). IEEE, New York, pp 93–102
20. Suh GE, Devadas S (2007a) Physical unclonable functions for device authentication and secret key generation. In: 2007 44th ACM/IEEE Design Automation Conference. IEEE, New York, pp 9–14
21. Suh GE, Devadas S (2007b) Physical unclonable functions for device authentication and secret key generation. In: 2007 44th ACM/IEEE Design Automation Conference, pp 9–14
22. Tehranipoor M, Peng K, Chakrabarty K (2011) Introduction to VLSI testing. In: Test and Diagnosis for Small-Delay Defects. Springer, New York, pp 1–19
23. Vashistha N, Hossain MM, Shahriar MR, Farahmandi F, Rahman F, Tehranipoor M (2021) eChain: A blockchain-enabled ecosystem for electronic device authenticity verification. IEEE Trans Consum Electron 68(1), 23–37
24. Yao Y, Kim M, Li J, Markov IL, Koushanfar F (2013) ClockPUF: Physical unclonable functions based on clock networks. In: 2013 Design, Automation and Test in Europe Conference and Exhibition (DATE). IEEE, New York, pp 422–427

Chapter 2
Intrinsic-Transient PUF

2.1 Introduction

Since the 1960s, metal oxide silicon (MOS) semiconductors have significantly contributed to integrated circuits (ICs). Following Moore's law, with the development of technological advancement in the past decades, ICs have an incredible performance on extreme processing speed, highly integrated capacity, and reduced nanometer-scale size, as compared to the earliest computer chips. However, hardware security problems have emerged as important data can be leaked through various attacks [1, 9, 10, 15, 26, 31].

PUF was first presented in [13]. A PUF plays a significant role in hardware security because it provides a unique "silicon fingerprint." The unique response of PUFs, which introduce casual but unique physical variability, satisfies the requirement of hardware security primitive, specifically for hardware cryptography applications. The unpredictable and uncontrollable features make a PUF's repeatability unrealizable hence it is unclonable. PUFs are crucial for preventing IC piracy, authentication, and data integrity for hardware security [27]. An Arbiter PUF utilizes the random path delay since the response variable is uncontrollable [13, 20]. A composition of identical ring oscillators with slight differences in oscillation frequency is known as a Ring-Oscillator (RO) PUF [2, 22]. Static random access memory (SRAM) PUF takes advantage of the SRAM-matrix in microelectronic fields [36]. A Butterfly PUF, alike to the SRAM-PUF, can generate a state that is floating to gain a random state [5]. Field-programmable gate array (FPGA), described by hardware description language (HDL), is an integrated circuit whose configuration can be designed after manufacturing. Due to the substantial resources of RAM blocks and logic gates, FPGAs can perform intricate computations. Compared with an application-specific integrated circuit (ASIC), FPGAs could be programmed to modify or update circuit level design after manufacturing [35]. Also, the cost of FPGAs is gradually decreasing, and their versatile usages make them suitable for quick design, development, and testing of PUF circuits [4]. While in the most common PUFs, RO-PUF's statistical properties are highly favored in FPGA implementation [6, 8]. However, the ring oscillators are not always stable due to changes in environmental parameters such as temperature, and their output can change with the variation in operating conditions. Thus, a new design of PUFs should meet the challenge of the variation in working conditions and retain an expected performance. Therefore, a glitch in a basic PUF design or a transient effect in ring oscillators is a big concern for circuit designers. To address the problem mentioned earlier, a variant of RO-PUF called Glitch PUF, or the transient effect ring oscillator (TERO)-based PUF, can be used. This PUF is based on metastability and utilizes a positive spike generated by path-delay or transitory oscillations.

In this chapter, the background of TERO-PUF is discussed in Sect. 2.2 In Sect. 2.3, we discuss the design strategies covering architecture, implementation, and application of Glitch/TERO-PUFs. Finally, we conclude in Sect. 2.4.

2.2   Background

Due to the horizontal business model of IC manufacturing adopted by fabless semiconductor companies today, most production is done overseas foundries to save the high cost of maintaining an advanced technology node fab [23, 29, 30]. Moreover, the foundry has access to design layout and a tightly controlled fabrication process with IP has owner has little to no information about the process [5, 32]. Therefore, IC production is vulnerable to trust issues, and appropriate preventative measures must be taken to ensure proper authentication and key generation is established for every IC developed [34, 39]. The foundry is supposed to be trusted with the complete layout design and is responsible for the requested number of chips. Since the IP owner supplies all the test patterns to the foundry and responses for every circuit design, if a foundry becomes adversarial, it can overproduce dies sell or intellectual property (IP) without the knowledge of the IP owner [11, 14]. Other security threats include invasive attacks, such as decapsulation and de-processing of an already produced chip to allow the attacker to reverse engineer the IC and resell the chip as a clone [27]. Without a unique identifier or watermark to prove authentication, it can be challenging for the semiconductor OEMs to manage operational risks and prevent attackers from overproducing or cloning ICs [3]. Active attacks such as electromagnetic fault inject, clock speedups, voltage, and temperature manipulations can cause unintended behaviors from the chip and cause the chip to fail [1, 33]. Passive attacks such as optical inspection, timing, and power analysis can provide secret information in memory that was not intended to be looked at [10, 19, 21]. Encryption keys and secret content can be extracted by an attacker and must be hidden at all costs to prevent companies from declining financially [31]. The cost of design for trust outweighs the millions of dollars in court fees and additional expenses that a company would have to spend if hardware security primitives were not implemented with the most up-to-date security. PUFs appear to be a promising method of ensuring authentication, traceability, and access control. They provide secure inexpensive authentication, and numerous of their architectures are groupable. Memory-based PUFs, such as the SRAM and the DRAM, are the type of PUFs that generate cryptographic keys for specific memory locations [12, 24]. Another type of PUF is delay-based PUF, which comprises arbiter PUF and RO-PUF., to prevent IP infringement and is presented in [38].

Previous solutions come with many challenges in security today, and new primitives are needed to ensure trust. For example, ring oscillators were once considered breakthrough security primitive for FPGAs until electromagnetic analysis was discovered to provide the PUF location. If an attacker knows the location of the PUF, they can use electromagnetic injections that can result in locked RO cells [17]. Therefore, PUFs need to be completely hidden and unknown to the attacker, which provides new challenges for the growing technology. Not only should a PUF be hidden, but it should also generate values based solely on process variation alone.

The quality of a PUF's response correlates to the random process variation within the die and can be categorized based on its reliability and uniqueness. Reliability assesses the PUF's capability of reproducing the same challenge/response pair (CRPs) across distinct environmental parameters over time. It measures the intra-device variation by placing identical PUFs in multiple locations within the die. Ideally, the intra-device hamming distance for a stable PUF is 0%, resulting in a zero-bit error rate. On the other hand, uniqueness measures the distinctive CRPs from die-to-die and is considered the inter-die variation. An ideal hamming distance of 50% results in a truly random PUF output between devices.

A PUF can also be categorized as weak or strong. A weak or strong PUF has nothing to do with its ability to ensure trust but is related to the number of challenge/response pairs. Weak PUFs require a small number of CRPs and are mainly used for key generation, while strong PUFs require many CRPs to produce a unique chip identifier. With the combination of strong and weak PUFs, trust is bolstered between the IP owner and the foundry.

2.3 Design Strategy and Applications

2.3.1 PUF Design for FPGA-Based Embedded Systems

As discussed earlier, an IC can generate its unique signature with broader applications in the area of hardware security, such as IP/IC piracy prevention, device authentication, and data protection by cryptographic key generation. This unique signature of an IC can be utilized as a PUF circuit for hardware security applications [4]. This PUF design, in contrast to earlier ones, may be seamlessly integrated into a design's HDL, using very little space, and does not necessitate "hard macros" with fixed routing.

The central hypothesis is that the PUF offers copies of combinational paths matched to the FPGA architecture. Thus, these paths can generate different delays

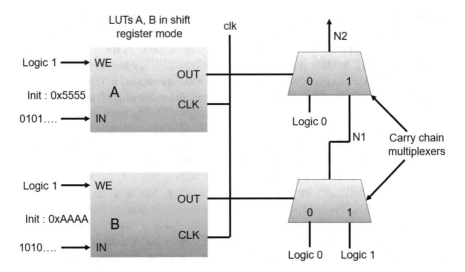

Fig. 2.1 Illustration of the PUF circuit implemented using FPGA LUTs to exploit variation in delay due to process variation

from manufacturing variations. There are two significant improvements over the prior PUF designs. First, a single FPGA logic block is used to signature of the PUF. Its architecture is more straightforward because it does not call for matched routing outside the logic block. Secondly, the PUF does not require using hard macros with fixed routing because the PUF is designed in VHDL. This PUF design is based on Xilinx Virtex-5 65nm FPGAs. Additionally, this PUF will generate the output of either 0 or 1, based on random process variations. The core of this design includes two look-up tables (LUTs), A and B, which are combined within the two slices of Virtex-5. A 16-bit shift register mode is applied to these LUTs. Below is the content of the 16-bit shift register:

LUT A: 0x5555 (In hexadecimal) 0101010101010101 (In binary)

LUT B: 0xAAAA (In hexadecimal) 1010101010101010 (In binary)

The string of initial LUT A is complementary to LUT B. Figure 2.1 indicates the basic design of two-LUT architecture. A sequence of 0101...a 1...will be produced by the shift register in LUT A. A sequence of 1010...a 1...will be produced by the shift register in LUT B. the respective carry chain multiplexers are driven by the OUT pin on the shift register. At the same time, the input of "0" on both multiplexers is fixed with logic-0. The input of "1" on the lower multiplexer is fixed with logic-1. The lower multiplexer's output drives the top multiplexer's "1" input.

The designers emphasized two scenarios: one in which the multiplexer is driven by LUT B and another in which LUT A is faster. In the first case, when LUT B shifts from logic 1 to 0, the N1 signal also moves from logic 1 to 0. The slower LUT A will then go from logic 0 to logic 1, whereas N2 will remain at logic 0. Because LUT A and its multiplexer are quicker than LUT B and its multiplexer in the second case, when LUT A flips from logic 0 to 1, N1 does not switch from logic 1 to 0. This case will generate a short positive glitch on N2 before N1 transitions from logic 1 to 0. This presence or absence of the glitch is determined by the delay between two LUTs and their multiplexers. And as a result, it can be used to generate a PUF signature. The primary benefit of this PUF design is that it is entirely specified in VHDL. The automation feature in synthesis, place, and route tools stops the manual intervention. Secondly, a "push-button" FPGA design flow can be applied to this PUF design. Thus, no external matched routing or hard macros are needed. Additionally, since only two slices are involved with the PUF bit generation, this design has a small area.

As with most delay-based PUF implementations, this PUF design is susceptible to side-channel attacks [18]. Although addition, the PUF design is reliable at high temperatures, there are still some signature changes, which add to its instability.

The signature uniqueness was analyzed through the Hamming distance between all PUF pairs. The ideal distribution of the hamming distance is expected at around the value of 64 bits. The average Hamming distance was found to be 61.8, which is closed to the expected value of 64. The smallest and largest distance was 43 and 79, respectively. The design was also analyzed for the reliability of the PUF design at low and high temperatures. The result shows that 72 % of the signatures altered by five or fewer bits when in the case of high temperature, and there is no indication that signatures change over 10 bits. The results also indicate that the PUF design has the advantage of uniqueness, and the degree to which it is affected by temperature is relatively small.

2.3.2 Transient Effect Ring Oscillator-Based PUF

This PUF is designed using transient effect ring oscillators (TERO) [6]. TERO-PUF, unlike previously designed RO-PUFs, is not prone to the locking situation. The locking situation limits the use of ROs in the design of PUFs and True Random Number Generators (TRNG). For security requirements, the ring oscillators must be running independently, and oscillation frequency must be hidden from attackers, which is no longer guaranteed due to electromagnetic attacks. These security vulnerabilities are addressed in TERO-PUF design by making the design insensitive to the locking phenomenon (because it does not consider the frequency of oscillations during the entropy extraction process). Instead, a TERO-PUF takes advantage of cross-coupled elements' oscillatory metastability by extracting entropy from their transient oscillatory cycles.

The architecture of this PUF is based on a TERO loop of an SR flip-flop. The TERO loop is made up of two AND gates and an odd number of NOT gates. The two branches of the TERO loop have the same number of inverters and can be extended to enhance transient oscillations. If there is positive feedback, the inputs of the SR flip-flop are connected to a control signal, which causes transient oscillations on every rising edge. As a result, the RC time constant is less than the total time delays of the loop's logic elements. Figure 2.2 [6] depicts a 26-bit accumulator, an 8-bit

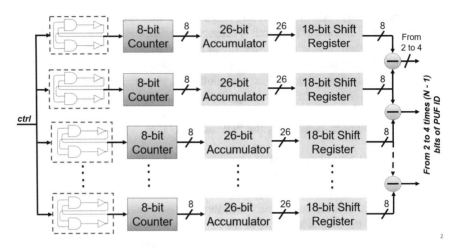

Fig. 2.2 Illustration of overall approach to TERO-PUF

counter, and shift register which is 18-bit, all of which are required to calculate the mean number of transient oscillations. The oscillations should never stop if the loop is symmetrical. However, due to the intrinsic asymmetry T_d defined in [28], the circuit usually oscillates for a small duration before stopping. The T_d denotes the difference in delays between the two branches of the loop and is related to random contributions to logic gate delays caused by intrinsic noise. Unlike the structure in [28], which controls the loop with XOR and AND gates, this architecture stimulates the oscillatory metastable state with AND gates and inverters. As a result, the loop used in [28] is smaller than the current design.

Deviations in the manufacturing process affect the TERO loop's behavior in various ways: the steady state of the S signal (Zero or One) and the amount of transient fluctuations. In nine Intel DE1 boards, the designers evaluated this PUF for different stable end values of the S signal at all potential locations of the TERO loop. When employing the final state of S as a PUF response, it was discovered that the oscillating state represents a considerable disadvantage. In some circumstances (roughly one-third of the TERO loops implemented), it can be assumed that the T_d values are pretty tiny and inferred that the final state of S cannot be the origin of entropy from the fabrication process. This observation contradicts prior TERO loop design outcomes [37]. The three least significant bits (numbers 0, 1, and 2) of the 8-bit counter output determine whether the TERO-PUF is a true random number generator or simply an RNG. These least significant bits (LSBs) passed the Federal Information Processing Standard (140–1 and 140–2) tests with a high margin [7]. According to AIS-31, this result is insufficient to ensure security of the TRNG. Therefore, an entropy extraction model is required for PUF and TRNG. More research is needed to optimize entropy extraction. This TERO-PUF implementation is completely scalable, with PUF ID sizes of 126, 189, or 252-bits, which are significantly larger than previously built PUFs (Arbiter, RO, SRAM, Butterfly, and Loop-PUF).

The experimental results demonstrate that the LSBs show instability due to electric noise and can be utilized as an entropy source for TRNG. The MSBs were stable, and the statistical properties were used to characterize the TERO-PUF. The PUF's intra-device and inter-device fluctuation can be characterized by placing four identical 64-Loop TERO-PUFs in each of the nine available FPGAs. Further, the bias, uniqueness, and reliability for ID sizes ranging from 126 to 252 bits were evaluated using 64-loop TERO-PUF. From the 128 samples taken, subtractions for the top two to four bits of the PUF response were used for variation comparison. It was found that the top two MSBs had less than 2% intra-device variation and could always be used as a PUF response, while the third and fourth bits could be used with error correction codes because they have a variation of less than 5%. The TERO-PUF architecture is more appealing than other cutting-edge PUF architectures due to its low intra-device fluctuation and increased bit response.

2.3.3 Glitch PUF

The purpose of this PUF architecture was to address the present Delay-PUF challenge of estimating the relationship between delay and produced information [25]. This architecture takes advantage of glitches that act in a non-linear manner due to gate delay variation and the pulse propagation characteristic of each gate, hence the name Glitch PUF.

This PUF exploits the transient state of output signals from a set of inter-connected combinational logic. This output signal is called the glitch signal. The glitch signal is generated by exploiting the different delays that signals incur while traveling through combinational logic. The glitch generator receives a selection signal SEL_{gc} comprising a combinational circuit and a v-1 selector, with the circuit performing $Y = f(X)$ for every value of X. The layout of the glitch generator can be seen in Fig. 2.3 [25]. The selector chooses one bit from v bits of Y based on the selection signal SEL_gc. The ping signal delay circuit comprises a buffer chain that outputs h_d, a delayed signal of h. The chain's depth is calculated during the design phase by simulating the occurrence timing of the glitch signals produced by the glitch generator.

This PUF contains glitches that change shape depending on delay variation. The problem here is determining how and when to capture the shape of an intermittent pulse precisely. Simultaneously, the acquisition procedure should be implemented as a digital circuit. One of the general solutions is the phase-shift method, which involves preparing several clocks with distinct stages to sample a tiny pulse. The sampling accuracy increases as the amount of distinct phased clocks increases. However, this method is impractical due to the large number of clock lines required. In addition, only a few low-jitter global clock lines use time division, which impacts acquisition speed suffers. As a result, a method is required in which the target data

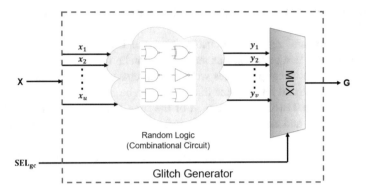

Fig. 2.3 Illustration of glitch generator circuit

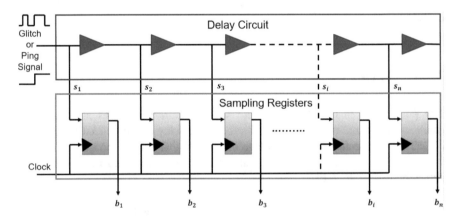

Fig. 2.4 Illustration of sampling circuit

is moved by a tiny amount of time while being sampled by the same clock. Figure 2.4 [25] depicts a sampling circuit that can be used to capture a sampled output from a glitch signal generated.

Due to the propagation delay within the sampling registers, the sampling registers will have varying latch times. This varying latch time causes sequentially placed registers to have a nonsequential time order and thus incur jitter on the sampled output. This jitter problem can be solved by using a ping signal to find the time order of the data of the registers. A delay circuit is used to manipulate the properties of this ping signal, and a variable delay circuit is used to manipulate the shared clock of the sampling registers for more thorough testing.

With the acquired data, jitter correction pre-processing can occur. After acquired, the sampled output from a given glitch signal is converted into a single bit of response. This conversion is done by the G2R (Glitch to Response) converter, which uses the parity of the amount of rising edges in the glitch waveform to determine the response bit. The Glitch PUF can be seen in Fig. 2.5 [25].

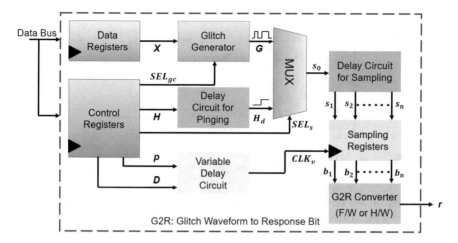

Fig. 2.5 Overall architecture of Glitch PUF

Due to the architecture's nature, the entropy and error rate can be simulated without analog SPICE simulations. The observed intra-device hamming distance has a mean error rate of 1.3% and a maximum error rate of 6.6% at 80°C, before any masking, and is acceptable. The observed inter-device hamming distance falls short yet is still sufficient in many cases.

The Glitch PUF creates a non-linear relationship between delay variation and the response. It has a good intra-device hamming distance variation. Additionally, the Glitch PUF is easy to evaluate without using a SPICE simulation, unlike the SRAM-PUF. Unfortunately, the Glitch PUF does not have an ideal inter-device hamming distance variation. As machine learning techniques advance, the efforts here will become outdated.

2.3.4 TERO-PUF on SRAM FPGAs

The TERO-PUF is based on extracting the entropy of process variations by comparing TERO cell characteristics, as previously discussed in Sect. 2.3.2. As a result, this TERO cell must be meticulously designed to construct a PUF. Furthermore, accuracy is required during the sizing of the logic gates and the interconnect delays within the logic cell. As a result, designing TERO-PUF in FPGA is extremely difficult.

The structure of the TERO cell consists of two equally mirrored branches. An AND gate for initialization followed by odd number inverters for the typical PUF cell. Since the TERO cell works on the oscillation properties of the circuity, the symmetry of the two branches becomes much more significant. If not designed correctly, the cell could have properties, such as bias, toward a particular value to

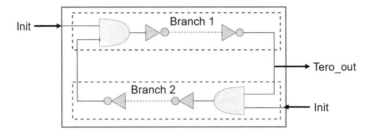

Fig. 2.6 Illustration of TERO cell structure

not oscillate at all. The number of NOT gates must be the same across two branches to achieve high symmetry and identity. The connection between elements has to be equal in length in the same position of the two branches. Last, the connection of the two branches has to be of equal length, as shown in Fig. 2.6 [16].

TERO works by first activating the raising the signal "init" to a specific value to initialize the cell. The initialization will cause two opposite sate to propagate inside the TERO cell and create the oscillating phenomenon. The oscillation will never stop in an ideal case where everything is perfectly symmetric. In reality, manufacturing/processing variation will cause enough asymmetry leading to a propagation speed difference. After a certain amount of time, one state will catch up with the other, thus ending the oscillation. This TERO-PUF has been implemented using Intel Cyclone V and Xilinx Spartan 6 FPGAs [16]. Both are SRAM-based FPGAs that make identical connections difficult with the design tool provided by their manufacturer, but extensive study on the configuration has made it possible.

For FPGA implementation of TERO cell, the following constraints must be considered:

1. The quantity of inverters in the cell's two branches must be the same.
2. All elements' connections must be pairwise equal. For example, in both branches, the delay between the initialization phase and the first NOT gate must be the same.
3. The connections between the two branches must have same delay.

The first constraint appears straightforward because a designer has complete control over the number of elements he/she implements, but this constraint does not apply to Intel FPGAs. Furthermore, the last two constraints are difficult to meet with any SRAM-based FPGA. Control of the place and route tool's connections, for example, is in reality impossible, and only element placement may be imposed. As a result, finding a configuration that fits all restrictions is tough.

The Xilinx Spartan 6 FPGAs structure consists of configurable logic blocks (CLB) with two circuit element slices in each block. Only one type of slice is used for this application, slice X, consisting of four LUTs. Due to the requirement of TERO cell, each LUT is used for only one gate. A total of four slices were used to create a branch with 1, 3, 5, or 7 inverters and interconnections with branches. This

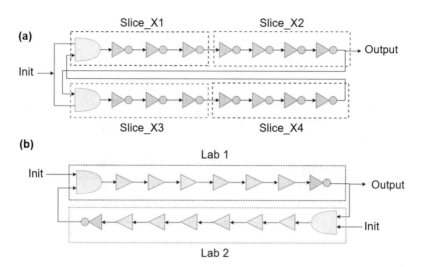

Fig. 2.7 TERO cell schematic implemented on (**a**) Xilinx Spartan 6 and (**b**) Altera Cyclone V FPGAs

approach also allows identical connections between two branches and eliminates any difference in routing delay. In this case, 7 inverters type was chosen for this design, shown in Fig. 2.7a [16]. An AND gate was created with the Xilinx LUT6 component for the initialization. To ensure the layout on the FPGA was correct and according to our design, an estimation of the delay is performed with the "FPGA editor" from the Xilinx tool, which indicates the identical delay between elements in two different branches. At this step, a hard macro is created for fixing the TERO cell's routing to avoid future changes. This step ensures one can replicate the TERO cell on other FPGAs.

Intel Cyclone V has many differences from the Xilinx Spartan 6. The technology note of Cyclone is 28 nm compared to 45 nm on Spartan 6, and the logic array blocks (LAB) in an array arrangement, with each LAB having ten adaptive logic modules (ALM). Each ALM is equipped with two LUTs, each with six inputs and two outputs. Implementing TERO cell on Altera Cyclone is much more challenging since the synthesis tool will optimize the design even when limitations are set. The delay element, or CELL, was used to combat this drawback because it will not affect the design tool. The TERO cell on the Cyclone FPGA has two AND gates, but instead of having inverters like the one on the Xilinx Spartan 6, it has two inverters and 12 LCELLs. A two-lab-placed side-by-side scheme is used to mitigate the delay of this layout, shown in Fig. 2.7b [16]. The Altera timing quest analyzer tool confirmed that the total delay difference is 0.035 ns.

Figure 2.8 [16] shows the system of the TERO-PUF, which consists of the FPGA part and the processing part. Two identical blocks (A&B) consist of 128 TERO cells, a selector, and a multiplexer for the FPGA part. The output of the PUFs is a 128-bit signal that is then passed to a counter. Two counter values are then passed to a

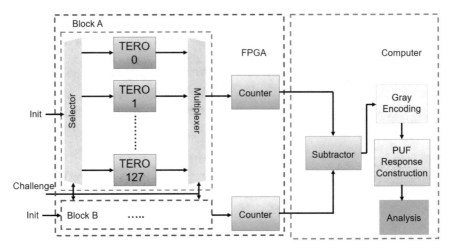

Fig. 2.8 Illustration of overall architecture of TERO-PUF architecture for Xilinx and Altera FPGAs

subtractor. Finally, the result is coded as Gray code, ready to be used to construct the response.

In total, 30 Xilinx Spartan 6 FPGAs and 18 Intel Cyclone V are used to generate results. For this PUF system, bitstream from subtracting two 16-bits counters is used with bit 0 to represent the least significant bit while 15 is the most significant one. Unlike the RO-PUF, where only one bit is used for comparison, TERO-PUF could use more than one bit to generate the response for each challenge. With this in mind, a robustness analysis study on bits 4, 5, and 15. There are also two combinations of bits under investigation: bits 5 and 15 and bits 4, 5, and 15. The two robustness tests were performed with temperature, and voltage variations [16]. The results indicate that all the bit responses have steadiness below 10% from 15°C to 35°C, which indicates a temperature fluctuation of 40% with the center temperature put at 25°C. The effect of voltage variation on the steadiness of bits shows that for the below 10% mark, the supply voltage range should be between 1.18 V and 1.22 V, and a 1.5% variation compared to 1.2 V. Bit 4 consistently performs the worst in temperature and voltage tests, while bit 15 performs the best in those tests.

2.3.5   TERO-PUF in IoTs

Today's world is becoming increasingly interconnected. The "Internet of Things" connects everything, from televisions to smartphones, vehicles, buildings, and household appliances (IoT). The Internet of Things is now a part of our daily lives. Over ten billion devices are already linked, with a fivefold increase expected in the next decade. While IoT integration and deployment are growing, one of the main challenges is providing practical solutions to IoT security, and privacy problems. Interoperability and scalability, traceability, authentication, and access control must all be included in protection and security mechanisms while remaining lightweight. PUFs that use some physical characteristics of similar but different integrated circuits to provide a unique identifier are one of the most promising methods for such security mechanisms. As a result, these functions may be used to validate integrated circuits, offer traceability, and limit access [17].

An extensive design study of TERO-PUF implementation on Xilinx and Altera FPGAs is covered in Sect. 2.3.4 [16]. For the IoT level design scenario, an almost identical layout was used to evaluate the performance of TERO-PUF further [17]. The layout and number of inverters in Xilinx's FPGAs are identical to the previous design, while the number of delay cells (LCELL) has increased from 12 to 16 in Altera's case. The architecture of the TERO-PUF structure implemented on both kinds of FPGAs is identical to as shown in Fig. 2.7, and the same architecture was used, as shown in Fig. 2.8. Also, the Xilinx tool and Altera TimeQuest analyzer tool were used to evaluate the delay and layout of the PUF. During testing, it has been observed that some cells do not seem to stop oscillating in the experiment, so deciding the acquisition time is challenging. Therefore, three different acquisition times were segmented and allocated in three cases to better evaluate the proper acquisition time frame. In case 1, the acquisition window is short compared to all TERO cell oscillations, which will lead to good steadiness, but a low uniqueness response. In case 2, we have enough time for most cells to reach a steady state. It is interesting since there is no way of predicting which cell will continue oscillating infinitely. Lastly, the acquisition time in case 3 is long enough that all the cells have reached a steady state.

The data system acquisition required a 16-bit counter to capture the number of oscillations for every cell. A total of 30 Xilinx and 16 Altera FPGAs were used to get the mean of uniqueness and steadiness result of the counter, starting from 0, which is the least significant bit, to 15, which is the most significant bit (MSB). The data acquired from Xilinx FPGAs is considered with a threshold of 10%, meaning that only the bits with less than 10% steadiness are considered. Another aspect is the fact that the uniqueness of the bit drops significantly after bit 5. For authentication purposes, bits 4, 5, and 15 were chosen due to their desirable properties. The same set of data was acquired with Altera FPGA, and the same set of bits demonstrates favorable characteristics. Another aspect of PUF is its ability to resist environmental variations. In this design, the test result from PUF on Altera Cyclone V with the acquisition window set as 30 clock cycles is analyzed for Bits 5, 6, 15, and bits with various voltages, where bit 15 had the lowest steadiness and bit 15 had the highest. For the threshold steadiness to be less than 10%, the voltage must be from 1.07 V to 1.13 V, which is 1.5% of the variation. The TERO-PUF allows the extraction of multiple bits per challenge while maintaining statistical characteristics. The speed and lightweight nature of this PUF make it a viable candidate for IoT applications.

2.4   Conclusions

This chapter analyzes multiple research studies for different types of delay-based PUFs on commercially available FPGAs. The first design utilized varying delays of combinational paths, the two shift registers, and their corresponding carry chain multiplexers to create a glitch spike response. The TERO-PUF architecture uses oscillatory metastability to create the response, and remove the precursor's susceptibility (the RO-PUF), to the locking phenomenon. An extensive study has examined the PUF's uniqueness and robustness under different environmental variations. The Glitch PUF utilizes a similar concept to the first architecture. Still, instead of utilizing the delay of its multiplexers and shift registers, the Glitch PUF utilizes the delays between its combinational logic. However, the PUF designs discussed in this chapter do have some limitations. The first PUF design is vulnerable to side-channel attacks and has signature changes at high temperatures. Furthermore, the TERO-PUF was not optimized for area variation in steadiness (reliability) and voltage and temperature differences. The Glitch PUF was the most limited of all PUFs reviewed in measured uniqueness. Nevertheless, the unique characteristics and requirements of PUFs show a promising solution for the security and authentication of IoT devices. Further research collaboration is required between the industry and academic community to bring more robust and advanced PUFs into our next-generation devices.

References

1. Ahmed B, Bepary MK, Pundir N, Borza M, Raikhman O, Garg A, Donchin D, Cron A, Abdelmoneum MA, Farahmandi F, et al (2022) Quantifiable assurance: From IPs to platforms. arXiv preprint arXiv:220407909
2. Amsaad F, Pundir N, Niamat M (2018) A dynamic area-efficient technique to enhance ROPUFs security against modeling attacks. In: Computer and Network Security Essentials. Springer, Berlin, pp 407–425
3. Anandakumar NN, Rahman MS, Rahman MMM, Kibria R, Das U, Farahmandi F, Rahman F, Tehranipoor MM (2022) Rethinking watermark: Providing proof of IP ownership in modern SoCs. Cryptology ePrint Archive
4. Anderson JH (2010) A PUF design for secure FPGA-based embedded systems. In: 2010 15th Asia and South Pacific Design Automation Conference (ASP-DAC). IEEE, New York, pp 1–6
5. Bhunia S, Tehranipoor M (2018) Hardware security: a hands-on learning approach. Morgan Kaufmann, Burlington
6. Bossuet L, Ngo XT, Cherif Z, Fischer V (2013) A PUF based on a transient effect ring oscillator and insensitive to locking phenomenon. IEEE Trans Emerg Top Comput 2(1):30–36
7. Brown KH (1994) Security requirements for cryptographic modules. Fed Inf Process Stand Publ, New York, pp 1–53
8. Choudhury M, Pundir N, Niamat M, Mustapa M (2017) Analysis of a novel stage configurable ROPUF design. In: 2017 IEEE 60th International Midwest Symposium on Circuits and Systems (MWSCAS). IEEE, New York, pp 942–945
9. Contreras GK, Rahman MT, Tehranipoor M (2013) Secure split-test for preventing IC piracy by untrusted foundry and assembly. In: 2013 IEEE International symposium on defect and fault tolerance in VLSI and nanotechnology systems (DFTS). IEEE, New York, pp 196–203

10. Dey S, Park J, Pundir N, Saha D, Shuvo AM, Mehta D, Asadi N, Rahman F, Farahmandi F, Tehranipoor M (2022) Secure physical design. Cryptology ePrint Archive

11. Farahmandi F, Huang Y, Mishra P (2020) Automated test generation for detection of malicious functionality. In: System-on-Chip Security. Springer, Cham, pp 153–171

12. Garg A, Kim TT (2014) Design of SRAM PUF with improved uniformity and reliability utilizing device aging effect. In: 2014 IEEE International Symposium on Circuits and Systems (ISCAS). IEEE, New York, pp 1941–1944

13. Gassend B, Clarke D, Van Dijk M, Devadas S (2002) Silicon physical random functions. In: Proceedings of the 9th ACM Conference on Computer and Communications Security, pp 148–160

14. Guin U, DiMase D, Tehranipoor M (2014) Counterfeit integrated circuits: Detection, avoidance, and the challenges ahead. J Electron Test 30(1):9–23

15. Lee J, Tehranipoor M, Patel C, Plusquellic J (2007) Securing designs against scan-based side-channel attacks. IEEE Trans Dependable Secure Comput 4(4):325–336

16. Marchand C, Bossuet L, Cherkaoui A (2016) Design and characterization of the TERO-PUF on SRAM FPGAs. In: 2016 IEEE Computer Society Annual Symposium on VLSI (ISVLSI). IEEE, New York, pp 134–139

17. Marchand C, Bossuet L, Mureddu U, Bochard N, Cherkaoui A, Fischer V (2017) Implementation and characterization of a physical unclonable function for IoT: a case study with the TERO-PUF. IEEE Trans Comput Aided Des Integr Circuits Syst 37(1):97–109

18. Merli D, Schuster D, Stumpf F, Sigl G (2011) Side-channel analysis of PUFs and fuzzy extractors. In: International Conference on Trust and Trustworthy Computing. Springer, Berlin, pp 33–47

19. Park J, Anandakumar NN, Saha D, Mehta D, Pundir N, Rahman F, Farahmandi F, Tehranipoor MM (2022) PQC-SEP: Power side-channel evaluation platform for post-quantum cryptography algorithms. IACR Cryptol ePrint Arch 2022:527

20. Pundir N, Amsaad F, Choudhury M, Niamat M (2017) Novel technique to improve strength of weak arbiter PUF. In: 2017 IEEE 60th International Midwest Symposium on Circuits and Systems (MWSCAS). IEEE, New York, pp 1532–1535

21. Pundir N, Park J, Farahmandi F, Tehranipoor M (2022) Power Side-Channel Leakage Assessment Framework at Register-Transfer Level. IEEE Trans Very Large Scale Integr VLSI Syst

22. Rahman MT, Rahman F, Forte D, Tehranipoor M (2015) An aging-resistant RO-PUF for reliable key generation. IEEE Trans Emerg Top Comput 4(3):335–348

23. Shi Q, Vashistha N, Lu H, Shen H, Tehranipoor B, Woodard DL, Asadizanjani N (2019) Golden gates: A new hybrid approach for rapid hardware Trojan detection using testing and imaging. In: 2019 IEEE International Symposium on Hardware Oriented Security and Trust (HOST). IEEE, New York, pp 61–71

24. Sutar S, Raha A, Raghunathan V (2016) D-PUF: An intrinsically reconfigurable DRAM PUF for device authentication in embedded systems. In: 2016 International Conference on Compliers, Architectures, and Synthesis of Embedded Systems (CASES). IEEE, New York, pp 1–10

25. Suzuki D, Shimizu K (2010) The glitch PUF: A new delay-PUF architecture exploiting glitch shapes. In: International Workshop on Cryptographic Hardware and Embedded Systems. Springer, Berlin, pp 366–382

26. Tehranipoor M, Koushanfar F (2010) A survey of hardware Trojan taxonomy and detection. IEEE Des Test Comput 27(1):10–25

27. Tehranipoor M, Peng K, Chakrabarty K (2011) Introduction to VLSI testing. In: Test and Diagnosis for Small-Delay Defects. Springer, New York, pp 1–19

28. Varchola M, Drutarovsky M (2010) New high entropy element for FPGA based true random number generators. In: International Workshop on Cryptographic Hardware and Embedded Systems. Springer, New York, pp 351–365

29. Vashistha N, Lu H, Shi Q, Rahman MT, Shen H, Woodard DL, Asadizanjani N, Tehranipoor M (2018a) Trojan scanner: Detecting hardware Trojans with rapid SEM imaging combined

with image processing and machine learning. In: ISTFA 2018: Proceedings from the 44th International Symposium for Testing and Failure Analysis, ASM International, p 256

30. Vashistha N, Rahman MT, Shen H, Woodard DL, Asadizanjani N, Tehranipoor M (2018b) Detecting hardware Trojans inserted by untrusted foundry using physical inspection and advanced image processing. Journal of Hardware and Systems Security 2(4):333–344

31. Vashistha N, Rahman MT, Paradis OP, Asadizanjani N (2019) Is backside the new backdoor in modern socs? In: 2019 IEEE International Test Conference (ITC). IEEE, New York, pp 1–10

32. Vashistha N, Lu H, Shi Q, Woodard DL, Asadizanjani N, Tehranipoor M (2021) Detecting hardware Trojans using combined self-testing and imaging. IEEE Trans Comput Aided Des Integr Circuits Syst 41(6):1730–1743

33. Wang H, Li H, Rahman F, Tehranipoor MM, Farahmandi F (2021) Sofi: Security property-driven vulnerability assessments of ICs against fault-injection attacks. IEEE Trans Comput Aided Des Integr Circuits Syst 41(3):452–465

34. Wang X, Tehranipoor M, Plusquellic J (2008) Detecting malicious inclusions in secure hardware: Challenges and solutions. In: 2008 IEEE International Workshop on Hardware-Oriented Security and Trust. IEEE, New York, pp 15–19

35. Wiśniewski R (2009) Synthesis of compositional microprogram control units for programmable devices. University of Zielona Góra, Poland

36. Xiao K, Rahman MT, Forte D, Huang Y, Su M, Tehranipoor M (2014) Bit selection algorithm suitable for high-volume production of SRAM-PUF. In: 2014 IEEE international symposium on hardware-oriented security and trust (HOST). IEEE, New York, pp 101–106

37. Yamamoto D, Sakiyama K, Iwamoto M, Ohta K, Ochiai T, Takenaka M, Itoh K (2011) Uniqueness enhancement of PUF responses based on the locations of random outputting RS latches. In: International Workshop on Cryptographic Hardware and Embedded Systems. Springer, Berlin, pp 390–406

38. Zhang J, Wu Q, Lyu Y, Zhou Q, Cai Y, Lin Y, Qu G (2013) Design and implementation of a delay-based PUF for FPGA IP protection. In: 2013 International Conference on Computer-Aided Design and Computer Graphics. IEEE, New York, pp 107–114

39. Zhang X, Tehranipoor M (2011) Case study: Detecting hardware Trojans in third-party digital IP cores. In: 2011 IEEE International Symposium on Hardware-Oriented Security and Trust. IEEE, New York, pp 67–70

Chapter 3
Direct Intrinsic Characterization PUF

3.1 Introduction

Secret cryptographic keys are an essential component of hardware security [18]. The shift of the electronic supply chain from vertical to horizontal and globalization have induced many vulnerabilities in the electronic supply chain [3, 4, 20]. Therefore, these secret keys are required in an integrated circuit (IC) to perform secure activation, authenticate the chip, encrypt a secure communication, protect against intellectual property (IP) piracy, and IP counterfeiting. The conventional secret keys are derived from the digital data usually residing on the IC in some non-volatile memory. As a result, these conventional secret keys are susceptible to probing attacks [21]. The attacks against these conventional keys can be mitigated if the secret keys are generated at runtime by exploiting the inherent process variation of the IC. This led to the development of physical unclonable functions (PUFs), specialized hardware that exploits process variation in the IC to generate a random sequence of secret keys for the given challenge [16].

PUFs represent the characteristics of the physical system and are a one-way security function [2, 6]. Due to process variation, no two ICs can be alike; they will always have some differences even though the manufacturer uses the exact same components and process. PUFs take advantage of these subtle differences between chips to generate a security model. Due to subtle differences between chips, each challenge input to the PUF will have a different PUF response. This challenge–response pair (CRP) is unpredictable, and the response generated by each challenge is irregular. Hence, the CRPs can be used as a security key to provide hardware security in the ICs.

In the previous chapters, we saw that PUFs could be built with delay-based circuits. These circuits are low complexity hardware-based digital solutions for existing security issues (e.g., IP piracy, tampering, and counterfeiting) [7, 11, 13, 16]. However, these simple PUFs suffer from aging and environmental perturbations and may often produce unreliable outputs [15]. To compensate for these erroneous

© The Author(s), under exclusive license to Springer Nature Switzerland AG 2023
M. Tehranipoor et al., *Hardware Security Primitives*,
https://doi.org/10.1007/978-3-031-19185-5_3

outputs, Error Correction Coding (ECC) or fuzzy extractors are added to the original circuit, increasing area, and power overheads. A unique Direct Characterization PUF can be utilized to mitigate this overhead issue. Therefore, this chapter focuses on a special class of PUFs, proposed in the literature, which relies on PUF authentication through direct characterization of electronic components.

Direct Characterization PUFs implement the intrinsic characteristics of the electronic components and authenticate through direct characterization of these components. Therefore, in contrast to racetrack PUFs, which require different delay paths to be fabricated, direct characterization PUFs do not require additional fabrication steps. On the other hand, this direct characterization can exploit several different intrinsic and electronic properties such as capacitance, current in response to voltage, or interconnect circuitry. Some researchers also call these kinds of PUF "analog electronic PUF" because most of the such PUFs explore the analog voltage-current features of an electronic component. The PUFs covered in the chapter include Cellular Neural Network PUF (CNN PUF) [1], Power Distribution PUF (PD-PUF) [5], Quasi-Adiabatic Logic-based PUF (QUALPUF) [10], VIA-PUF [9], and Threshold Voltage (TV-PUF) [12].

3.2   Cellular Neural Network PUF

Inspired by the human brain's functionality, researchers in the machine learning community have tried recreating the biological neuron for the automated decision-making process. The mathematical model of this biological neuron is called perceptron. Like the biological neuron, which activates for specific input and passes the output to the next neuron, a perceptron calculates the input value's weighted sum, generates an output using the activation function, and then passes this output value as an input to the next layer of perceptrons. This kind of multi-layer perceptron, where all the perceptrons are connected, and one perceptron can communicate with all other perceptions, is called a neural network. A cellular neural network (CNN) is a parallel computing model comparable to a neural network in computer science. The only difference is that the perceptron/neuron in CNN is only authorized to communicate between neighboring neurons. Addabbo et al. [1] proposed to utilize the concept of CNN to build analog PUFs. It proposes a 2-neuron CNN model, laying the foundation for a 1-bit PUF core. Figure 3.1 shows the two neuron CNN model used as the foundation [1]. For this CNN model, the states x and y can be mathematically represented as:

$$\dot{x} = \frac{1}{\tau}(-x + k(g(x) - g(y)) + u(t)\xi)$$

$$\dot{y} = \frac{1}{\tau}(-y + k(g(y) - g(x)) + u(t)\xi),$$

(3.1)

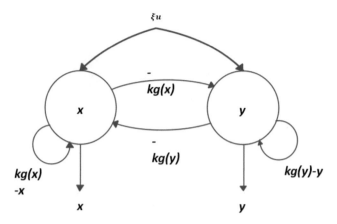

Fig. 3.1 Simple structure illustrating two neuron cellular neural network

where k is greater than 1, τ is a time constant, ξ is the excitation parameter, and $g(\cdot)$ is the activation function defined as follows.

$$g(\rho) = \begin{cases} -1, & \text{if } \rho < -1 \\ \rho, & \text{if } -1 \leq \rho \leq 1 \\ 1, & \text{if } \rho > 1. \end{cases} \tag{3.2}$$

This CNN system can have two configurations (the excited and non-excited case) depending on the value of the digital control signal function u, which activates the excitation ξ. For example, the excited case is achieved when $u = 1$ activates excitation ξ, whereas $u = 0$ gives the non-excited case where excitation ξ remains deactivated.

When $u = 0$, the CNN system will have three equilibrium points: two of these are stable and can be expressed as $p_1 = (2k, -2k)'$ and $p_2 = (-2k, 2k)'$, the other one p_0 is an unstable point which is located at the origin. Any initial condition that resides on the line $x = y$ makes the system create an asymptotically approaching trajectory toward the unstable point p_0. Any other initialization will make the system settle down in either point p_1 or p_2 with a probability of 1. Furthermore, if the initial condition is symmetrical to the line $x = y$, the system will have equal (0.5) probability to settle down on either p_1 or p_2. This dynamic behavior of reaching a particular equilibrium point depending on the initialization is exploited in designing CNN PUF.

When $u = 1$, the system enters active mode. If $\xi > 2k - 1$ the system will only have one stable point $p_{init} = (\xi, \xi)'$. p_{init} is located inside the region of the positive saturation of the activation function. This activated system also conserves its symmetry concerning the line $x = y$. As the CNN has only one global stable point, any arbitrary initialization close to p_{init} point will result in a trajectory toward the unique, attractive sink. Deploying control signal u, this dynamic behavior

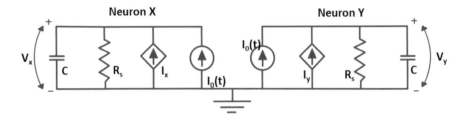

Fig. 3.2 Illustration of circuit for initializing the CNN PUF

initializes the CNN. It is critical to examine the excited system's rate of convergence in order to anticipate how near to the point p_init the CNN state will be after it has been excited for a long time. The converging rate is set not smaller than $\frac{1}{2\tau}$ to make the excited system globally exponentially stable [8]. Please refer to [1] for detailed velocity vector fields for both excited and non-excited cases of the CNN.

As discussed above, CNN may performs differently for different u. Thus, [1] defined a function to control u for two time instants such that $0 < t_0 < t_1$:

$$u(t) = \begin{cases} 1, & \text{if } t_0 \le t < t_1 \\ 0, & \text{otherwise.} \end{cases} \tag{3.3}$$

Here, the control impulse signal u starts at $t = t_0$ time and has a pulse width of $t_1 - t_0$. If the initial condition of $q = (x_0, y_0)'$ fulfills the $x_0 > y_0$ requirement during initialization ($t = 0$), the system state will be attracted toward p_1 during lag time $[0, t_0\}$. After activating excitation ξ at $t = t_0$, the CNN state starts to move from the p_1 state into the sink p_{init}. Finally, when the impulse signal u turns off the activation ξ at $t = t_1$, the CNN state goes back to the equilibrium point of p_1. This trajectory never crosses $x = y$ line and remains in the same half plane. So, the final equilibrium point is set by the initial condition q during $t = 0$. This final point p_1 will work as the signature that can be implemented through PUF. Here, the initialization phase of CNN has no contribution to find an ideal set for the system.

Exploiting this non-linear dynamic behavior of CNNs, PUFs can be implemented using physical differential relation between current and voltage across a capacitor. Figure 3.2 shows this 1-bit PUF circuit inspired by the two neuron CNN model proposed in [1]. Using Kirchhoff's current law, it can be shown that

$$\begin{aligned} C\frac{dV_x}{dt} &= \frac{-V_x}{R_s} + I_x + I_0(t) \\ C\frac{dV_y}{dt} &= \frac{-V_y}{R_s} + I_y + I_0(t) \end{aligned} \tag{3.4}$$

$$I_x = -I_y = \frac{k}{R_s}(g(V_x) - g(V_y)) \tag{3.5}$$

and

$$I_0(t) = u(t)\frac{\xi}{R_s}.\tag{3.6}$$

If we rearrange Eq. (3.4) we would get a system equivalent to Eq. (3.1), i.e.,

$$\frac{dV_x}{dt} = \frac{1}{R_sC}(-V_x + k(g(V_x) - g(V_y)) + R_sI_0(t))$$
$$\frac{dV_y}{dt} = \frac{1}{R_sC}(-V_y + k(g(V_y) - g(V_x)) + R_sI_0(t)).\tag{3.7}$$

For any case of $V_x = V_y$, the currents become $I_x = I_y = 0$ A and depending on the control signal u, voltages across each capacitor changes exponentially (either increases or decreases) with time constant R_sC. This physical behavior is a representation of system described in Eq. (3.1) while the system is in excited state and the CNN state resides on the line $x = y$.

The described circuit will be used as a one-bit PUF module if first powered on in the excited state. After the circuit has balanced around the p_{init} state, the control signal is switched off. This causes the circuit to go into a non-excited state. In this state, the circuit can converge to any one of the equilibrium states, i.e., p_1 or p_2. Depending on the state in which the circuit converges, the PUF bit is set to 0 or 1.

The experiments performed in [1] showed that CNN-based PUFs have high robustness against environmental changes. However, CNN PUFs showed limitations against delay-based PUFs due to high complexity in the designing process. For example, the system needs a high computational resource to find an ideal initialization. For this reason, CNN PUFs are only suited for simple cases, such as 1-bit PUF, but work poorly when the system's complexity rises.

3.3 Power Distribution PUF

It was seen that the PUF designs based on the active components of the ICs are most susceptible to environmental variations. Therefore, for developing robust PUFs resilient to environmental variations, one can take advantage of passive elements in the ICs. And since such passive elements are an inherent part of the IC design, they incur the least overhead. Helinski et al. [5] proposed one such PUF design which takes advantage of the resistance fluctuations in the metal lines of the Power Distribution System (PDS) of the IC, called Power Distribution PUF (PD-PUF). Since every chip needs power and ground connections to be activated, the power/metal lines in a grid formation are a very well-established design component in chip design. However, the challenge in implementing the PD-PUF is that the architecture that produces the key should not incur a large area overhead. The area overhead of a PD-PUF is confined to this challenge–response circuitry because the

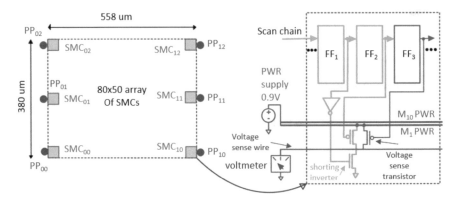

Fig. 3.3 (Left) Overview of location of stimulus measure circuits in power distribution PUF. (Right) Illustration of circuit architecture of stimulus measure circuit

power grid is a well-established component in every chip design. Furthermore, resistance changes of variable magnitudes are introduced by dispersed process variation effects across power grid regions. This property enhances the resilience of the PUF in such a way that PUFs from different ICs produce distinct responses.

In an IC, metal lines at the same metal layer are placed parallel to each other, either vertically or horizontally. From one metal layer to the next one, the metal lines run perpendicular to each other. Additionally, vias are placed in the intersections between different metal layers for routing. In a typical commercial chip design, multiple metal layers exist, and the width and granularity of these layers vary along with the layer depth. Usually, the width is wider, and the granularity is coarser in the upper metal layers than in the lower metal layers. This is because the metal layer width keeps increasing, and the granularity gets coarser as we go up in layers. The grid is connected to some Power Ports (PP) in the top metal layer. These ports allow power supply to be connected to the grid. As no two manufactured chips are the same owing to the process variation, the resistance and power distribution will vary for each chip across different power grid regions. PD-PUF uses this intrinsic power and resistance variance of a chip to identify the chip uniquely.

To measure the resistance of a point in the power grid, one needs to measure the voltage and current at that point. To measure the current, [5] used the Global Current Source Meter (GCSM), which provides 0.9 V to the power grid of the test chip. And to measure the voltage, [5] used a voltmeter connected to the pin, which is then internally connected to a global voltage sense wire. The IC used in [5] has six power ports, and under each power port, a measure circuit is inserted. Figure 3.3 shows the architecture of the measure circuit proposed in [5]. Each measure circuit consists of a shorting inverter, three scan flip-flops connected to the transistors, and a voltage sense transistor. The shorting inverter in the measure circuit is used as a stimulus to measure the voltage of the power grid. When flip-flops 1 and 2 are set to 0, the voltage sense transistor measures the power grid's voltage by enabling flip-flop three by setting it to 0 [14].

The generation of the secret signature using the PD-PUF can be done in two ways. The first strategy relies on measuring the voltage drop at the six measuring circuits embedded below the power pads. Voltage drop can be measured as a difference between the nominal supply voltage and the measured voltage at each measure circuit. However, since the measured voltage is the function of the current flowing through the shorting inverter, it is most susceptible to changes in environmental conditions. This affects the reliability of responses generated and thus making the PUF suitable only as a random number generator.

Therefore, the second method, which relies on the equivalent resistance, is proposed by Helinski et al. [5]. By dividing the recorded voltage drop by the global current, this approach eliminates the reliance on current and makes it resistant to environmental fluctuations. The PD-PUF proposed using only six measure circuits; however, multiple measure circuits can be embedded in a large commercial IC. For assessing the quality of PUF responses, [5] used 36 test chips and Euclidean distance between the pairs to find the probability of aliasing across chips. The results showed that the voltage drop method has an aliasing probability of 1 in 28 billion, whereas the equivalent resistance method has an aliasing probability of 1 in 15 million.

3.4 QUALPUF

QUALPUF is an abbreviation for Quasi-Adiabatic Logic-based PUF. The term adiabatic comes from thermodynamics, where adiabatic processes do not exchange heat with their surroundings. Similarly, in electronics, adiabatic logic effectively recycles the charge held in the load capacitor throughout each clock cycle. This logic is used to create ultra-low-power circuits. However, constructing a fully adiabatic circuit is complex, and the logic necessitates more area overhead than a conventional CMOS circuit. As direct intrinsic characterization PUFs aim at lowering power and area overhead, using a quasi-adiabatic circuit to design PUF makes more sense [19].

Adiabatic logic is utilized in ultra-low-power circuit design techniques to lower the overall power required by the circuit by efficiently re-purposing the charge stored in the design's load capacitance. Power clocks are used to recuperate the energy, which is then employed in the next computing cycle. The concept of the adiabatic circuit is illustrated in Fig. 3.4 (adapted from [10]). If a constant current source supplies the charge of the load capacitance, then the power dissipated in an adiabatic circuit will be denoted by

$$E_{diss} = \frac{RC}{T} C V_{dd}^2, \tag{3.8}$$

where C denotes the load capacitance, T indicates the discharging time of C, and V_{dd} is the power clock's full swing. The power dissipated by the adiabatic circuit will be lower than the CMOS circuit if $T > 2RC$.

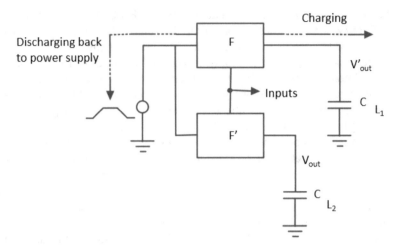

Fig. 3.4 Simple block diagram illustrating the concept of adiabatic circuit

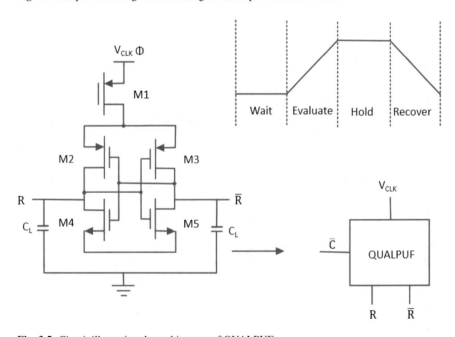

Fig. 3.5 Circuit illustrating the architecture of QUALPUF

Figure 3.5 shows the concept of Quasi-adiabatic circuit that can be implemented as a PUF, as proposed in [10]. The power clock that recovers energy is represented here by V_clk. The architecture employs five transistors, with transistor $M1$ serving as the sleep transistor and transistors $M2$, $M3$, $M4$, and $M5$ constituting the latch structure. The random bits are generated as a result of the threshold mismatch in this

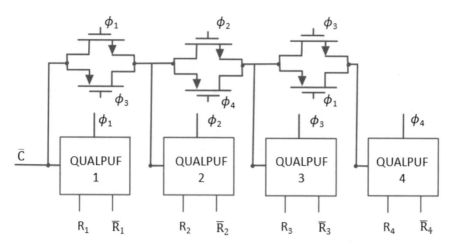

Fig. 3.6 Cascaded structure illustrating the architecture of PUF cells used to form local PUF

latch construction. The single challenge input is used to query the QUALPUF cell, which excites with two answer bits, R and $\bar{R}$.

Figure 3.5 shows the four different phases of V_{CLK}, i.e., the wait, evaluate, hold, and recover stage. The challenge bit $C = 0$ means that the QUALPUF cell is not queried with any challenge, and the cell remains inactive. Sleep transistor will be switched OFF when there is no challenge in the QUALPUF. The challenging bit will become 1 ($C = 1$) when it gets changed. The wait, evaluate, hold, and recover phase will sequentially occur when the cell is queried with this challenging bit.

The wait phase starts with V_{CLK} at GND. After that, $\bar{C}$ will slowly decrease from V_{DD} to 0. For PMOS to be turned on, the voltage across the source and gate (V_{sg}) must be greater than the PMOS threshold voltage (V_{tp}), meaning $V_{sg} > V_{tp}$. In the evaluate phase, V_{CLK} will increase from GND to V_{DD}. After $M1$ turns on, node 1 in Fig. 3.5 will increase from GND to V_{DD}. However, when node 1 rises to V_{tp}, $M2$ and $M3$ will be turned on. Although $M2$ and $M3$ are built to have the same V_{tp}, their manufacturing process variation will cause slight differences in their threshold voltages. The PMOS, which has the lower V_{tp}, will turn on first, making the other one flip (as the gate of the later one will increase to 1). This phenomenon creates the output where R is logic 1, and $\bar{R}$ is logic 0. The clock will hold the current output value during the hold phase. The stored charge in the load capacitor is recovered to V_{CLK} during the recovery phase via either $M2$ or $M3$ and $M1$.

To extract the unique IDs from any system, four QUALPUFs are cascaded together to form a local PUF (LP), which will create four random bits, as shown in Fig. 3.6. The QUALPUF will be set to 90 degrees phase difference with the next one. The 90 degrees difference in this part will be set as a single-phase difference of the V_{CLK} cycle. When QUALPUF 1 is in the waiting phase, QUALPUF 2 will be in evaluating phase, QUALPU3 will be in the hold phase, and the QUALPUF4 will be in the recovery phase (meaning QUALPUF1 will be in a 180 phase difference

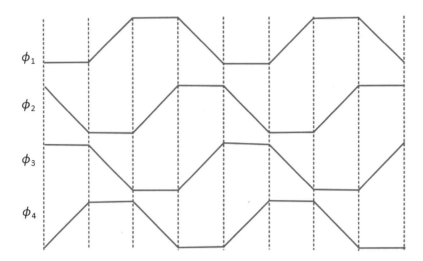

Fig. 3.7 Waveform illustrating the timing diagram of four-phase clocks

from QUALPUF3). It will create a 4 bits output. Because the QUALPUF 1 could be four phases, the output could be four different outputs, which has enough diversity than other PUF. Each local PUF is queried with a single challenge bit to determine the system's unique ID. Each local PUF is managed by four clocks, each having a 90-degree phase variation from the next. Figure 3.7 depicts the timing diagram of the four-phase clocks used to operate the LP cell.

The assessment results in [10] showed that the uniqueness and reliability of the QUALPUF are 40.5% and 96.2%, respectively, which is relatively lower than the other proposed delay-based PUFs in the previous chapter. However, QUALPUFs consume lower power than the others. It may also be used in low-power portable electronic devices like RFIDs (Radio Frequency Identification), wireless sensor nodes, and so on. Furthermore, QUALPUFs do not generate the constant bit in each phase, so the attacker cannot analyze its output by simply modeling it.

3.5 Via-PUF

Physical-based PUFs use responses obtained by a specific physical electric layer (metal, via, poly, etc.) as the unique ID. The one-bit PUF answer (0 or 1) is determined by the presence of an open or short circuit between distinct conducting layers. These responses are particularly dependable since the open and short circuit current and voltage characteristics are readily differentiated. Furthermore, due to external noises, these properties do not vary with time. To confirm open or short states between conducting layers, every chip manufacturer has predefined design guidelines. However, these design principles can be intricately violated to ensure

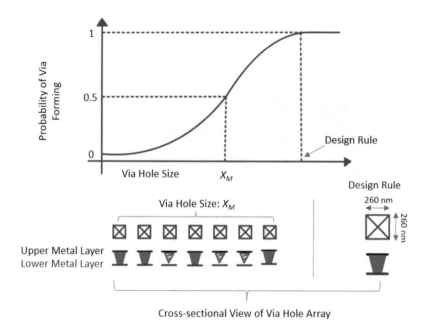

Fig. 3.8 Graph illustrating the probability of via as a function of via hole size

that the conducting layers will be open (or short) due to process changes. Via-PUF uses physical via in a probabilistic method to manipulate design rules so that process variation will result in specific open or short conducting layers.

A vertical connection between two different metal layers in ICs is called a via. When the connection is formed between two metal layers using via, the layers become short. To confirm via formation during fabrication, the manufacturer has standard design rules like regulating a minutest via hole size. If the via hole size during fabrication becomes less than the standard design rule, the likelihood of via formation becomes less than 1 (as shown in Fig. 3.8). A PUF can be generated, which will produce a random bit sequence that is based on the likelihood of via formation from an array of vias. Uniformity means that 50% of the response bit should be 1, whereas 50% should be 0. Further post-processing step is necessary to increase the uniformity of the response bit sequence. For uniformity improvement, 8-bit sequences are divided into two groups of 4-bit sequences, and a comparator compares the two groups to get an output response. However, this introduces overhead as eight via holes are required to generate one bit.

A response via-PUF can be obtained through ground PMOS, which works as the read transistor. A via is placed between a resistive load and the read transistor array. Figure 3.9 shows the overall architecture of response bit generation using the Via-PUF, as proposed in [9]. The row selection logic is connected to a row of transistors. The column selection logic is connected to the switches and sequentially activates them to connect PMOS memory cells to the load. When using the VIA-PUF, the

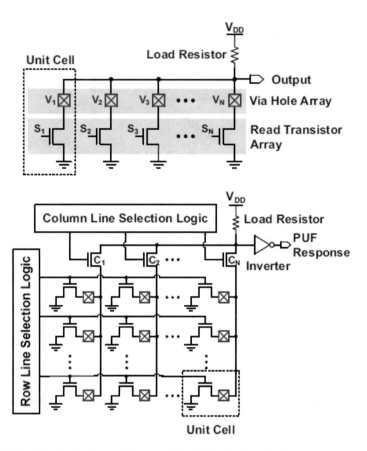

Fig. 3.9 Circuit illustrating the architecture of Via-PUF extraction circuit

response is determined by the inverter. For example, if the via is in an open state, the inverter outputs 0. On the other hand, when the via hole is in a short state, the inverter gives an output of 1. These outputs can be utilized as the response pair for PUF.

Via-PUF is built on the physical connection probability between two conducting layers. For the Via-PUF, the bit error rate is really low, and reliability is high. However, the post-processing on the original design is needed to achieve 50% uniformity, resulting in increased overhead. Therefore, Via-PUF can be used in device identification, authentication, or RFID (Radio Frequency Identification) key generation for its high uniformity, reliability, and low bit error rate.

3.6 Threshold Voltage PUF

In modern ICs, there can be billions of MOSFETs. The threshold voltage of these MOSFETs depends on the manufacturing process, which is usually the same for all in a die. However, one can observe slight variations in the threshold voltage of these MOSFETs due to process variation. It was observed that the quantity of impurity dopants could vary across MOSFETs, leading to variations in threshold voltage [17]. Lofstrom et al. [12] proposed to use the variations in threshold voltages of the MOSFETs as a source of entropy to develop PUF, called threshold voltage PUF (TV-PUF). TV-PUF can be implemented with an array of addressable MOSFETs, as shown in Fig. 3.10.

All the MOSFETs in TV-PUF have a common gate and source. The gates of the MOSFETs can be sequentially switched on, and while on, the gates drive the resistive load. Because of the process variation, the threshold voltage, as well as the drain current of individual MOSFETs, will be different. This results in a sequence of random voltage across the load resistance. These analog voltages are read sequentially (first E1, then E2, E3, and so on.) and then converted into a binary identification sequence with an auto zeroing comparator. This comparator compares successive random voltages to each other and produces a unique binary ID.

Hamming distance between two binary IDs refers to the number of bits that are different between them. Intra-hamming distance indicates how many bit differences a single device produces in multiple runs. A single device should always produce the same binary ID, and that is the reason the intra-hamming distance must be as small as possible (ideally 0%). This intra or "self" Hamming distance helps to uniquely identify a device. Inter-hamming distance indicates how much bit difference multiple devices produce in a single run. Different devices should have a different unique ID, so the inter-hamming distance must be close to 50%. If the inter-

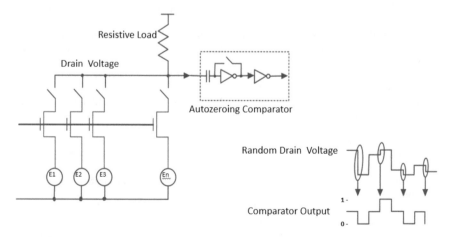

Fig. 3.10 Simple circuit illustrating the structure of threshold voltage PUF

hamming distance is less than or greater than 50%, that means the output binary ID is biased toward either 0 or 1. Lofstrom et al. [12] reported that the intra or "self" Hamming distance of the TV-PUF is low, whereas the inter or the "other" Hamming distance is averaging 55%, which makes TV-PUF reliable with high uniqueness.

Environmental noise, mobile ion contamination, and aging can change the threshold voltages of the MOSFET, thus affecting the reliability of TV-PUF. Some random drift can also change the sign of the bit, resulting in changed identification bits. Another issue with this method is that the IC sequence is not deterministic. Even though the intra and inter-Hamming distance of TV-PUF is very close to the ideal PUF, communication error can introduce a higher bit error rate. Error correction code is needed in that case scenario.

3.6.1 Conclusions

This chapter covered the basic working principles, architecture, applications, and limitations of different direct intrinsic characterization PUFs. This category of PUFs includes cellular neural network PUF, power distribution PUF, quasi-adiabatic logic-based PUF, VIA-PUF, and threshold voltage PUF. Direct intrinsic characterization PUFs do not require the external circuitry as they can rely on the chip's intrinsic properties to extract the source of entropy, thus, reducing the area overhead required to build security primitives.

References

1. Addabbo T, Fort A, Di Marco M, Pancioni L, Vignoli V (2013) Physically unclonable functions derived from cellular neural networks. IEEE Trans Circuits Syst I Regul Pap 60(12):3205–3214. DOI 10.1109/TCSI.2013.2255691
2. Amsaad F, Pundir N, Niamat M (2018) A dynamic area-efficient technique to enhance ROPUFs security against modeling attacks. In: Computer and Network Security Essentials. Springer, Berlin, pp 407–425
3. Guin U, DiMase D, Tehranipoor M (2014a) Counterfeit integrated circuits: Detection, avoidance, and the challenges ahead. J Electron Test 30(1):9–23
4. Guin U, Huang K, DiMase D, Carulli JM, Tehranipoor M, Makris Y (2014b) Counterfeit integrated circuits: A rising threat in the global semiconductor supply chain. Proc IEEE 102(8):1207–1228
5. Helinski R, Acharyya D, Plusquellic J (2009) A physical unclonable function defined using power distribution system equivalent resistance variations. In: 2009 46th ACM/IEEE Design Automation Conference. IEEE, New York, pp 676–681
6. Herder C, Yu MD, Koushanfar F, Devadas S (2014) Physical unclonable functions and applications: A tutorial. Proc IEEE 102(8):1126–1141
7. Hossain MM, Vashistha N, Allen J, Allen M, Farahmandi F, Rahman F, Tehranipoor M (2022) Thwarting counterfeit electronics by blockchain
8. Hu S, Wang J (2002) Global asymptotic stability and global exponential stability of continuous-time recurrent neural networks. IEEE Trans Autom Control 47(5):802–807

9. Kim T, Choi B, Kim D (2014) Zero bit error rate id generation circuit using via formation probability in 0.18 μm CMOS process. Electron Lett 50(12):876–877
10. Kumar SD, Thapliyal H (2016) QUALPUF: A novel quasi-adiabatic logic based physical unclonable function. In: Proceedings of the 11th Annual Cyber and Information Security Research Conference, pp 1–4
11. Lim D, Lee JW, Gassend B, Suh GE, Van Dijk M, Devadas S (2005) Extracting secret keys from integrated circuits. IEEE Trans Very Large Scale Integr VLSI Syst 13(10):1200–1205
12. Lofstrom K, Daasch WR, Taylor D (2000) IC identification circuit using device mismatch. In: 2000 IEEE International Solid-State Circuits Conference. Digest of Technical Papers (Cat. No. 00CH37056). IEEE, New York, pp 372–373
13. Maiti A, Kim I, Schaumont P (2011) A robust physical unclonable function with enhanced challenge-response set. IEEE Trans Inf Forensics Secur 7(1):333–345
14. Rahman MS, Nahiyan A, Amir S, Rahman F, Farahmandi F, Forte D, Tehranipoor M (2019) Dynamically obfuscated scan chain to resist oracle-guided attacks on logic locked design. Cryptology ePrint Archive
15. Rahman MT, Forte D, Fahrny J, Tehranipoor M (2014) ARO-PUF: An aging-resistant ring oscillator PUF design. In: 2014 Design, Automation and Test in Europe Conference and Exhibition (DATE). IEEE, New York, pp 1–6
16. Suh GE, Devadas S (2007) Physical unclonable functions for device authentication and secret key generation. In: 2007 44th ACM/IEEE Design Automation Conference. IEEE, New York, pp 9–14
17. Tang X, De VK, Meindl JD (1997) Intrinsic MOSFET parameter fluctuations due to random dopant placement. IEEE Trans Very Large Scale Integr VLSI Syst 5(4):369–376
18. Tehranipoor M, Peng K, Chakrabarty K (2011) Introduction to VLSI testing. In: Test and Diagnosis for Small-Delay Defects. Springer, New York, pp 1–19
19. Teichmann P (2011) Adiabatic logic: future trend and system level perspective, vol 34. Springer, New York
20. Vashistha N, Lu H, Shi Q, Woodard DL, Asadizanjani N, Tehranipoor M (2021) Detecting hardware Trojans using combined self-testing and imaging. IEEE Trans Comput Aided Des Integr Circuits Syst 41(6):1730–1743
21. Wang H, Forte D, Tehranipoor MM, Shi Q (2017) Probing attacks on integrated circuits: Challenges and research opportunities. IEEE Des. Test 34(5):63–71

Chapter 4
Volatile Memory-Based PUF

4.1 Introduction

Globalization, cost, and competitive market have forced global semiconductor companies to switch from vertical supply chain to a horizontal supply chain [27]. Earlier, a single entity was responsible for the chip's design, fabrication, and packaging in a vertical supply chain. However, in current days, most tasks after the completion of the design phase are outsourced overseas to reduce fabrication and packaging costs. The integrated circuit (IC) design house sends the final design, also called intellectual property (IP), to the fabrication house, which fabricates, packages, and sends it to the distributor. Then printed circuit board (PCB) or system integrator sources different ICs from distributors and put all of them together to build a system and sells it in the market. In this entire supply chain, the fabrication house has complete access to the design IP and the test vectors to ensure the chip functions correctly after fabrication [8, 29, 30, 32]. Therefore, a fabrication house is fully capable of performing adversarial practices to have financial gains or to have/provide a competitive advantage. For example, it can modify an IP to insert hardware Trojans to cause safety and reliability issues in the systems where IC is used [9, 15, 23, 35]. It can overproduce ICs or do IP piracy to have financial gains [13, 14, 31]. And in extreme cases, it can sell low-grade or faulty, or counterfeit ICs in the market through unauthorized channels [28, 33]. The adversarial activities by a malicious foundry could cause significant earning and reputation loss to the IP owners and IC design houses. Therefore, IP owners resort to several security measures to protect against overproduction, IP piracy, and counterfeiting. One of the solutions is to verify the authenticity of the IC using a secret encrypted key stored in the EEPROMs (non-volatile memory (NVM)) and Flash [12]. However, such keys which are memory-based are at risk to invasive attacks (for example, optical and nano probing attacks) [19, 34]. Moreover, such invasive attacks can also be performed while the device is not operating [2].

© The Author(s), under exclusive license to Springer Nature Switzerland AG 2023
M. Tehranipoor et al., *Hardware Security Primitives*,
https://doi.org/10.1007/978-3-031-19185-5_4

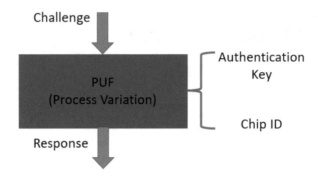

Fig. 4.1 Use of Physical Unclonable Functions for Chip Authentication and Chip ID generation

Prominent security measures against hardware Trojans, overproduction, piracy, and counterfeiting is a physical unclonable function (PUF) [4, 5, 18, 35]. Generation of unique challenge–response pairs (CRPs) takes place by leveraging process variations. These unique pairs can be used for authentication and generating unique chip IDs, as shown in Fig. 4.1. Due to the fabrication process, random features of the transistors, such as delay, are used to generate unique volatile secret keys. The keys are generated only at the time of check, thus, protecting against probing attacks. Furthermore, since the key generation exploits the process variation in the fabrication, it is difficult to predict or clone the keys, thus presenting a promising hardware-based security solution [24].

Based on the process variation of the chip, PUFs map a unique collection of challenges to a unique set of responses [24]. This set of CRPs for each device can be stored in a database after fabrication and used by IC design houses for chip verification anytime. The design house can validate the authenticity of the ICs by sending a challenge to the chip/user and asking for a response generated by the chip. To authenticate the chip, the received response is compared to the responses saved in the design house's database [1, 17]. PUFs have unique characteristics, making them a suitable security primitive against various supply chain threats. For example, they are immune to cloning and replay attacks. Due to random process variation, an attacker cannot predict their response before or after production. Some strong PUF candidates can generate a large sample space of CRPs; therefore, to carry out a successful attack, an attacker would need to leak all the CRPs. And finally, in PUF, all secret keys are generated at the time of check. This prevents against traditional non-volatile memory-based attacks where an adversary tries to leak the key when device is not in use [24].

Literature review describes many different types of PUFs. This chapter specifically discusses the volatile memory-based PUFs such as Bistable Ring PUF [3], SRAM PUF [10], MECCA PUF [17], DRAM PUFs [25], and Rowhammer PUFs [21]. It focuses on the novelty of these security primitives, the pros and cons of each proposed design, the efficiency and efficacy of the proposed methods, and, finally, the possible solutions to design challenges. The chapter is organized as follows. Section 4.2.1 discusses the PUFs classification and the metrics of the PUFs' performance. Section 4.3 describes the analysis of the following PUFs in order: Bistable Ring PUF, DRAM PUFs, MECCA PUFs, Intrinsic Rowhammer PUFs, and SRAM Random Address Error-based Chip ID Generation. Finally, Sect. 4.4 concludes with a summary of PUFs discussed.

4.2 Background

4.2.1 PUF Performance Evaluation Metrics

In this section, we discuss metrics based on performance used to determine the quality of PUFs. These metrics include uniqueness, randomness, and reliability.

Uniqueness refers to how distinct CRPs generated by a PUF is across different chips. Since CRPs from a PUF are used to identify the chip, CRPs from different chips should be very different from each other to allow easy authentication of chips. Calculating the inter-PUF/chip Hamming distance (HD) ratio is a common way to assess a PUF's uniqueness. Multiple PUFs/chips are used to perform such assessment and the formula is given by Scholz et al. [22]

$$HD_{inter} = \frac{2}{n(n-1)} \sum_{i=1}^{n-2} \sum_{j=i+1}^{n} \frac{HD(R_i, R_j)}{k} \times 100\%, \tag{4.1}$$

where n stands for number of PUFs under assessment, k is the response bit length, R_i and R_j are the response from PUF_i and PUF_j. $HD(R_i, R_j)$ is the hamming distance of response of PUF_i and PUF_j. For the ideal case, the inter-PUF hamming distance should be 50%, considering the maximum differences between the PUF responses of the two chips.

Randomness or Uniformity refers to the unpredictability of a PUF by ensuring that the response bits generated are completely random. The statistical tests should prove that the response of a PUF is not biased and should have an equal distribution of "1" and "0" distributed randomly [11].

Reproducibility or Reliability of PUF refers to the generation of the same CRP across different environmental conditions and times. Typically, measurement is performed concerning intra-Hamming distance, which is defined as follows:

$$HD_{intra} = \frac{1}{m} \sum_{y=1}^{m} \frac{HD(R_i, R'_{i,y})}{k} \times 100\%, \tag{4.2}$$

where m represents the number of samples of a PUF, k is the response bit length, $HD(R_i, R'_j)$ refers to the hamming distance between the response R_i and the yth response of the response bit $R'_{i,y}$ of the PUF under assessment. Ideally, a PUF should generate the same challenge pair response, meaning the ideal intra-hamming distance should be 0%.

4.3 Comparative Analysis of Volatile Memory-Based PUF

This section elaborates on different volatile memory-based PUFs proposed in the literature and presents their performance metrics.

4.3.1 Bistable Ring PUF

Since a bistable ring consists of inverters that are even in number [3], it has two possible stable states. For example, in a 64-bit bistable ring, the two stable states are $0xAAA...AA$ (101010...10) and $0x555...55$ (010101...01). On the ring's power-up, all inverters try to take the high voltage "1" state; however, it is not possible due to the inverter structure. Therefore, even if one-bit stables, then because of inverter chain structure, all bits stables into an alternating pattern.

A bistable ring in itself is of no use because it has only two stable states. Therefore, to create a large amount of CRPs and create a bistable ring PUF (BR-PUF), [3] proposed changes to the bistable ring, as shown in Fig. 4.2. First, each inverter in the ring is replicated, and a MUX and DEMUX are added to select the inverter. Then, each inverter is replaced with a 2-input NOR or NAND gate to accommodate for reset signal. This way, the challenge to the BR-PUF can select different combinations of NOR/NAND gates. Therefore, for an n-stage BR-PUF, 2^n different rings can be created with the help of challenges.

Settling time is very crucial for the BR-PUFs. The time needed for a BR-PUF to reach a stable response state after handling challenges is called the settling time. Figure 4.3 shows the concept of a settling time in a nutshell. There are two ways to measure the evaluation time for PUF response. External circuitry is one option, while setting a standard evaluation period for all responses is another. It was found that the average and max settling time for a 64-bit BR-PUF was 10.7 μs and 22.2

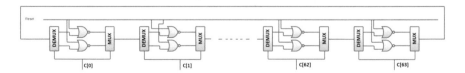

Fig. 4.2 Illustration of architecture of a bistable ring PUF with 64-bit stages

Fig. 4.3 Block diagram illustrating the concept of settling time in BR-PUF

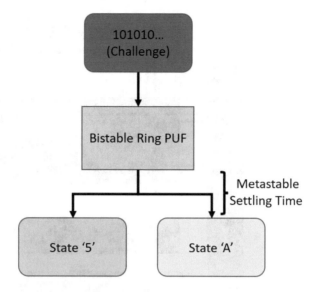

μs, respectively. Therefore, the maximum settling time was chosen as the standard evaluation period for calculating the PUF response.

In [3], it was noted that challenges with short settling time were the most reliable but were also constant across chips. Therefore, if these CRPs are used for analysis, it can significantly bring down the inter-HD metric, affecting the uniqueness of the BR-PUF. On the contrary, challenges with longer settling were less reliable; however, they had more uniqueness, closer to 50%. Therefore, the CRPs with longer settling times should be used for the identification and authentication of chips. In summary, it was noted that BR-PUFs could generate an exponentially large number of CRPs with high uniqueness and reliability metrics. And, no modeling attacks are yet demonstrated against BR-PUFs because they would be difficult to model due to the complex nature compared to standard delay-based PUFs. Similarly, variations in the settling time allow for separating different CRPs for different purposes. And, long settling time causes longer read-out times, which makes it difficult for an attacker to exploit the PUF [20].

4.3.2 DRAM-Based Intrinsic PUF

Tehranipoor et al. [25] proposed the use of DRAM for the generation of PUFs. Since there is a large availability of bits in a DRAM, a large sample space of CRPs can be generated. Moreover, the cost of DRAM is significantly lower than SRAMs and found abundantly in modern computing systems, thus, making them a perfect candidate for PUFs.

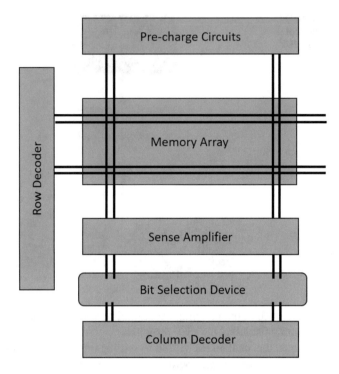

Fig. 4.4 Block diagram illustrating one-transistor DRAM array structure

Before [16, 25] proposed the use of DRAM PUFs-based on disabling and altering the refresh cycle of DRAMs. The basic idea behind refresh DRAM PUFs was to initialize all cells to "1" and then disable the refresh cycle. However, since DRAM cells leak charge; therefore, after some time, some cells will discharge to "0." However, the challenging issue is that sometimes it may take several minutes to hours for a significant amount of cells to discharge to "0."

In [25], it was observed that DRAMs show similar start-up properties as that of SRAM. That is, not all DRAM cells initialize to "0," and this randomness in start-up values can be exploited to create PUF. Figure 4.4 depicts a typical DRAM array structure and the peripheral circuitry of the memory structure. The process of generating the PUF bits starts by initializing the memory cells, or in other words, the voltage stored in the capacitor. The cells are precharged to VDD/2 (PEQ is the signal to precharge the bit lines). When the word line is active, the DRAM cells are read by setting the bit line to VDD/2, and the bit-line voltage changes slightly based on the storage capacitor's capacitance. This is because the amount of stored voltage in a capacitor will change depending on inherent process variations. The sense amplifier detects this slight deviation from the precharge voltage as a "1" or "0," depending on how much the capacitor has discharged. As a result, the sense amplifier is fairly likely to register a "1" or a "0." This characteristic permits the DRAM's initial values to act as a PUF.

To evaluate the metrics based on performance of the DRAM PUF, [25] used eight different 1 MBit Hitachi HM51100AL CMOS DRAMs in a DIP package. The uniformity of the bits produced is close to ideal. However, it was seen that slightly more than half the bits were "1" [25]. Furthermore, the results showed poor uniqueness when random PUF bits were selected because of spatial bias in the DRAM array. However, a selection bit algorithm can be used to address the issue. If only stable bits are selected using the algorithm, an inter-chip hamming distance close to 50% could be achieved. It was shown that strong randomness with 99% confidence could be achieved while testing the generated CRPs with the NIST suite. The reliability of the proposed PUF scheme was evaluated under different conditions: normal condition (NC), accelerated aging (AA), low voltage (LV), high voltage (HV), low temperature (LT), and high temperature (HT) [25]. It was seen that the effect of voltage variation has more impact than changes in other operating conditions, like aging or temperature.

4.3.3 MECCA PUF

(ME)emory (C)ell-based (C)hip (A)uthentication PUF exploits the failure mechanism in the SRAM cell arrays to create the PUF [17]. The proposed PUF controls the word line duty cycle to determine the failure probability of the cells during reading and write access. The control over the word line duty cycle allows the generation of a large sample space of CRPs, thus qualifying as a strong PUF. Random process variation can influence the failure probability of the cells, which is then translated into a digital response in MECCA PUF.

Figure 4.5 depicts the architecture of the MECCA SRAM PUF in the form of a block diagram. The PUF contains a programmable delay generator and other SRAM arrays and peripherals. The programmable world line duty controller exists at the clocking and control part. In Fig. 4.6 a detailed circuit diagram of the duty controller is shown. By selecting different multiplexer switches, the circuit creates the different duty cycles for the word line (WL).

During memory cell access, four types of failures can be observed:

- Write Failure: This occurs when an SRAM cell fails to discharge during word line activation.
- Read Failure: During the read operation the SRAM cell data is flipped.
- Access Failure: Difference between the bit lines in terms of voltage is less than the offset voltage of the sense amplifier.
- Hold Failure: When standby mode has low supply voltage, it leads to the internal node voltage to fall below the inverter's switching threshold.

The reliability of a single SRAM cell is assessed in MECCA PUF by causing a write failure. This is achieved by modifying the word line duration. For operating normally, the duration of the word line is selected in a way that it passes across all process variations. However, when used as a PUF, the stable cells may not function

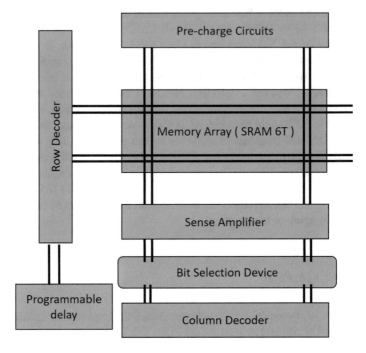

Fig. 4.5 Block diagram illustrating the architecture of a MECCA PUF with programmable delay to control the word line duration

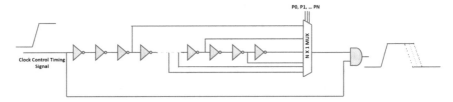

Fig. 4.6 Illustration of programmable controller circuit used to control the word line duty cycle

properly in modified word line duration due to process variation. For this, [17] uses a programmable word line delay that shortens the word line causing random failures of the cells. Furthermore, read and hold failures are not used by the MECCA PUF as they are static in nature and hence, cannot be controlled using word line duration. Following is the summary of steps performed to create a unique signature:

- Address of n-cells is used as a challenge.
- To ensure cells are stable, normal write operation is performed. It also helps to ensure the cells are in a known initial state.
- Word line is shortened using programmable delay, and write operation is performed.
- N-bit response is generated by reading out the values at n-cells.

Overall, the method is inexpensive as it has less area overhead compared to other PUF methods because in the SoC it uses the existing SRAM cells. Moreover, the uniqueness of the MECCA PUF is found to be close to 49.9% from 1000 chips. And the worst-case reliability (intra-HD) at 0.8 V was close to 18% (ideal 0%). Reliability during varied temperatures showed that 93.3% responses flipped with less than 3 bits for 1000 chips. Only 6% of all cells fall within the guard band for the aging effect, producing an inaccurate output during the product's lifetime. In comparison to ROPUFs (delay-based PUFs), MECCA PUF has 16.6% less area overhead.

4.3.4 Intrinsic Rowhammer PUF

Schaller et al. [21] proposed to use the bit flips caused in DRAMs due to the Rowhammer effect as a source of randomness to create PUF. The Rowhammer effect occurs when a memory row is accessed quickly and repeatedly, causing neighboring cells to discharge quickly, generating bit flips. However, the effect has been primarily used to attack a computer system's security by modifying sensitive memory contents [6]. The proven stable spatial distribution of the effect has motivated its use as reliable unclonable security primitive. Multiple rows, referred to as hammer rows, are accessed repeatedly, and changes in the adjacent rows and termed PUF rows, are observed for bit flips.

DRAM cells are most susceptible to the Rowhammer effect because it stores bit as a charge on a capacitor. The charge in the capacitor leaks over time, causing the data to flip. To prevent this data loss, the word lines of each row of DRAM should be periodically accessed so that sense-amplifiers can recharge the capacitors of that row. This entire phenomenon is known as DRAM refresh and is periodically done every 32 to 64 ms. If the discharge rate due to repeated writes/reads exceeds the DRAM refresh rate, one can notice the bit flips due to the Rowhammer effect.

However, the Rowhammer effect can be limited to higher refresh rates, thus providing low entropy for a strong PUF. To circumvent this, [21] proposed three techniques. First, to disable the DRAM refresh of the memory cells which are to be used a PUF. Second, to use multiple rows as rowhammer PUF. And finally, to control the initial values of DRAM cells and the hammering time to induce maximum bit flips. Figure 4.7 shows two Rowhammer types to achieve maximum bit flips.

As the charge state-logic representation may differ across DRAM (True Cells represent "1" in the charged state, whereas anti-cells represent "0" in the charged state), the initializing value of the rows (to ensure a charged state of the cells) is a pre-requisite to ensure the maximum number of bit flips. Based on the location of the bit flips (which varies across modules due to process variation-induced coupling among DRAM cells), a unique PUF fingerprint can be extracted. The uniqueness of this spatial distribution of the flipping cells can be evaluated using the Jaccard index [26], a widely used statistic used to understand the similarities between sample sets in computer vision. This metric focus on the similarity of flip bit locations between

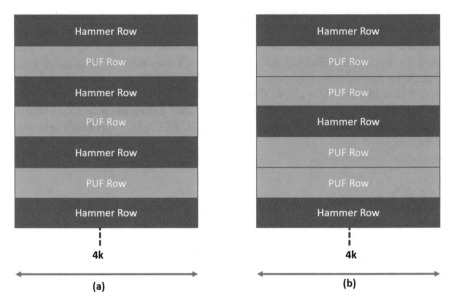

Fig. 4.7 Illustration of different Rowhammer types (**a**) Double-sided Rowhammer (**b**) Single-sided Rowhammer

the set of indices of flipped bits can be calculated as $J(S_1, S_2) = |S_1 * S_2|/|S_1 + S_2|$, where S_x denotes the set of indices for flipped bit attributed to a Rowhammer PUF measurement.

Schaller et al. [21] evaluated for uniqueness, reliability, and entropy, following an exploratory approach (as true-cell and anti-cell distribution were not known) by assessing all potential parameters that yielded the most significant impact with rowhammering. The number of bit flips were primarily influenced by the initial values of the rows, meaning more manifestation/exposition of a hammering generated distinctive bit flip. As the PUF draws its characteristics from bit flip locations, the discernible histogram plot of the Jaccard index of the inter-chip and intra-chip PUF measurements substantiates Rowhammer PUF's feasibility in Device Authentication. Also, the PUF showed a similar Jaccard index (less than 2%) across temperature variations, proving its robustness.

4.3.5 SRAM Random Address Error-Based Chip ID Generation

SRAM cells can be used to generate a unique fingerprint of an IC, i.e., a chip ID [10]. Furthermore, these SRAM cells can generate a random bit pattern for the chip ID by causing random bit failures by lowering static-noise-margin (SNM). Although there have been many designs for SRAM PUFs, [10] design is based on the efficient

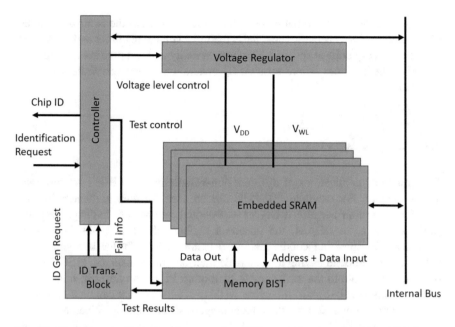

Fig. 4.8 Block diagram illustrating the generation of ID translation from the chip's fail information

retrieval of stable and reliable data from an SRAM array as a chip ID during an authentication protocol. Since the threshold voltage of the six transistors that form the SRAM varies due to process variations, only specific cells show a fault at a particular word line voltage. Therefore, a chip ID is generated by identifying these faulty bits and their addresses.

Figure 4.8 shows a test chip comprising of an SRAM, an MBIST (memory built-in self-test), an ID translation block, a voltage regulator, and a controller. The test voltage is first initialized by the voltage regulator, following which the memory built-in self-test detects failing addresses, and the number of fail bits (FB) is counted. If the number of fault bits is not in the desirable range, the voltage regulator changes the voltage level and re-reads the fault bits. A suitable test voltage is determined using iteration, and the word line voltage is set to that value [7]. After setting the word line voltage to a suitable value, several faults are triggered in the SRAM cells in random addresses. Finally, from the memory built-in self-test translation of the faulty bits of information is performed by the ID translation block to the 128-bit chip ID.

A modified ID was proposed to enhance the uniqueness of the ID creation process by incorporating both the failure address and the failure data input/output bit location. The measurement results indicated that for 512 test iterations, the stability of the generated chip ID was 99.9%. And the test time was also found to be 737 μs, which is considerably small to be eligible for practical purposes [10].

In the proposed method there are area overheads (such as the controller and the ID translation block) because most SoCs feature a voltage regulator and SRAM IPs with the memory built-in self-test. However, stability and test time are not mutually exclusive. In instances where a quick ID generation is necessary, the scheme is unsuitable.

4.4 Conclusions

The limits and applications of different non-volatile memory PUFs are discussed in this chapter. For example, BR-PUF can be used for chip authentication and chip ID generation because it has an exponential number of challenge–response pairings. Again, in DRAM, the proposed method of creating bits is sensitive to changes in operating conditions. As a result, it is unfit for key generation. Furthermore, although having a reasonable inter-hamming distance, MECCA PUF is more susceptible to the environment than other PUFs. As a result, it should be utilized in chips with a short life cycle. Making it age-resistant can be a good idea, but it will probably cost more than the non-age-resistant version. When comparing RO (delay-based PUFs) and MECCA PUF, it is clear that MECCA PUF has 16.6% less area overhead than RO-PUFs. Finally, there is a trade-off between stability and test time in SRAM Random Address Error-based Chip ID Generation. In some instances where a quick ID generation is necessary, the scheme is unsuitable. As a result of these non-volatile PUF ideas, specialized PUFs can be created for specific applications. Since memory makes up most of today's SoCs, intrinsic non-volatile memory has the most potential for chip authentication and security among all other PUFs due to its low cost and efficiency.

References

1. Amsaad F, Pundir N, Niamat M (2018) A dynamic area-efficient technique to enhance ROPUFs security against modeling attacks. In: Computer and Network Security Essentials. Springer, Berlin, pp 407–425
2. Asadizanjani N, Rahman MT, Tehranipoor M (2021) Optical inspection and attacks. In: Physical Assurance. Springer, Cham, pp 133–153
3. Chen Q, Csaba G, Lugli P, Schlichtmann U, Rührmair U (2011) The bistable ring PUF: A new architecture for strong physical unclonable functions. In 2011 IEEE International Symposium on Hardware-Oriented Security and Trust, HOST 2011 (August 2019), pp 134–141. https://doi.org/10.1109/HST.2011.5955011
4. Chowdhury S, Xu X, Tehranipoor M, Forte D (2017) Aging resilient RO PUF with increased reliability in FPGA. In: 2017 International Conference on ReConFigurable Computing and FPGAs (ReConFig). IEEE, New York, pp 1–7
5. Cruz J, Farahmandi F, Ahmed A, Mishra P (2018) Hardware Trojan detection using ATPG and model checking. In: 2018 31st international conference on VLSI design and 2018 17th international conference on embedded systems (VLSID). IEEE, New York, pp 91–96

6. Du M, Liu N, Hu X (2019) Techniques for interpretable machine learning. Commun ACM 63(1):68–77. https://doi.org/10.1145/3359786
7. Farahmandi F, Huang Y, Mishra P (2020a) Automated test generation for detection of malicious functionality. In: System-on-Chip Security. Springer, Cham, pp 153–171
8. Farahmandi F, Huang Y, Mishra P (2020b) Hardware Trojan detection schemes using path delay and side-channel analysis. In: System-on-Chip Security. Springer, Cham, pp 221–271
9. Farahmandi F, Huang Y, Mishra P (2020c) Trojan detection using machine learning. In: System-on-Chip Security. Springer, Cham, pp 173–188
10. Fujiwara H, Yabuuchi M, Nakano H, Kawai H, Nii K, Arimoto K (2011) A chip-ID generating circuit for dependable LSI using random address errors on embedded SRAM and on-chip memory BIST. IEEE Symposium on VLSI Circuits, Digest of Technical Papers 5:76–77
11. Georgescu C, Simion E, Nita AP, Toma A (2017) A view on NIST randomness tests (in) dependence. In: 2017 9th International Conference on Electronics, Computers and Artificial Intelligence (ECAI). IEEE, New York, pp 1–4
12. Guin U, DiMase D, Tehranipoor M (2014a) Counterfeit integrated circuits: Detection, avoidance, and the challenges ahead. J Electron Test 30(1):9–23
13. Guin U, Huang K, DiMase D, Carulli JM, Tehranipoor M, Makris Y (2014b) Counterfeit integrated circuits: A rising threat in the global semiconductor supply chain. Proc IEEE 102(8):1207–1228
14. Hossain MM, Vashistha N, Allen J, Allen M, Farahmandi F, Rahman F, Tehranipoor M (2022) Thwarting counterfeit electronics by blockchain
15. Karri R, Rajendran J, Rosenfeld K, Tehranipoor M (2010) Trustworthy hardware: Identifying and classifying hardware Trojans. Computer 43(10):39–46
16. Keller C, Gürkaynak F, Kaeslin H, Felber N (2014) Dynamic memory-based physically unclonable function for the generation of unique identifiers and true random numbers. In: 2014 IEEE international symposium on circuits and systems (ISCAS). IEEE, New York, pp 2740–2743
17. Krishna AR, Narasimhan S, Wang X, Bhunia S (2011) Mecca: A robust low-overhead PUF using embedded memory array. In: Preneel B, Takagi T (eds) Cryptographic Hardware and Embedded Systems—CHES 2011. Springer, Berlin, pp 407–420
18. Pundir N, Amsaad F, Choudhury M, Niamat M (2017) Novel technique to improve strength of weak arbiter PUF. In: 2017 IEEE 60th International Midwest Symposium on Circuits and Systems (MWSCAS). IEEE, New York, pp 1532–1535
19. Rahman MT, Tajik S, Rahman MS, Tehranipoor M, Asadizanjani N (2020) The key is left under the mat: On the inappropriate security assumption of logic locking schemes. In: 2020 IEEE International Symposium on Hardware Oriented Security and Trust (HOST). IEEE, New York, pp 262–272
20. Rührmair U, Jaeger C, Bator M, Stutzmann M, Lugli P, Csaba G (2010) Applications of high-capacity crossbar memories in cryptography. IEEE Trans Nanotechnol 10(3):489–498
21. Schaller A, Xiong W, Anagnostopoulos NA, Saleem MU, Gabmeyer S, Katzenbeisser S, Szefer J (2017) Intrinsic rowhammer PUFs: Leveraging the rowhammer effect for improved security. In: 2017 IEEE International Symposium on Hardware Oriented Security and Trust (HOST). IEEE, New York, pp 1–7
22. Scholz A, Zimmermann L, Sikora A, Tahoori MB, Aghassi-Hagmann J (2020) Embedded analog physical unclonable function system to extract reliable and unique security keys. Appl Sci 10(3), 759. https://doi.org/10.3390/app10030759. https://www.mdpi.com/2076-3417/10/3/759
23. Shi Q, Vashistha N, Lu H, Shen H, Tehranipoor B, Woodard DL, Asadizanjani N (2019) Golden gates: A new hybrid approach for rapid hardware Trojan detection using testing and imaging. In: 2019 IEEE International Symposium on Hardware Oriented Security and Trust (HOST). IEEE, New York, pp 61–71
24. Suh GE, Devadas S (2007) Physical unclonable functions for device authentication and secret key generation. In: 2007 44th ACM/IEEE Design Automation Conference, pp 9–14

25. Tehranipoor F, Karimian N, Xiao K, Chandy J (2015) DRAM based intrinsic Physical Unclonable Functions for system level security. Proceedings of the ACM Great Lakes Symposium on VLSI, GLSVLSI 20-22-May-2015:15–20. https://doi.org/10.1145/2742060.2742069

26. Tehranipoor F, Karimian N, Xiao K, Chandy J (2015) Dram based intrinsic physical unclonable functions for system level security. In: Proceedings of the 25th Edition on Great Lakes Symposium on VLSI, Association for Computing Machinery, New York, NY, USA, GLSVLSI '15, p 15–20. https://doi.org/10.1145/2742060.2742069. https://doi.org/10.1145/2742060.2742069

27. Tehranipoor M, Koushanfar F (2010) A survey of hardware Trojan taxonomy and detection. IEEE Des Test Comput 27(1):10–25

28. Tehranipoor MM, Guin U, Forte D (2015c) Counterfeit test coverage: An assessment of current counterfeit detection methods. In: Counterfeit Integrated Circuits. Springer, Cham, pp 109–131

29. Vashistha N, Lu H, Shi Q, Rahman MT, Shen H, Woodard DL, Asadizanjani N, Tehranipoor M (2018a) Trojan scanner: Detecting hardware Trojans with rapid SEM imaging combined with image processing and machine learning. In: ISTFA 2018: Proceedings from the 44th International Symposium for Testing and Failure Analysis, ASM International, p 256

30. Vashistha N, Rahman MT, Shen H, Woodard DL, Asadizanjani N, Tehranipoor M (2018b) Detecting hardware Trojans inserted by untrusted foundry using physical inspection and advanced image processing. Journal of Hardware and Systems Security 2(4):333–344

31. Vashistha N, Hossain MM, Shahriar MR, Farahmandi F, Rahman F, Tehranipoor M (2021a) eChain: A blockchain-enabled ecosystem for electronic device authenticity verification. IEEE Trans Consum Electron 68(1):23–37

32. Vashistha N, Lu H, Shi Q, Woodard DL, Asadizanjani N, Tehranipoor M (2021b) Detecting hardware Trojans using combined self-testing and imaging. IEEE Trans Comput Aided Des Integr Circuits Syst 41(6):1730–1743

33. Villasenor J, Tehranipoor M (2013) Chop shop electronics. IEEE Spectr 50(10):41–45

34. Wang H, Forte D, Tehranipoor MM, Shi Q (2017) Probing attacks on integrated circuits: Challenges and research opportunities. IEEE Des Test 34(5):63–71

35. Wang X, Tehranipoor M, Plusquellic J (2008) Detecting malicious inclusions in secure hardware: Challenges and solutions. In: 2008 IEEE International Workshop on Hardware-Oriented Security and Trust. IEEE, New York, pp 15–19

Chapter 5
Extrinsic Direct Characterization PUF

5.1 Introduction

A counterfeit chip is an electronic component that deviates from set standards in terms of features, performance, or material quality while being advertised and sold as the genuine version [9, 10]. Examples of counterfeit include unauthorized copies, remarked or recycled dies, cloned designs, or failed real parts [5, 31]. Counterfeit chips are a big concern due to the size of the global electronic industry. According to [26], it is valued at about 5.5 trillion dollars. Of this number, the counterfeit industry accounts for a huge 75 billion dollars [26]. Due to the recent global chip shortage, [19], one of the many effects of the COVID-19 pandemic, it is expected that the counterfeiting chip market share will be on the rise. It is imperative to make sure that counterfeit chips do not find their way into consumer electronics, and if they do, it will be a breach of the privacy of the individuals, as well as a threat to national security. However, there are ways to identify counterfeit electronic parts [2, 8, 9, 39]. But the average consumer may not be able to go to such lengths to verify the authenticity of electronic components. Apart from the lack of awareness, equipment unavailability reduces the likelihood of identifying a compromised device. The concept of physical unclonable function (PUF) comes into the picture as a countermeasure to counterfeit ICs. PUFs confirm the authenticity of the electronic ICs as they have a unique and strong correlation with the IC's physical structures [18].

One major feature in the working principle of PUFs is how randomness arises. It can be natural or explicitly added [23]. Characterization of PUFs is based on the source of entropy, i.e., internal or external randomness. Extrinsic characterization is done on the PUFs built with extrinsic components for evaluation. Direct extrinsic characterization attempts to exploit non-intrinsic components in the production of PUFs. The success of this method expands the possibilities of PUF-based security for devices that do not produce the amounts of variation required intrinsically for PUF creation.

© The Author(s), under exclusive license to Springer Nature Switzerland AG 2023
M. Tehranipoor et al., *Hardware Security Primitives*,
https://doi.org/10.1007/978-3-031-19185-5_5

The characterizations described in the chapter are inherently different from the conventional PUFs. The first evaluation involves the potential of diode-based PUFs [29]. Here, a novel design of PUF using ALILE diodes is demonstrated. These diodes are fabricated in special ways that introduce process variation, suitable for a PUF. The second characterization showcases nano-electromechanical (NEM)-based PUFs [16]. NEMs have strong static friction (stiction) that allows for strong adhesion in the switch, thus making the PUF more robust to harsh environmental conditions. The third characterization presented in this chapter is done on carbon nanotube-based PUFs [15]. A ternary bit map is created first instead of a binary bit map. This PUF can cancel the need for error-correcting hardware, common in silicon-based security primitives. The nanotubes' unique characteristics have the potential to make the ICs side-channel attack(SCA) resilient. The fourth characterization technique utilizes the MEMS accelerometer sensor and its unique characteristic for PUFs. The MEMS accelerometer sensor is wildly available and has a reasonable price which makes it suitable for low-cost system integration [4]. Finally, the fifth evaluation technique discusses PUFs using capacitance in printed circuit boards(PCBs). Capacitor banks are embedded in the PCBs, and their variations are utilized for PUF key generations [36].

5.2 Background

This section provides some of the basic background information associated with PUFs and relevant challenges associated with direct extrinsic characterization PUFs.

5.2.1 PUF Preliminaries

Electronic products are common in every facet of life with the prevalence of the Internet of Things (IoT). It is imperative that these devices remain secure, and research continues to progress in this area from academic and industrial efforts [1]. One of the areas of increasing growth involves PUFs. PUFs are physical structures meant to be embedded in a device's physical structure. Ideally, they must be easy to evaluate and hard to predict. Some of the most common types of PUFs include: arbiter PUFs [25], ring-oscillator (RO) PUFs [21], and Static Random Access Memory (SRAM) PUFs [30]. Arbiter PUF produces a unique response based on path delay caused by fabrication variation in two paths on the chip. An RO PUF uses the difference in frequencies between a pair of ROs. SRAM PUF exploits the cross-coupled inverter composition of an SRAM cell initializes to a random value due to power mismatches [22].

The most desirable qualities of a PUF are uniqueness, reliability, and uniformity. The ability of a PUF to identify between various chips within a collection of the same type of chips using Hamming distance is known as uniqueness (HD). The recommended percentage should be 50%. The inter-chip Hamming distance (HD) measures uniqueness and is calculated using the following formula.

$$\text{Uniqueness} = \frac{2}{k(k-1)} \sum_{i=1}^{k-1} \sum_{j=i+1}^{k} \frac{HD(R_i, R_j)}{n} \times 100\%, \tag{5.1}$$

where for a given challenge k and arrays "i" and "j," R_i and R_j represent the n-bit responses of the arrays. k represents total amount of arrays, and n represents array size. Uniformity involves the distribution of '0's and '1's in the response bits [22]. Uniformity is calculated as the percentage Hamming weight. Ideally, the uniformity value is 0.5 (50%) because it indicates that the response is highly random, making its output unpredictable and unclonable. The uniformity of response is computed using:

$$\text{Uniformity} = \frac{1}{n} \sum_{i=1}^{n} b_i \times 100\%, \tag{5.2}$$

where b_i represents the i-th bit of the response. n represents the size of the PUF array. Reliability analyses how well a PUF can repeat the response bits under varying conditions. Operating conditions like temperature, voltage, and aging can create different bit responses. Under different conditions, the ability to produce the same responses indicates the robustness of the PUF. The intra-Hamming Distance is the measure of the reliability of the PUF.

$$\text{Reliability} = \frac{1}{k} \sum_{i=1}^{k} \frac{HD(R_{i,1}, R_{i,2})}{n} \times 100\%, \tag{5.3}$$

where $R_{i,1}$ and $R_{i,2}$ represents the n-bit responses of the "i" challenge in two different times. n represents the PUF array size, and k represents the amount of calculated arrays. The ideal intra-chip HD (reliability) is 0. The priority for utilizing direct extrinsic characterization is ensuring that the above-stated qualities are adequately met. They need to be as good as, if not better, than their intrinsic counterparts. Also, extrinsic direct characterization PUFs require additional electronic components to be added during the fabrication process to generate responses.

5.2.2 Challenges of Direct Extrinsic Characterization

As mentioned earlier, a good PUF must satisfy the qualities of uniqueness, reliability, and uniformity. Although products of direct extrinsic characterization may satisfy these needs, some may require supplementary assets for improvement. Finding these assets may present a set of problems. To begin with, extrinsic characteristics, if not properly protected, provide avenues for malicious parties to obtain key information [3]. The supplementary features could also create avenues for exploitation. The right combination of extrinsic components and extra security components that do not make the device less secure may be the biggest challenge in direct extrinsic characterization. Another issue to consider is the possible overhead from utilizing extrinsic characteristics.

5.3 Extrinsic Direct Characterization PUF

This section presents five direct extrinsic characterization-based PUFs, namely ALILE Diode-based, nano-electrochemical-based, carbon nanotube-based, MEMs accelerator sensor-based, and capacitor-based, in detail.

5.3.1 ALILE Diode-Based PUF

In [29], proposed the usage of unique current-voltage characteristics of a diode to build PUF. It was shown that diodes with irregular I–V curves could be used for cryptography and security applications. For these specific characteristics, the diodes are created using a special crystallization-based fabrication technique that is known as the aluminum-induced layer exchange (ALILE) process [27].

ALILE process is different from the regular diode fabrication process because it introduces process variation. In regular diode fabrication, the diode maker can control the amount of doping materials that get mixed to create the PN junction structure. By contrast, in the ALILE process, the maker has no control over the amount of doping materials that end up in the mix. During the ALILE process, silicon (Si) and aluminum (Al) layers are placed on top of each other, and the whole structure is placed at a temperature below the eutectic temperature of Al-Si system. Under this condition, the formation of the crystallization (p-type structure) will happen automatically and randomly. Figure 5.1 shows a schematic diagram of the ALILE diode. The formation can neither be predicted nor controlled, meaning that every ALILE diode will have different amounts of doping and different I–V responses.

The process variation from the ALILE process can be used for physical cryptography. There are three possible applications for this ALILE diode: Certificates of Authenticity (COAs), Physically Obfuscated Keys (POKs), and Strong Physical Unclonable Functions (strong PUFs). COAs requires to generate a non-imitable signature that could be measured by an external device to prove the device's authenticity. In the case of ALILE diodes, the complex and varied range of I–V curves successfully satisfy this given objective. Therefore, they can be used as COAs, and cheap readers can read out their electrical outputs to obtain the unique signature to authenticate the chip.

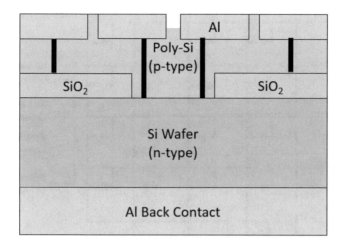

Fig. 5.1 Illustration of a simplified architecture of the ALILE diode

Random characteristics of ALILE diodes can be used as non-volatile memories/storage to store the secret keys. The traditional way of storing the keys in EEPROMs is susceptible to invasive attacks to extract the key. In contrast, the random characteristics of ALILE diodes can make it significantly difficult to perform invasive attacks. ALILE diodes as POKs rely on extracting the key from the varied set of I–V curves. And since the bits are not stored as 1's or 0's like in traditional memories, it is resilient to invasive attacks.

Similarly, ALILE diodes can be used as strong PUFs. The main characteristic of a strong PUF is to generate an exponential number of challenge-response pairs (CRPs), so that even if certain CRPs are leaked to the attacker, it is impossible to guess the entire CRP sample space. For ALILE diodes to be used as a strong PUF, they can be arranged in a crossbar structure, as shown in Fig. 5.2. The principle is similar to a read-only memory function. Challenges are applied on word lines, and unique identification values are read on the bit lines. When a different challenge is applied to the system, the system will respond with a different unique value. When the same challenge is applied, the system will give the same value. Following are some characteristics of the designed crossbar structure:

- It is not feasible to read out different memory units (i.e., diodes) in parallel.
- Writing will be destroyed if an attacker does a faster read-out than the present limit of the structure, thus not allowing read-out of the remaining structure.

To prove the effectiveness of the ALILE diode for the COAs, [29] collected data from 16 different individual ALILE diodes on one chip. Repeated data were taken for each diode, and the mean value was calculated to find the I–V curves and the average/maximum deviation. For experiment conducted in [29], 4 helper data were selected for reliable diode identification: -1.3 V, -0.65 V, 0.65 V and 1.3 V. Every 16 diodes gave unique current readings at these four voltage points, showing good

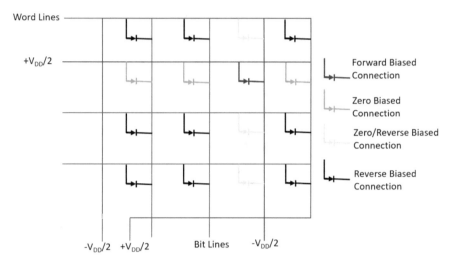

Fig. 5.2 Illustration of the use of ALILE diodes in crossbar structure to be used as a strong PUF

uniqueness. And the measurement of each diode at these four voltage points also showed good reliability with a max deviation of less than 10%. The results showed that it is easy to distinguish between all 16 ALILE diodes due to the random process inserted in the I-V characteristics.

To apply the ALILE diodes for POKs, the main focus is to generate a unique key for the diodes. For experimental purpose in [29], 4 helper data: -1.3 V, -0.56 V, 0.56 V, and 1.3 V were used for key generation. The current axis was divided into sections, and each section was assigned with 0 or 1. One bit from the recent value is generated at each helper data point, four bits per diode and 16 diodes in total; as a result, 64 bits unique key is generated for these ALILE diodes. Similarly, the strong PUF generated using ALILE diodes showed high uniqueness, reliability, and resiliency against machine learning attacks [29].

5.3.2   Nano-Electro-Mechanical-Based PUF

The structure of NEM-based PUF is typically inspired by the dual-gate FinFET [32]. However, the gate oxide layer is removed from the structure of NEM-based PUFs. In this NEM structure, an air gap surrounds the region below the fin region, carved using wet etching of the oxide. Because of this carving, the fin can move between the two gates, Gate 1 and Gate 2. The illustration of this movement is shown in Fig. 5.3. In ideal conditions, the NEM should have a symmetric structure; however, due to inevitable random process variation, stiction can happen, causing fins to attach to either gate with equal probability. This random event is a source of randomness in the NEM-based PUF and a key element of its success. The value of bits in the

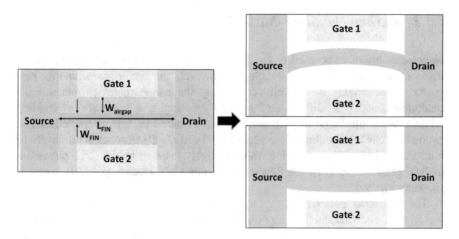

Fig. 5.3 Illustration to show the structure of Nano electromechanical-based PUF switch. The fin of the switch can stick to Gate 1 or Gate 2 due to the random variation from stiction

NEM switch is determined using this mechanical state of the fins. If the fin touches Gate 1, it is in state "0," and if it touches Gate 2, it is in state "1." Though the initial state of the NEM switches is determined by the stiction phenomenon, van der Waals force dominates in later stages. This force ensures that the switch remains in its state, overcoming the tough environmental conditions. That is why NEM-based PUFs have very high uniqueness, and reliability in extreme environmental variations [16].

Advantages of the NEM-PUF include nearly ideal uniformity, uniqueness, and robustness. Experimental data in [16] shows that the average uniformity result is 49.6% and the standard deviation is 3.16% (ideal uniformity should be 50%). The average uniqueness value is 50.3% with a standard deviation of 7.4% (ideal uniqueness should be 50%). And when testing robustness with varying harsh conditions, the intra-chip HD was 0, which is the ideal value. This shows that NEM-PUF has great robustness, uniformity, and uniqueness. NEM-PUF's strong robustness is what gives it the edge over silicon-based PUFs in security, military, and aerospace applications. The NEM-PUF is significant because it can be used with CMOS just like regular silicon PUFs. Additionally, due to the NEM design, the NEM-PUFs have improved reliability and robustness in harsh environmental conditions.

However, in terms of fabrication, NEM-PUF can be expensive to fabricate because, unlike a silicon PUF, NEM-PUF requires extra steps involving evaporated drying process of BOX to get the random contact between fin and G1 or G2. So NEM-PUF could be more challenging and expensive to fabricate for lower technology nodes.

5.3.3   Carbon Nanotube-Based PUF

Nanotechnology can be used to create robust and tamper-resistant security primitives compared to CMOS technology. However, the challenge with this new emerging technology is difficulty controlling the nanoparticles to enhance the randomness and reliability. Therefore, [15] proposed a new method of creating reliable and controllable random nano-structures-based on self-assembled carbon nanotubes (CNTs). Random placement of CNTs in HfO_2 trenches is the source of randomness in CNT-based PUFs, which is highly inspired by the CNT placement method proposed in [28]. An illustration of this CNT placement in 2D arrays is shown in Fig. 5.4.

The 2-D array uses CNTs, or self-assembled carbon nanotubes(CNTs). They are aligned and deposited on HfO_2 trenches that are modified with an NMPI monolayer, ranging from 70 nm to 300 nm wide. The sidewalls are made of negatively charged SiO_2, which repels the CNTs wrapped in negatively charged SDS. The SDS-CNT is attracted to the monolayer, and the competing attractive and repulsive forces create a random pattern of CNTs in the array.

In CNT-based PUF, since the stored bits rely completely on the CNTs placement, the PUF is immune to environmental variations, thus improving its reliability. In [15], experiments show that it is possible to fabricate trench dimensions on a Multiphysics model and create the arrays using electron-beam lithography and chemical methods. The goal is to create an array with maximum randomness, which

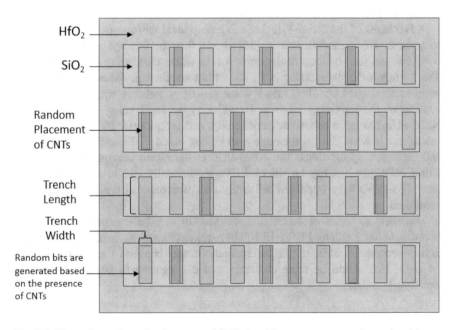

Fig. 5.4 Illustration to show the placement of CNTs in a 2D array to generate the random bits

means that 50% of the bits in the trench array would be connected with a carbon nanotube. In the experiment, the trench width was chosen to be 80 nm because of its near-perfect alignment of the attractive and repulsive forces. The I–V curves of 2560 devices with an 80 nm trench width showed high randomness.

A given CNT can either be open (disconnect) or closed (connected), leading to a binary bit array. However, CNTs can also be either semiconducting or metallic, which can be used to create ternary bit arrays. The CNT being semiconducting or metallic can only be differentiated after the synthesis process by setting the on-off ratio between currents at gate voltages of −3 and −0.5 V to be larger than 20. The use of CNTs as a ternary bit array doubles the key size. In [15], it is claimed that a bit combination of 3.43×10^{30} can be generated out of a 64-bit key, which is many orders of magnitude higher than a binary key and therefore more secure.

The reliability and repeatability of the arrays are satisfactory according to the conducted experiment in [15]. Inter-HD randomness of devices was centered at 0.5 with a variance of 0.0039 in the binary keys and 0.6219 for ternary keys with a variance of 0.0042. They both have a very low intra-HD of about 0.03, which points to high reliability and repeatability. Compared to the 30% BER of common CMOS devices, the CNT-based device may not need error-correcting hardware for reliability. The randomness is further justified by showing that the random bits generated using the proposed method pass the NIST statistical randomness test suite [17].

However, the CNT-based PUF has some limitations [15]. In the reliability temperature test results, the arrays were tested at 25° and 85 °C, and current leakage was found. The large noise margin at high temperatures needs improvement as well. It is not easy to envision the production of the CNT-based PUFs because of the high-temperature unreliability. Also, there is no power analysis on the devices created from the arrays to see if they can prevent power-based side-channel attacks [15].

The arguments for using CNTs in security primitives are plenty. Its randomness and reliability without using expensive and insecure error-correcting hardware around the primitive are compelling arguments. The arrays are created with industry-understood methods at low temperatures, which can be scaled up at a low cost. Duplication of the CNT-PUF will be extremely difficult due to the randomness of the nanotube placement using the drift processes. Invasive attacks would be seriously hindered as well. Electron microscope imaging will be impossible to use in attacks because when the CNTs are exposed, the plasma etching or polishing will destroy them. In addition, the CNT interconnections are on a nanoscale and extremely difficult to identify, let alone measure in a common non-invasive and invasive manner.

5.3.4 MEMs Accelerometer Sensor-Based PUF

Different micro-electro-mechanical systems (MEMS) have become a frequent feature of electronic devices to acquire data from the surroundings as the Internet of Things (IoT) has progressed. These sensors allow the IoT devices to read environmental conditions such as temperature, altitude, speed, noise, humidity, etc. The device can process this environmental information and perform accordingly. However, such sensors can be a source of entropy and thus an excellent candidate for PUFs. Building up on this observation, [4] proposed the use of one such MEMS sensor, i.e., accelerometer, to build PUF.

Many digital fingerprint (security authentication key) techniques have been developed for the low-cost system, and each has its shortcoming. For example, the SRAM approach utilizes power-up values of the SRAM cells by power cycling the entire SRAM [7, 13]; however, this method is not practical during runtime. Another method using the microprocessor-intrinsic variation to generate unique result when executing instruction tend to be slow because it takes time for the unique result to come out [20]. The method using delays of dedicated logic has the same drawback as the microprocessor method of being slow. The advantage of this MEMS sensor method compared to the others is that it is fast and can be utilized during runtime. And since the MEMS sensors are very commonly used in low-power electronic devices and are very cheap and easy to obtain, the MEMS sensors fit the criteria of low cost very well.

The MEMS sensor can be used as PUF because of the process variations during manufacturing introduced due to the small fabrication dimension error and the complexity of the mechanical parts. One particular MEMS sensor, 3-axis accelerometer, is described in [4] as the PUF. Because of process variation, each accelerometer will have a different static offset value when 0g or no acceleration is applied. Therefore, the first way to generate a unique key is by utilizing this 0g offset calibration value. Another way to generate a unique key is to utilize the self-test feature of the accelerometer. Figure 5.5 shows the block diagram illustrating the authentication protocol based on MEMS sensor [4]. In the experiment, during the self-test, the accelerometer exerted an electrostatic impulse to all its axis, and this

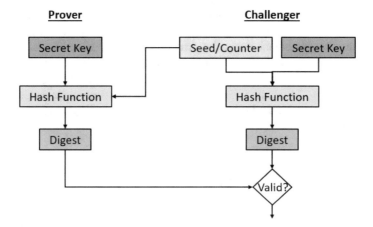

Fig. 5.5 Block diagram illustrating the authentication protocol using MEMS sensor

Fig. 5.6 Block Diagram to
illustrate the easy
MEMS-PUF setup using only
microcontroller and a
accelerometer

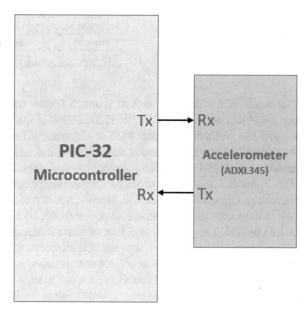

force caused a shift in the x-y-z-axis. Due to the process variation, the shift value will be different from sensor to sensor, and this shift value can be used to generate keys.

A simple experiment is conducted in [4] to test the viability of this accelerometer sensor as a PUF. A simple 3-axis accelerometer called ADXL345 and a development board with a PIC32 microcontroller is used for the experiments. Figure 5.6 shows the illustration of the serial communication link between the accelerometer and the microcontroller.

The experiment tested the two methods (0 g offset calibration value and self-test feature) across 20 different accelerometer sensors. The development board accumulated 1000 samples of the output values from each sensor using the 0g offset calibration method and 1000 simples sample values from each sensor using the self-test feature method. To evaluate the performance of the sensor, two metrics for the assessment of physical authentication keys are used: uniqueness and reliability. Uniqueness measures how different each bit of a sample from one device compares to each bit of a sample from another device. The ideal uniqueness is 50%. Reliability measures how similar each bit of a sample compares to another sample of the same device. The deal reliability is 100%. The result showed that when applying the 0g offset calibration method, the sensor achieves uniqueness of 30.2% and 86.23% reliability. When applying the self-test method, the sensor achieves a uniqueness of 42.64% and reliability of 92.17%, as shown in Table 5.1

This technique makes an effort to use an accelerometer sensor as a physical authentication on a low-cost device. As can be seen in Table 5.1, the 0 g offset calibration method is much less unique and less reliable than the self-test method.

Table 5.1 Reliability and uniqueness results of each authentication methods

Proposed Methods	Uniqueness (%)	Reliablity (%)
Calibration	30.2	86.23
Self-test	42.64	92.17

The result of the self-test method is much better, with a uniqueness of 42.64% and reliability as high as 92.17%. It appears that the accelerometer sensor using the self-test method is a decent PUF application for the low-cost system; however, there are still some limitations of this system. First, only 20 different accelerometer sensors were tested in the aforementioned experiment. It is not a big enough sample size. There is a low-power mode for the accelerometer sensor also, but this mode introduces more noise to measure. However, it is reasonable to assume that a low-cost system may also operate in low-power mode. So more experiments are needed in this regard. The process variation of the MEMS accelerometer sensor can be utilized for PUF, and the self-test method produces acceptable results. Due to the wide availability of MEMS accelerometer sensors, it is cheap. It is also easy to integrate into an existing system. However, one glaring drawback of this experiment is the relatively small sample size. It is difficult to justify the MEMS accelerometer sensor-based PUF being a reliable alternative for the PUF with small sample sizes, although they show great potential.

5.3.5 Capacitor-Based PUF

Similar to ICs, printed circuit boards (PCBs) can also be susceptible to counterfeiting, reverse engineering, hardware Trojans, and other supply chain attacks [6, 11, 24, 33, 35]. The use of counterfeit PCBs can lead to loss of sensitive information or degradation in system performance when used in critical applications [14, 34]. Therefore, solutions for counterfeiting prevention and detection of PCBs should also be the main focus of the community. However, in the past majority of supply chain solutions based on hardware security primitives (PUFs) were designed for ICs. Very limited work has been done in the past that targets the development of solutions to prevent counterfeiting of PCBs [12, 36, 38].

Wei et al. [36] proposed BoardPUF in an attempt to address the need for PCB authentication using PUFs. The solution relies on fabricating a number of capacitor banks in the PCB and a secure chip connected to these capacitors to generate a unique key. The source of entropy in the proposed PUF is the capacitors embedded within the PCB layers. These capacitors will also be exposed to process variation during PCB manufacturing and generate a random frequency when read by the secure chip. The frequency count of the capacitors is stored internally in the chip, and response bits are generated by comparing the counts of two capacitors. Figure 5.7 shows the overall architecture of BoardPUF proposed in [36].

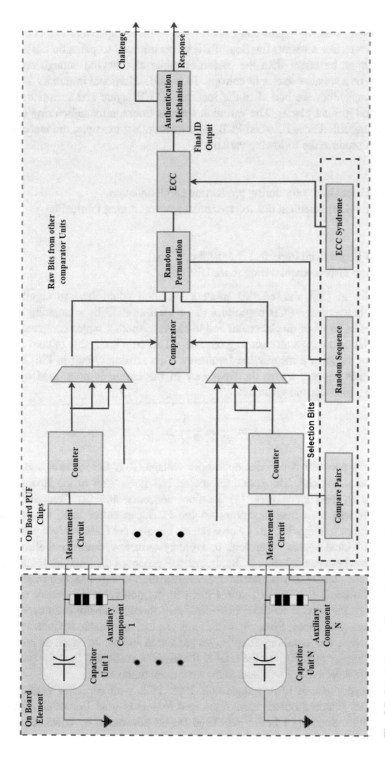

Fig. 5.7 Block diagram illustrating the overall architecture of BoardPUF

The parasitics of the board can cause noise to these capacitors embedded in the PCB layers. As a result, for BoardPUF to be immune to parasitic noise, the capacitors must be larger than the parasitics while also having enough process variation to be a suitable source of entropy. To provide additional immunity against noise, the capacitors are buried in the innermost PCB layers and surrounded by uninterrupted ground planes. The variation sources leverage manufacturing errors to expand the differences between PCB capacitances. For example, the inaccurate process can produce the following variations.

1. Global:

    (a) Thickness may vary during production and lamination.
    (b) Pattern Displacement due to layer misalignment during fabrication.

2. Local:

    (a) Burr Edges from imperfect chemical etching.
    (b) Pattern displacement during mask fabrication.

The study in [36] showed that process variation could lead to significant differences between the PCB capacitors to be used as PUF by simulating these capacitances in terms of displacement and thickness. Another major component of the BoardPUF is the capacitor sensing unit/chip. This chip is capable of converting the capacitor value to a measurable frequency. This sensing circuit in [36] based on the Schmitt trigger, where the frequency of the circuit is controlled by the capacitance under test and given by:

$$f = \frac{1}{K} \times \frac{1}{R \times C}, \tag{5.4}$$

where K is a constant dependent on supply voltage, R is the resistor, and C is the capacitor under test. The response/random bits generation in the BoardPUF is governed by two techniques, (1) Random sequence and (2) Compare Pair Selection. It was seen that capacitors on the PCB can suffer from systematic variation similar to chips/ICs [37]. Some boards are also cut from a larger panel of repeated individual boards. Therefore, to avoid systematic variation and eliminate the correlation between different PCBs, BoardPUF proposed the Random sequence. In this method, the PCB manufacturer generates the random bits before slicing the PCB into smaller PCBs and stores them in the non-volatile memory of the BoardPUF. In the second method, PUF designers should measure each capacitor in two extreme conditions. First, capacitor units producing unstable frequencies are eliminated. Next, the remaining frequencies are sorted and paired based on ranking differences greater than N/2 for the total N frequencies available. Then the pairs of capacitors are selected that will lead to the maximum uniqueness.

In the experiment of [36], a quadruple layer PCB with capacitor banks was designed. One hundred boards were fabricated to demonstrate BoardPUFs effectiveness. Xilinx Spartan- 6 FPGA was used for bit generation by connecting the output of the Schmitt trigger to it. The ID length chosen was 24 bits. The tests were run at 25°, 35°, 45°, 55° and 65 °C. 4.5 V, 5.0 V, and 5.5 V were used as source voltage. Like all PUFs, the metrics used are inter-chip Hamming Distance (Ideally 50% of total bit length) and intra-chip Hamming Distance (0%).

In addition to the two, the inter-board-intra-chip (IBIC) was also used to check the difference between two PCBs soldered with the identical authentication chip. The ideal value is 50%. The mean Intra-distance percentage for the experiment is reported to be 3.63%. This is reasonably close to the ideal value, especially considering the somewhat non-ideal experimental setup. Inter-chip HD also averages at 47.21%. The one result that was subpar was the IBIC distance. The value stood at 39.7%, but reproducing an exact ID would still be costly to an attacker. The percentage of bits flipped at various environmental conditions was addressed to demonstrate intra-chip HD.

Attackers can also use various means to attack PUFs. These include machine learning algorithms, side-channel attacks, and invasive attacks. The basis of these attacks is the availability of internal structure information to the adversary. Therefore, compare pairs and random sequences should also be securely stored to prevent against key-recovery attacks. In BoardPUF, one source of vulnerability can be that an attacker gets access to the capacitor units. In this case, an oscilloscope can be used to measure signals and possibly retrieve keys. However, there is a high probability that the probe can encounter parasitics capacitance and read out the wrong frequencies.

5.4 Conclusions

In this chapter, five different direct characterization techniques were thoroughly investigated, focusing on applying various PUFs with extrinsic sources of randomness. This chapter also summarizes how these PUFs can be applied to hardware security. The chapter also forecasts the potential and future application of diode-based PUFs, NEM-based PUFs, carbon nanotube-based PUFs, MEM sensor-based PUFs, and PCB capacitor-based PUFs. These non-conventional extrinsic PUFs offer advantageous qualities (uniqueness, reliability, and uniformity), supplementing the conventional silicon-based PUFs. There is a strong indication that PUFs created from Direct Extrinsic randomness have the potential to overtake silicon PUFs and make their way into everyday electronics.

References

1. Ahmed B, Bepary MK, Pundir N, Borza M, Raikhman O, Garg A, Donchin D, Cron A, Abdel-moneum MA, Farahmandi F, et al (2022) Quantifiable assurance: From IPs to platforms. arXiv preprint arXiv:220407909

2. Amsaad F, Pundir N, Niamat M (2018) A dynamic area-efficient technique to enhance ROPUFs security against modeling attacks. In: Computer and Network Security Essentials. Springer, Berlin, pp 407–425
3. Armknecht F, Maes R, Sadeghi AR, Standaert FX, Wachsmann C (2011) A formalization of the security features of physical functions. In: 2011 IEEE Symposium on Security and Privacy. IEEE, New York, pp 397–412
4. Aysu A, Ghalaty NF, Franklin Z, Yali MP, Schaumont P (2013) Digital fingerprints for low-cost platforms using mems sensors. In: Proceedings of the Workshop on Embedded Systems Security, pp 1–6
5. Basak A, Zheng Y, Bhunia S (2014) Active defense against counterfeiting attacks through robust antifuse-based on-chip locks. In: 2014 IEEE 32nd VLSI Test Symposium (VTS). IEEE, New York, pp 1–6
6. Cruz J, Farahmandi F, Ahmed A, Mishra P (2018) Hardware Trojan detection using ATPG and model checking. In: 2018 31st international conference on VLSI design and 2018 17th international conference on embedded systems (VLSID). IEEE, New York, pp 91–96
7. Guajardo J, Kumar SS, Schrijen GJ, Tuyls P (2007) FPGA intrinsic PUFs and their use for IP protection. In: International workshop on cryptographic hardware and embedded systems. Springer, Berlin, pp 63–80
8. Guin U, Forte D, Tehranipoor M (2013) Anti-counterfeit techniques: From design to resign. In: 2013 14th International workshop on microprocessor test and verification. IEEE, New York, pp 89–94
9. Guin U, DiMase D, Tehranipoor M (2014a) Counterfeit integrated circuits: Detection, avoidance, and the challenges ahead. J Electron Test 30(1):9–23
10. Guin U, Huang K, DiMase D, Carulli JM, Tehranipoor M, Makris Y (2014b) Counterfeit integrated circuits: A rising threat in the global semiconductor supply chain. Proc IEEE 102(8):1207–1228
11. Harrison J, Asadizanjani N, Tehranipoor M (2021) On malicious implants in PCBs throughout the supply chain. Integration 79:12–22
12. Hennessy A, Zheng Y, Bhunia S (2016) JTAG-based robust PCB authentication for protection against counterfeiting attacks. In: 2016 21st Asia and South Pacific Design Automation Conference (ASP-DAC). IEEE, New York, pp 56–61
13. Holcomb DE, Burleson WP, Fu K (2008) Power-up SRAM state as an identifying fingerprint and source of true random numbers. IEEE Trans Comput 58(9):1198–1210
14. Hossain MM, Vashistha N, Allen J, Allen M, Farahmandi F, Rahman F, Tehranipoor M (2022) Thwarting counterfeit electronics by blockchain
15. Hu Z, Comeras JMML, Park H, Tang J, Afzali A, Tulevski GS, Hannon JB, Liehr M, Han SJ (2016) Physically unclonable cryptographic primitives using self-assembled carbon nanotubes. Nat Nanotechnol 11(6):559–565
16. Hwang KM, Park JY, Bae H, Lee SW, Kim CK, Seo M, Im H, Kim DH, Kim SY, Lee GB, et al (2017) Nano-electromechanical switch based on a physical unclonable function for highly robust and stable performance in harsh environments. ACS Nano 11(12):12,547–12,552
17. Kim SJ, Umeno K, Hasegawa A (2004) Corrections of the NIST statistical test suite for randomness. arXiv preprint nlin/0401040
18. Kumar SS, Guajardo J, Maes R, Schrijen GJ, Tuyls P (2008) The butterfly PUF protecting IP on every FPGA. In: 2008 IEEE International Workshop on Hardware-Oriented Security and Trust. IEEE, New York, pp 67–70
19. Laschat D, Ehrmann T (2021) Systemic risk in supply chains: a vector autoregressive measurement approach based on the example of automotive and semiconductor supply chains. Available at SSRN 3882809
20. Maiti A, Schaumont P (2012) A novel microprocessor-intrinsic physical unclonable function. In: 22nd International Conference on Field Programmable Logic and Applications (FPL). IEEE, New York, pp 380–387
21. Maiti A, Casarona J, McHale L, Schaumont P (2010) A large scale characterization of RO-PUF. In: 2010 IEEE International Symposium on Hardware-Oriented Security and Trust (HOST). IEEE, New York, pp 94–99

22. Maiti A, Gunreddy V, Schaumont P (2013) A systematic method to evaluate and compare the performance of physical unclonable functions. In: Embedded systems design with FPGAs. Springer, Berlin, pp 245–267
23. McGrath T, Bagci IE, Wang ZM, Roedig U, Young RJ (2019) A PUF taxonomy. Appl Phys Rev 6(1):011303
24. Mehta D, Lu H, Paradis OP, MS MA, Rahman MT, Iskander Y, Chawla P, Woodard DL, Tehranipoor M, Asadizanjani N (2020) The big hack explained: detection and prevention of PCB supply chain implants. ACM J Emerg Technol Comput Syst 16(4):1–25
25. Morozov S, Maiti A, Schaumont P (2010) An analysis of delay based PUF implementations on FPGA. In: International Symposium on Applied Reconfigurable Computing. Springer, Berlin, pp 382–387
26. Morrell PS (2021) Airline finance. Routledge, New York
27. Nast O, Wenham SR (2000) Elucidation of the layer exchange mechanism in the formation of polycrystalline silicon by aluminum-induced crystallization. J Appl Phys 88(1):124–132
28. Park H, Afzali A, Han SJ, Tulevski GS, Franklin AD, Tersoff J, Hannon JB, Haensch W (2012) High-density integration of carbon nanotubes via chemical self-assembly. Nat Nanotechnol 7(12):787–791
29. Rührmair U, Jaeger C, Hilgers C, Algasinger M, Csaba G, Stutzmann M (2010) Security applications of diodes with unique current-voltage characteristics. In: International Conference on Financial Cryptography and Data Security. Springer, Berlin, pp 328–335
30. Schrijen GJ, Van Der Leest V (2012) Comparative analysis of SRAM memories used as PUF primitives. In: 2012 Design, Automation and Test in Europe Conference and Exhibition (DATE). IEEE, New York, pp 1319–1324
31. Tehranipoor MM, Guin U, Forte D (2015) Counterfeit test coverage: An assessment of current counterfeit detection methods. In: Counterfeit Integrated Circuits. Springer, Cham, pp 109–131
32. Toh EH, Wang GH, Samudra G, Yeo YC (2007) Device physics and design of double-gate tunneling field-effect transistor by silicon film thickness optimization. Appl Phys Lett 90(26):263507
33. Vashistha N, Rahman MT, Shen H, Woodard DL, Asadizanjani N, Tehranipoor M (2018) Detecting hardware Trojans inserted by untrusted foundry using physical inspection and advanced image processing. Journal of Hardware and Systems Security 2(4):333–344
34. Vashistha N, Hossain MM, Shahriar MR, Farahmandi F, Rahman F, Tehranipoor M (2021) eChain: A blockchain-enabled ecosystem for electronic device authenticity verification. IEEE Trans Consum Electron 68(1):23–37
35. Wang X, Han Y, Tehranipoor M (2019) System-level counterfeit detection using on-chip ring oscillator array. IEEE Trans Very Large Scale Integr VLSI Syst 27(12):2884–2896
36. Wei L, Song C, Liu Y, Zhang J, Yuan F, Xu Q (2015) BoardPUF: Physical unclonable functions for printed circuit board authentication. In: 2015 IEEE/ACM International Conference on Computer-Aided Design (ICCAD). IEEE, New York, pp 152–158
37. Yin CE, Qu G (2013) Improving PUF security with regression-based distiller. In: Proceedings of the 50th Annual Design Automation Conference, pp 1–6
38. Zhang D, Han Y, Ren Q (2019) A novel authorization methodology to prevent counterfeit PCB/equipment through supply chain. In: 2019 IEEE 4th International Conference on Integrated Circuits and Microsystems (ICICM). IEEE, New York, pp 128–132
39. Zhang X, Tuzzio N, Tehranipoor M (2012) Identification of recovered ICs using fingerprints from a light-weight on-chip sensor. In: Proceedings of the 49th Annual Design Automation Conference, pp 703–708

Chapter 6
Hybrid Extrinsic Radio Frequency PUF

6.1 Introduction

Radio Frequency Identification (RFID) was first used by British Airforce to identify surrounding aircraft as friend or foe during World War II [18]. Not until 1999 did RFID technology prevail as a viable option for hardware security measures [9]. This shift toward implementing RFID technology was prompted by the ability for the devices to store more information than traditional barcodes, be fabricated at a low cost, and the ability for the technology to remain accessible even when the device is not exposed. RFID tags are prevalent in everyday life, without consumers even knowing they are implemented on their devices. Retail stores use RFID tags to prevent theft, and some clothing and fashion designers even use them to authenticate their designs, like its application in the hardware security domain. IoT devices are becoming used more regularly in the consumer domain, creating a need to secure their data [25]. Cell phones now contain RFID scanners to transfer payment information from the phone to credit card reading devices in stores. RFID tags are also used in technology tracking and identification. RFID systems are contactless, non-line-of-sight (non-LOS), and undetectable identification systems [21] that operate in untrusted environments. Thus, anyone within range of the tag can intercept the device's data and ID, creating a need to secure these devices to prevent data leakage, losses in revenues, damaged reputation, and loss of human life [12]. Since RFID tags have tight area and power constraints, it is challenging to implement cryptographic algorithms to transfer data to the user securely. Previous works have studied the ability to counterfeit and the safety of RFIDs. For example, a study conducted in 2005 displayed how a transponder produced by Texas Instruments was easily cloneable [9]. Another study exposed the ability to cheaply produce RFID readers that could scan passports and track travelers [3]. This exposes RFID devices as a potential threat to consumers and companies. Leveraging the functionality of physical unclonable functions (PUFs) and certificate of authentication (CoA), RFIDs can become a secure way to transmit sensitive data in an untrusted environment and prevent counterfeit products from entering the consumer market [24].

A PUF is a physical device implemented on System-on-Chip (SoC) designs to authenticate products and prevent counterfeiting [10, 11]. PUFs uses Challenge–Response Pairs (CRPs) to uniquely identify the reaction (response) to a stimulus (challenge) [1, 19]. The ability for a PUF to repeatedly generate an identical response for the given the same challenge is known as reliability. Ideal PUFs return the same response when provided the same challenge. The ability of multiple PUFs to generate unique responses for the given the same challenge is also essential and is known as uniqueness. Ideal PUFs provide a different response for different challenges to ensure that multiple challenges cannot invoke the same response. Intellectual Property (IP) Owners select challenges and record individual PUF responses to them to authenticate their IP [2]. PUFs can be implemented by many technology methods such as Ring Oscillators (RO), Arbiter, SRAM PUFs, etc., to exploit material process variations and random physical properties to "randomize" the responses to challenges [2, 17, 19, 23]. The device technology that will be discussed is Radio Frequency (RF) PUFs. These PUFs utilize RF principles that maintain the physical characteristics to return a reliable output when provided an input challenge. The structure of these specific PUFs is helpful because the devices are passively powered and can be scanned to authenticate the IP. A scanner challenges the PUF device using electromagnetic (EM) radio waves to provoke a response from the PUF. Feedback from the EM stimulus will be unique and truly random compared to all other instances. A spectrum analyzer can also be used as a tool to extract the attenuation properties that make up the RF-PUF.

Certificates of Authenticity (CoA) are devices used to verify authorship and are not limited to devices such as PUFs implemented by IP owners [5]. The following requirements must be fulfilled to classify a product as a CoA officially.

1. The cost to produce the certificate shall be minimal, relative to the desired security confidence.
2. The cost to manufacture the certificate shall be several orders of magnitude less than the cost to reverse engineer and replicate the certificate to produce copies.
3. The cost to verify the certificate shall be small, as it shall be easily accessible and inexpensive.
4. When reverse engineering the known characteristics of a functioning certificate, construction of the certificate shall be computationally challenging to replicate [6].

These requirements ensure the usefulness, applicability, and security of the CoAs. If any of these requirements are not fulfilled, the CoA becomes expensive or useless and is not a valid solution. A secure and robust RFID system can be created by applying the requirement for a CoA to the design and manufacturing of RFID systems.

RFID is a wireless way of automatically retrieving and transmitting data to uniquely identify a device using the radio spectrum [26]. It automatically uses RF signals' transmission properties and space coupling to identify a device [27]. RFIDs can be broken into two parts, the tag and the reader. The tag is a passive transponder attached to or implemented in a device. The reader reads and writes data to the tag. RFID systems operate in the near-field and the far-field range [6]. First, the reader transmits a carrier signal through the transmission antennas in RFID systems. Then, when the tag is in range of the working area of the reader's transmitter antenna, the tag is activated and transmits its data to the reader. The reader's receiver antenna receives the transmitted data from the tag and demodulates and decodes the received signal [27]. Then it sends the fingerprint and data to a user. In this chapter, we elaborate on the use of RF-PUFs and RF-CoAs to create a robust and secure RFID system for devices.

6.2 Background

This section discusses the threat model and challenges faced to protect against privacy and security attacks.

6.2.1 Threat Model

Many security threats exist based on the way RFID systems work. The RFID threat model follows the STRIDE model [21]. STRIDE is an acronym that combines different threat categories: spoofing identity, tampering of data, repudiation, information disclosure, denial of service (DoS), and elevation of privilege [21].

1. **Spoofing Identity:** This happens when an adversary can successfully pose as an legitimate user of the RFID system. An adversary can do this by replacing the authentic RFID reader to read the RFID tags [21]. This can lead to tampering or observing the RFID tag's sensitive data.
2. **Data Tampering:** Tampering with Data occurs when the attacker modifies the data by adding, deleting, or restructuring the original data [21]. For example, an attacker can modify a vehicle's information in a toll transponder to reflect false information about a vehicle or make it look like another vehicle passed through the toll booth.
3. **Repudiation:** Repudiation happens when a legitimate personnel denies an activity and there is no evidence that the action occurred [21]. In theory, this can be done by anyone, but insider threats pose the highest risk. The insider threat can deny receiving sensitive data from a system if that system does not use a non-repudiation protocol.
4. **Information disclosure:** This happens when the RFID receiver or tag information is leaked to an unauthorized user [21]. For example, an adversary country uses a powerful directional reader to locate the military's types of weapons on its base since the military uses RFID tags to track and identify assets.

5. **Denial of Service:** An attacker can perform attacks efficiently since RFIDs are non-contact, non-LOS systems [21]. An adversary can use a powerful jammer to jam a patient's tag in a hospital to prevent their medication from automatically dispensing.
6. **Elevation of Privilege:** This happens when a lower privileged user gains unauthorized access to a higher privilege level in a system. An adversary can miniplate the RFID tag embedded in passports to change the status of a person from a criminal to a lawful citizen [21].

6.2.2 Challenges

According to the STRIDE model, RFID systems face numerous privacy and security challenges, making them vulnerable to attacks. Combating these threats is a difficult issue given the tag's small area and power overhead constraint. The use of a cryptographic algorithm to encrypt the data and fingerprint communication between the reader and tag is a simple way to solve most of the threats posed by an RFID system. Because it does not have the same space or power constraints as tags, this solution is relatively simple for the reader to implement. However, because of the limited area of these tags, cryptographic algorithms cannot be integrated to encrypt transmitted data and ID [13].

Furthermore, an attacker can easily digitally forge the ID and data of the tag [13, 15]. RFIDs are also vulnerable to reverse engineering and side-channel attacks such as timing and EM side-channel analysis. Reverse engineering and side-channel attacks in the reader can be mitigated using cutting-edge countermeasures; however, area constraints and a lack of power within passive RFID tags make implementation difficult. This means that RFID tags can be easily physically forged [13]. Given that an RFID tag and reader are encrypted, replay attacks are a significant threat to RFID [4]. Because RFID systems are non-LOS, contactless, and undetectable identification systems, anyone within range of the tag can intercept the message. Even if the message is encrypted, the ID remains constant and present in every message. As a result, an adversary can use brute force to extract the IDs and use them to conduct DoS, tampering, spoofing, and other attacks. Attacks on an RFID system can be mitigated by combining the ideology of a CoA or a PUF's parameters, such as uniqueness and reliability, with the RF properties of the device. RFID systems have the potential to become a secure, robust system for identifying devices and transmitting data from those devices in an untrusted environment.

6.3 Radio PUF in IoT Security

With the growing number of Internet of Things (IoT) devices, many of these devices operate in untrusted environments and are vulnerable to malicious attacks.

At the moment, the security of these devices is based on symmetric key encryption. However, the key is typically stored in nonvolatile memory or SRAM, which is vulnerable to critical hacking and consumes significant power and area. This is problematic because the size of these devices is constantly decreasing [4].

Various researchers over the years have proposed RF-PUF variants. Chatterjee et al. [4] created a novel RF-PUF for IoT nodes by leveraging the concept of PUFs and RF fingerprints. Tekbas et al. [20] used RFIDs to identify wireless nodes in a network by utilizing the devices' transient properties. However, this necessitates high oversampling rates, resulting in costly receiver (Rx) architectures [4]. Gerdes et al. [8] identified transmitter (Tx) nodes using fixed random channel access (RACH). This method uses the steady-state portion of the signal to obtain the device signature, which an attacker can easily identify [4]. DeJean and Kirovski [5] also proposed RF-DNA for IoT device authentication; however, unlike RF-PUF, RF-DNA requires additional hardware at the Tx end.

The RF-PUF proposed in [4] makes use of transmitter process variation (Tx). It also employs Machine Learning (ML) to extract entropy and generate a robust PUF in order to securely identify the receiver (Rx) end nodes. Tx refers to the data transmitted by the IoT device, and Rx refers to the gateway where data from all nodes is received. The RF-PUF in [4] is preamble-less, does not require Rx oversampling, and makes use of higher dimensionalities in the feature space. It also internally compensates for nonideal Rx signatures [4]. Preamble-less approaches provide real-time device authentication without knowing the predicted bitstream. The RF-PUF requires no additional hardware at the Tx nodes, which is significant because IoT nodes are already resource-constrained. It is assumed that the IoT nodes are not controlled during the fabrication process. Thus, RF-PUFs can securely exploit process variation to identify each node at the Rx end.

Furthermore, using a lightweight ML framework in Rx enables the ML engine to compensate for Rx shortcomings. Furthermore, it takes into account data and channel variability. This is accomplished through the use of an Artificial Neural Network (ANN). The ANN enables the connection of asymmetric small to medium IoT systems to a single gateway. RF-PUF adds a physical layer of security that can be used as a stand-alone authentication security feature. RF-PUF can be combined with Open Systems Interconnection (OSI) to provide multifactor authentication. The RF-PUF can be embedded in various network devices, such as monitoring patients' vital signs in a hospital or a fleet of vehicles in a warehouse. RF-PUFs can also be expanded to include the Internet of Military Things (IoMT), monitoring soldiers' biometrics and securely sync weapon systems.

The overall RF-PUF system framework can be divided into two parts: receiver and transmitter. A system-level diagram of the RF-PUF implementation is shown in Fig. 6.1. Three features are used to verify a node's authenticity securely. It is important to note that these features do not necessitate the installation of additional circuitry in devices. The first characteristic is frequency. Because of process changes in the local oscillators, each Tx signal has a distinct frequency offset from the ideal carrier frequency (LO). Using a minimal jitter with zero mean reference clock in the Rx, this approach may determine frequency offsets from multiple Tx(s) using

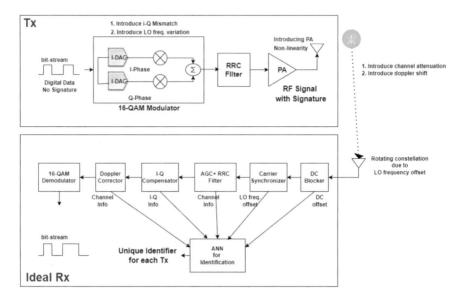

Fig. 6.1 Block diagrams illustrating the Tx and Rx blocks of RF-PUF during system-level setup

the carrier synchronizer module. This module is already included in the Rx module for LO offset compensation [4]. As shown in Fig. 6.1, the carrier synchronizer determines the signal's parts per million (ppm) frequency offset and passes the ppm value to the ANN. The second feature is the mismatch between in-phase (I) and quadrature (Q). The signal's I-Q is unique to each Tx. Using the I-Q in conjunction with the frequency offsets allows each Tx in the network to be uniquely identified. The communication channel is the final feature. When a signal travels through space, it experiences attenuation, distortion, and Doppler shift, which must be compensated for in the Rx to ensure the RF-PUF operates reliably. The Rx employs automatic gain control (AGC) and root-raised cosine (RRC) filters, as shown in Fig. 6.1. The combination of AGC and RCC aids in the reduction of intersymbol interference (ISI) and provides the ANN with the appropriate channel attenuation [4]. A Doppler corrector is also used to approximate and correct Doppler shift caused by physical movement of the transmitter and receiver. The Rx's ML framework consists of a three-layer ANN that accepts as inputs LO frequency offset, I-Q mismatch, channel information, and DC offset. Based on the training data, the ANN can correctly identify the Tx(s). The ANN is trained using various pseudo-random bit-streams in various channel conditions. This accounts for the data and channel variations produced by the Tx in the field and allows for preamble-less operation.

The RF-RUF has significant advantages when comparing the RF-PUF to other RFID solutions, such as RF-DNA or other PUF types. As previously stated, RF-PUF is a low-cost method that eliminates the need for node identification via preambles or keys. To send its signature to the Rx, the IoT device requires no additional circuitry.

This enables the use of RF-PUFs in devices with limited resources. Furthermore, it consumes no additional power in the Tx to embed the device's signature, allowing RF-PUF to be used in devices with limited power or battery-powered devices. An ANN must be integrated to implement the solution on the Rx side. In [4], the ANN only consumes an additional 5% of the power. RF-PUFs, unlike other strong PUFs, are immune to the malicious PUF model because they do not store any digitally encoded signatures [4]. However, the RF-PUF does have limitations, such as being vulnerable to replay attacks. The Tx repeatedly sends identical digital data streams: this cannot be avoided. Furthermore, an external attack on the ML engine is possible. An adversary could create a training set for the ANN by modeling the RF-PUF responses and monitoring the Tx data to access the unique identifier. However, the adversary's ANN modeling accuracy is significantly affected by the ratio of CRPs accessed by an adversary to total CRPs [4]. The RF-PUF was simulated in MATLAB using a 16-QAM modulation scheme, and 10,000 Tx(s) were simulated in [4] under various channel conditions. The false detection probability of Tx increases as the number of transmitters increases. Furthermore, having a Hidden Layer of 50 or more does not reduce error significantly. When the Intra-HD (Hamming Distance) and Inter-HD results are compared, the worst-case Inter-HD occurs with 1000 Tx(s), where the Inter-HD is 3.9 ppm, and the average Intra-HD is 2.8 ppm [4].

Two software-defined radios (SDRs) were used to demonstrate the hardware feasibility of the RF-PUF [4]. Both SDRs were set up as Tx(s), with one as an Rx. Unique Tx(s) were artificially emulated by changing the ambient temperature in a controlled environment to validate the ANN learning engine. The temperature has been adjusted in discrete 5C steps between 0C and −25C citesp8. The Rx uniquely identified all Tx(s), demonstrating that the ANN can detect Tx(s) with minor RF feature differences. Furthermore, the RF-PUF's security has been compared to other PUFs by examining the false rejection rate (FRR) and false acceptance rate (FAR). The FRR and FAR were very close to the pairwise-comparison ring-oscillator PUF, one of the most secure and robust PUFs, when the Tx(s) was around 50. Due to the smaller inter-HD, the RF-PUF's robustness and security decreased significantly as the number of Tx(s) in the network increased. Therefore, when using the RF-PUF in larger network systems, a trade-off between security and scalability must be considered.

6.4 PUF-Embedded RFID

Kang et al. [13] evaluated the first commercial PUF-embedded RFID tags that allowed consumers to authenticate them using a smartphone. These devices use near-field communication (NFC) technology, which is compatible with the majority of modern smartphones, making this PUF more accessible to a broader consumer base. The PUF in these commercial tags draws power from the smartphone that captures the PUF ID via electromagnetic induction. Kang et al. [13] conducted environmental testing to stress the device under varying temperatures to characterize and diagnose reliability, safety, and overall build quality due to the instability of this power source compared to direct voltage sources from power supplies. These tests are carried out because the previous testing concentrated on experimental PUF devices implanted on Application-Specific Integrated Circuit (ASIC) or Field Programmable Gate Array (FPGA) platforms for research purposes rather than for the consumer market [7, 13]. This test provided useful data for a device that is already on the market and is not in the research and development stage. In this case, CRPs were chosen for ten PUF devices and iterated ten times to characterize reliability and uniqueness. The device temperature is varied from −45° C to 95 °C during the iterations to identify any temperature anomalies. PUF device reliability is determined by comparing Intra-HD within responses to other responses from the same device with the same challenges. Uniqueness is measured by comparing the same challenge Intra-HD to the same challenge Inter-HD to determine differences in response across the ten devices.

When each RFID PUF was tested in a temperature chamber at various temperatures, the results show that temperature has no effect on the reliability of Verayo's product. The results in [13] demonstrated that the peaks from the same challenge Intra-HD of PUF 1 differ from the peaks from a different challenge Intra-HD. This demonstrated that when distinct challenges are applied to the same PUF, the PUF responds differently. Similarly, the PUF maintains the same response when the same challenge is applied, making the PUF reliable in varying temperatures. Temperature variation does not affect performance when characterizing uniqueness. The same challenge Intra-HDs were compared to InterHDs across all 10 PUF devices for uniqueness. There is no identification error because there is no overlap between peaks. This demonstrates that each device has its own unique identifier for the same challenge. Temperature tests revealed that the RFID PUF's reliability and uniqueness are not significantly affected. This test validates Verayo's device's capabilities under temperature stress. These devices can be reliably integrated into systems requiring stringent environmental qualifications. This device can be used in defense systems to reduce counterfeiting of classified technology.

6.5   RFID Tags as Certificates of Authenticity

RFID tags with digital encoding are used for applications ranging from inventory management to personnel identification in centers, warehouses, customs offices, checkout counters, and any other location where the authenticity of a product must be verified. The RFID reader would extract, disjoin, decrypt, and compare the encrypted and plaintext versions of the responses to ensure that the private key owner had, in fact, programmed the RFID. Figure 6.2 shows the CoA programming steps. Finally, the CoA fingerprint would be read, as described in Fig. 6.3, and compared to the stored RFID responses to authenticate the target item. Compared to far-field observations, the described system was said to have higher response uniqueness and lower power requirements, and measuring in the near-field had the

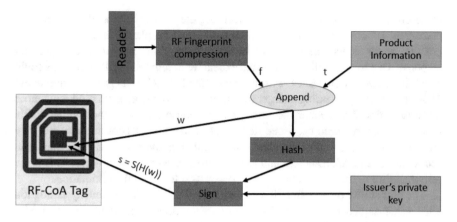

Fig. 6.2 Block diagram illustrating the certificates of authenticity programming steps

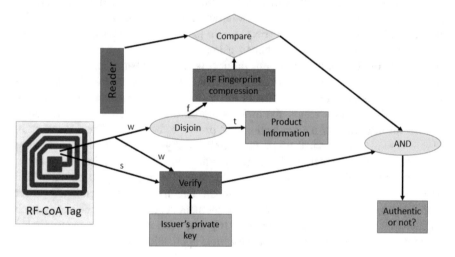

Fig. 6.3 Block diagram illustrating the certificates of authenticity verification steps

added benefit of not being easily eavesdropped, spoofed, or jammed in the far-field [15]. Furthermore, the CoA instance and reader may be small enough to be unnoticed in many applications. The proof-of-concept CoA instances were 1" × 1", but the actual implementation could be much smaller.

The system's drawbacks included measurement noise and inaccuracy, CoA placement positional tolerance, RFID storage size limits, and potential insider threats. Even when measurements were taken in rapid succession, the analog to digital converters produced digital results that varied by up to 20%. Even after averaging a group of 20 recordings for the same signal, there were still deviations of up to 8% between average measurements. Despite this measurement limitation, each CoA fingerprint could be uniquely identified with nearly 100% confidence, as described in the results in [15]. Concerning the above-mentioned limitations of placement accuracy and precision, each CoA must be placed in nearly the exact position as its actual measurement in order to reproduce its unique fingerprint response. Due to prohibitively large and expensive automation equipment, this could cause implementation issues. Furthermore, standard RFID tags only have about 1KB of memory, preventing a larger transmitter/receiver array from producing more response pairs. The greater the number of response pairs, the more quantifiably unique a response could be. Some RFID tags have up to 4 KB of memory, but they are not supported by most standard equipment. Finally, insider threats may gain access to issued CoA instances and use them to create counterfeit items. This is the responsibility of the owner and is not necessarily a technological limitation.

The probability density function was estimated using a kernel density estimation after measuring each of the 47 CoA responses ten times. The probability of a false negative or false positive occurring during CoA authentication was the primary metric of interest. The probability that a fake CoA was verified as authentic or an actual CoA was rejected as fake was determined by the intersection of false positive (FP) and false-negative (FN). A non-parametric, conservative estimate of the response distribution was obtained using kernel density estimation [15]. The probability of the described system producing a false positive or false negative was demonstrated to be on the order of 10-200, which is inconceivably small. The probability was even lower in some of the more significant CoA instances, on the order of 10-300. As a result, the authors' claim that their CoA was sufficiently unique for high-fidelity authentication was proven to be correct. Previous work in [14] had exploited physical properties based on physically unclonable functions; [15] was the first to create an RFID tag solution that did not require power input. The previous PUF solution necessitated the use of a microchip to house a PUF and generate a response to a challenge input. This method had significant power and area overhead, making it difficult to implement at scale. Other techniques, such as chemical CoAs on paper documents [16], had been used, but the measurement equipment was deemed too expensive for most applications.

6.6   PUF in Ambient World

RFIDs have risen to a high level of pervasiveness in modern society due to their low cost, high information content, and non-line-of-sight nature, especially with the recent growth of IoT devices. In 2004, the counterfeit market was estimated to be worth more than $500 billion, with RF IC authenticity concerns threatening major companies such as Texas Instruments, various automobile manufacturers, and even the designers of passport-embedded RFIDs [9]. Due to the significance of threats against RF ICs and emergence of hardware attacks, [9] centered their work around a new methodology for deploying a cryptographic key onto the nodes of a network. Guajardo et al. [9] used the Message-In-a-Bottle (MIB) protocol as the building block of the network. The MIB protocol is composed of five entities: the base station, which controls the wireless network; the sensor node, which shares the secret data with the base station upon key activation; the keying device, which enters the Faraday cage with the node in order to send the initial key facts and derive key; the keying beacon, which jams communication outside the Faraday cage and provides status information to the user; and the user, who performs the key deployment.

Guajardo et al. [9] argued that malicious hardware tampering could be easily solved in both this protocol and the alternative case of manufacturer-installed keys by introducing PUFs with fuzzy extractor schemes (algorithms that improve the reliability and uniformity of the PUF to cryptographic standards). The key-generating PUFs served the additional benefit of providing unclonability and tamper detection in their first proposed low-trust model. Their second model (described in Fig. 6.4), which assumed the presence of a Trusted Third Party (TTP) such as a manufacturer, relaxed the level of trust slightly in exchange for significantly lowering the complexity and cost of implementation. There is no need for a key beacon, keying device, or Faraday cage in the second model. However, one must accept that the TTP will be aware of the deployed keys for specific nodes and should proceed if the employees installing the modules can correctly follow the procedure [9].

Another intriguing proposal in [9] was the creation of a new PUF known as an "LC-PUF." This was based on a simple LC circuit consisting of a capacitor connected in series with an inductor. When an EM-field in the RF range is produced in the vicinity of the PUF, the circuit absorbs an unusual amount of power due to the random microscopic properties of the components. A response curve of the LC-PUF can be generated as its fingerprint by performing a frequency sweep. The LC-PUF outperformed previous work because it was ultra-low-cost, had strong resonance peaks, and did not require a high level of positional accuracy like the RF-CoA PUF. Assuming that the LC-PUF adheres to the assumptions of tamper-evidence, challenge–response pairs yielding negligible information about other pairs, and the impossibility of generating a challenge–response pair without physical access, this has broad applications for any design that requires a PUF at a low cost (perhaps even the key deployment scheme described previously) [9].

Overall, the benefits of this protocol stem from the presence of tamper evidence, tamper resistance, and unclonability in the introduced PUFs. Furthermore, the protocol is simpler and more efficient, with a lower implementation cost. Furthermore, even in the stricter model of no TTP, it is more difficult for the manufacturer to determine the factory key if it is used, as opposed to a simple NVM that stores the key. The main disadvantage, on the other hand, stems from the fact that, with the exception of the stricter model, the security assumptions are relaxed to the point where a third-party manufacturer could theoretically exploit knowledge of key-node pairs [9]. The limitations stem from the standard PUF assumptions, as well as the fact that the LC-PUF appears to be a weak-PUF (with a small challenge–response space) [9].

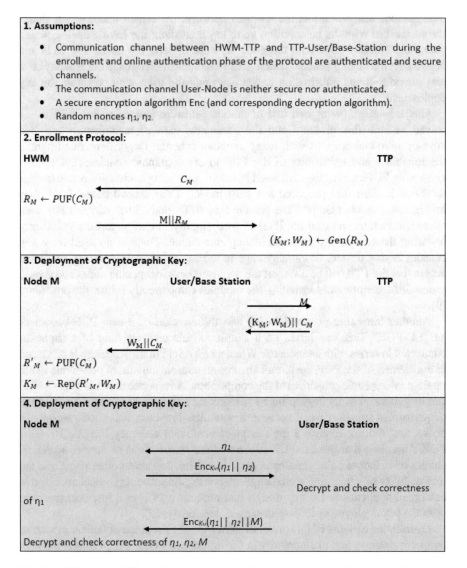

Fig. 6.4 Different steps illustrating the author's key development protocol for trusted third party

6.7 Radio Frequency-DNA

Another RF-PUF technique we'll discuss is based on two near-field properties of electromagnetic fields. The first property is that dielectric and conductive objects with surface features comparable to the wavelength of an electromagnetic wave are significant EM scatterers. The second property is that reflection and refraction at the boundary of two mediums produce near-field effects that are difficult to predict accurately. These two properties aided in developing a PUF that functioned as a CoA, proposed in [6]. To meet the requirements for a CoA, it must be relatively low cost to create and digitally sign, low cost to manufacture in comparison to (theoretical) exact-replication cost, low cost to authenticate an instance, sufficiently robust to "wear and tear," and computationally difficult to construct a new object with a similar "fingerprint." The "fingerprint" is an extracted set of features that reliably and uniquely represent a CoA, as defined by proximity bound and distance metric. The fingerprint for RF-DNA is the near-field EM response measured by an RF transceiver matrix [6]. The PUF/CoA response in [6] was measured using a matrix of antennas that transmitted and received EM waves in the near-field.

The isotropic nature of these CoAs allows for mathematical linearity, making the response simple to calculate. However, reverse engineering and cloning of a specific arrangement will be beyond reasonable computation challenges. This satisfies the PUF property of "easy to measure, difficult to replicate," as well as the requirements of CoA [6]. The RF-CoA can be used for personnel authentication, product identification, or any other situation where physical authenticity is required. At the same time, we must remember that there is computational complexity in the inverse linear solution to the PUF response if there is sufficient quantity and randomness in the orientation of conductive/dielectric items such that the PUF cannot be cloned. Even if an attacker has no computational complexity, there will be manufacturing complexity (i.e., the difficulty of manufacturing precise 3D replicant objects). This appears true for end user attackers but not necessarily for government or corporate-level adversaries [6].

The uniqueness of the CoA's response has met the PUF requirements in this technique. Another advantage of the technique is the ease with which new RF-DNA CoAs can be designed and manufactured, as well as the near-field EM benefits of low-power design and hard-to-jam communication [6]. The RF-DNA work, on the other hand, is not without limitations. Designing and implementing the back-end circuitry and antenna matrix are not easy tasks. Furthermore, the area impact of RF-DNA should not be overlooked, especially in today's age of microscopic RFIDs. Mitigation techniques such as dimension reduction and linearity abuse can reduce computational difficulty. Finally, it should be noted that it has had little impact compared to other existing techniques.

6.8 Conclusions

Using RF technology to operate as PUFs is a reliable and accessible method of reducing piracy and counterfeit output. As demonstrated by the discussion, RF-PUFs meet the complexity requirements for classification as CoAs. The uniqueness and reliability show that the devices are effective as identifiers. Furthermore, in order to qualify as proper CoAs, the cost of attempting to clone the device must be several orders of magnitude greater than the cost of reverse engineering the PUF itself. Although RF-PUFs have proven to be as robust as RO-PUFs, the technology has limitations. A significant limitation for RF-PUFs is their decreasing effectiveness as IoT network sizes grow. As IoT connectivity grows, this limitation may become a hindrance as technology advances. The fact that LC-PUFs are weak-PUFs rather than strong PUFs means that they provide fewer CRPs and cannot be used in critical applications. Measurement noise and inaccuracy, CoA placement positional tolerance, RFID storage size limits, and potential insider threats are all disadvantages of RF-PUF systems. RF-PUFs, despite their major limitations, has proven to be a cost-effective and accessible method for businesses to confirm authenticity. This will greatly improve the security of the electronic device supply chain, resulting in increased product revenue for companies as piracy declines [22].

References

1. Amsaad F, Pundir N, Niamat M (2018) A dynamic area-efficient technique to enhance ROPUFs security against modeling attacks. In: Computer and network security essentials. Springer, Berlin, pp 407–425
2. Bhunia S, Tehranipoor MH (2019) Hardware security: a hands-on learning approach. Morgan Kaufmann Publishers, Burlington
3. Carluccio D, Lemke-Rust K, Paar C, Sadeghi AR (2006) E-passport: The global traceability or how to feel like a ups package. In: International workshop on information security applications. Springer, Berlin, pp 391–404
4. Chatterjee B, Das D, Sen S (2018) RF-PUF: IoT security enhancement through authentication of wireless nodes using in-situ machine learning. In: 2018 IEEE international symposium on hardware oriented security and trust (HOST), pp 205–208, https://doi.org/10.1109/HST.2018.8383916
5. DeJean G, Kirovski D (2007) RF-DNA: Radio-frequency certificates of authenticity. In: International workshop on cryptographic hardware and embedded systems. Springer, Berlin, pp 346–363
6. Dejean G, Kirovski D (2007) RF-DNA: Radio-frequency certificates of authenticity. In: Proceedings of the 9th international workshop on cryptographic hardware and embedded systems, CHES '07. Springer, Berlin, pp 346–363. https://doi.org/10.1007/978-3-540-74735-2_24
7. Farahmandi F, Huang Y, Mishra P (2020) Automated test generation for detection of malicious functionality. In: System-on-chip security. Springer, Cham, pp 153–171
8. Gerdes RM, Daniels TE, Mina M, Russell S (2006) Device identification via analog signal fingerprinting: a matched filter approach. In: NDSS, CiteSeer
9. Guajardo J, Škorić B, Tuyls P, Kumar SS, Bel T, Blom AH, Schrijen GJ (2009) Anti-counterfeiting, key distribution, and key storage in an ambient world via physical unclonable functions. Inf Syst Front 11(1):19–41. https://doi.org/10.1007/s10796-008-9142-z
10. Guin U, DiMase D, Tehranipoor M (2014) Counterfeit integrated circuits: Detection, avoidance, and the challenges ahead. J Electron Test 30(1):9–23
11. Guin U, Huang K, DiMase D, Carulli JM, Tehranipoor M, Makris Y (2014) Counterfeit integrated circuits: a rising threat in the global semiconductor supply chain. Proc IEEE 102(8):1207–1228
12. Juels A (2006) RFID security and privacy: a research survey. IEEE J Sel Areas Commun 24(2):381–394
13. Kang H, Hori Y, Satoh A (2012) Performance evaluation of the first commercial PUF-embedded RFID. In: The 1st IEEE global conference on consumer electronics 2012, pp 5–8. https://doi.org/10.1109/GCCE.2012.6379926

14. Kirovski D (2010) Anti-counterfeiting: Mixing the Physical and the Digital World, pp 223–233. https://doi.org/10.1007/978-3-642-14452-3_10
15. Lakafosis V, Traille A, Lee H, Gebara E, Tentzeris MM, DeJean G, Kirovski D (2011) RFID-CoA: The RFID tags as certificates of authenticity. In: 2011 IEEE international conference on RFID, pp 207–214. https://doi.org/10.1109/RFID.2011.5764623
16. Preradovic S, Karmakar NC (2009) Design of fully printable chipless RFID tag on flexible substrate for secure banknote applications. In: 2009 3rd international conference on anti-counterfeiting, security, and identification in communication, pp 206–210. https://doi.org/10.1109/ICASID.2009.5276935
17. Pundir N, Amsaad F, Choudhury M, Niamat M (2017) Novel technique to improve strength of weak arbiter PUF. In: 2017 IEEE 60th international midwest symposium on circuits and systems (MWSCAS). IEEE, New York, pp 1532–1535
18. Rieback MR, Crispo B, Tanenbaum AS (2006) The evolution of RFID security. IEEE Pervasive Comput 5(01):62–69
19. Suh GE, Devadas S (2007) Physical unclonable functions for device authentication and secret key generation. In: 2007 44th ACM/IEEE design automation conference. IEEE, New York, pp 9–14
20. Tekbas O, Serinken N, Ureten O (2004) An experimental performance evaluation of a novel radio-transmitter identification system under diverse environmental conditions. Can J Electr Comput Eng 29(3):203–209
21. Thompson DR, Chaudhry N, Thompson CW (2006) RFID security threat model. In: In Proceedings Acxiom Laboratory for Applied Research (ALAR) Conference
22. Vashistha N, Hossain MM, Shahriar MR, Farahmandi F, Rahman F, Tehranipoor M (2021) eChain: A blockchain-enabled ecosystem for electronic device authenticity verification. IEEE Trans Consum Electron 68(1):23–37
23. Xiao K, Rahman MT, Forte D, Huang Y, Su M, Tehranipoor M (2014) Bit selection algorithm suitable for high-volume production of SRAM-PUF. In: 2014 IEEE international symposium on hardware-oriented security and trust (HOST). IEEE, New York, pp 101–106
24. Yang K, Forte D, Tehranipoor M (2015) ReSC: RFID-enabled supply chain management and traceability for network devices. In: International workshop on radio frequency identification: security and privacy issues. Springer, Cham, pp 32–49
25. Yang K, Forte D, Tehranipoor MM (2015) Protecting endpoint devices in IoT supply chain. In: 2015 IEEE/ACM international conference on computer-aided design (ICCAD). IEEE, New York, pp 351–356
26. Yang K, Botero U, Shen H, Woodard DL, Forte D, Tehranipoor MM (2018) UCR: An unclonable environmentally sensitive chipless RFID tag for protecting supply chain. ACM Trans Des Autom Electron Syst (TODAES) 23(6):1–24
27. Zhai C, Fu H, Zhang Q, Cao X (2020) Research on military supply support application based on RFID technology. In: 2020 IEEE 20th international conference on communication technology (ICCT), pp 1620–1624. https://doi.org/10.1109/ICCT50939.2020.9295713

Chapter 7
Optical PUF

7.1 Introduction

Product counterfeiting has been a long-standing issue [15, 34]. When an idea or a product makes a good profit, some people will try to copy it because it is a lucrative way to make more money with less effort. Recent globalization has significantly reduced the number of barriers that exist between nations and business entities [32]. International trade has grown in popularity, necessitating the need to track, monitor, or authenticate the product throughout the supply chain [18, 35]. This is especially important in sensitive industries such as the military, medical, energy, and space. One direct approach to addressing this issue is to identify and authenticate each product with a unique, unclonable fingerprint [33, 39, 41]. PUFs, or physical unclonable functions, generate unique identifying signatures through physical manufacturing variability. Since [27] introduced the concept of physical one-way function in 2002, researchers have designed and proposed new forms of PUFs every year [2, 25, 26, 28, 40]. PUF provides a specific response to a given challenge that is tied to the PUF's physical unclonable properties. PUF has a one-of-a-kind set of challenge–response pairs (CRPs) that can be used for secure authentication, object identification, anti-counterfeiting applications, cryptography, and other security applications. PUF shot to prominence in 2010 when silicon fingerprints were embedded in smartcards using PUF [1]. PUFs are widely used as a one-of-a-kind cryptographic key generator for commercial FPGAs. To be useful in a security application, a PUF must possess certain characteristics. For example, a PUF should be unique, which means that it can be distinguished simply by analyzing its CRPs. PUF should be reliable, which means that its response to a specific challenge should not deteriorate with time and environmental change. PUF should have uniformity, which means that the response should be completely random and only depend on process variation, making it impossible to model or clone the PUF behavior by observing the response bias.

© The Author(s), under exclusive license to Springer Nature Switzerland AG 2023
M. Tehranipoor et al., *Hardware Security Primitives*,
https://doi.org/10.1007/978-3-031-19185-5_7

Optical PUFs are PUFs whose challenges and responses are highly dependent on the optical properties of the material as well as light wavelength. Optical PUFs of this type can be utilized as anti-counterfeit labels for currency and strategic weapons, according to [4]. In addition to software licensing and object identification, optical PUFs can generate fingerprints. Using optical PUFs, fingerprints can be generated and authenticated using optical methods. Manufacturing variability can be used to model the PUF in the case of optical disks such as CDs and DVDs, and authentication can be done by the photodiode while reading CDs and DVDs [17]. In the case of paper currencies, optical properties of fiber, color-changing ink, hologram, nano-optical fibers mixed in paper slurry, etc., can be used to generate optical fingerprints and then authenticated by an appropriate light source, viewing angle, filters, and microscopes [9]. Nanoscale fingerprints can be produced by randomly distributed silver nanowires or AgNWs [21], randomly distributed plasmonic nanoparticles such as Gold Nanoparticles, AuNPs, and Silver Nanoparticles, AgNPs [29], randomly distributed Phosphor particles [7]. Such nano-fingerprints can be used on different products or on the packaging of the products to provide anti-counterfeit measures. To facilitate identification Bragg's reflection from Cholesteric Liquid Crystals (ChLC) can be used to produce Security-in-the-shell (SSh) tags [23]. Also, Rayleigh Scattering in Optical Fibers can be utilized to generate a unique ID for each item leading to an IDoT (ID of Things) [6].

This chapter examines various optical PUFs. The benefits and drawbacks of using such PUFs in practical situations are weighed against well-established anti-counterfeit and identification methods. A threat model is also considered, which includes counterfeiting in the overall supply chain in electronics or Integrated Circuits and any other industry such as medicine, software, entertainment, heavy machinery, valuable packaged products, expensive paintings, etc. The goal is to find a comprehensive solution to the counterfeit problem and provide clarity on whether optical PUF is a viable solution.

7.2   Background

This section discusses the threat model and challenges faced to protect against piracy and counterfeiting.

7.2.1   Threat Model

Counterfeiting is not limited to electronic devices and software; it can also be seen in everyday commodities, including currency, throughout the global supply chain [11, 16]. With the expansion of global trade, the scope and threat of counterfeiting in every industry grow by the day. For years, brand owners have suffered huge losses as a result of the prevalent counterfeit industry, [20]. Software piracy and counterfeiting are major issues, and the industry is losing billions of dollars [13]. Often, pirated software is copied and distributed using CDs and DVDs, putting software copyright at risk. To maintain trust between nations, the current economy relies on the reliability and authenticity of currency. In other words, a nation's sovereignty is dependent on the credibility of its currency. There have been numerous instances of counterfeit currency, necessitating the development of counterfeit deterrent features for currency. Almost all industries face a threat from counterfeiting expensive items and selling them as genuine articles. In the medical field, such forgery can be fatal. Counterfeit paintings by famous artists have long been a source of concern for art collectors and museums. Overall, there is always the possibility of counterfeiting as long as the counterfeit cost is significantly less than the price of the product being counterfeited. RFID is widely used for identification purposes, but it is vulnerable to cloning, reverse engineering, and tampering attacks [38]. As a result of these factors, the demand for an anti-counterfeiting system is increasing on a daily basis. The anti-counterfeiting system necessitates the use of a physical identifier, which should be attached to the product as a PUF [14]. A consumer can easily authenticate a product by matching its challenge and responses with those on the server [35].

7.2.2 Challenges

To prevent software piracy, security measures to detect counterfeit CDs and DVDs should be implemented. SecuROM technology addresses this issue by linking generated identifiers to executable files. However, this technology appears to be fragile, as used CDs quickly become unidentifiable. As a result, a more durable solution is required. Visual recognizability, inherent resistance to copying, resistance to simulation, ease of machine-readability, manufacturability, recurring cost, durability, capital cost, and other factors must be considered when selecting counterfeit deterrent features for currencies. In the case of currency, multiple anti-counterfeit features must be used concurrently to ensure maximum confidence in authenticity. Anti-counterfeit labels, for example, must be simple to make, detect, and reproduce. This type of label is widely used in the electronics and pharmaceutical industries. Furthermore, sensors and anti-counterfeit labels must be included on the devices to protect them from tampering, or material aging [30]. The security flaws of RFID and other non-optical PUFs in object identification force researchers to develop a more comprehensive solution. To address such challenges, researchers are developing various types of optical PUFs that provide a high level of security while overcoming the shortcomings of non-optical PUFs.

7.3 CD Fingerprint as Optical PUF

Each CD has a unique fingerprint that varies depending on manufacturing variation, and software licenses can be associated with the CDs using the fingerprints [17]. The main advantage of this method is that no expensive scanner is required to extract fingerprints from the CD surface. Electrical signals can also be used that are generated by a photodiode inside a CD reader. Data in series in the form of lands and pits is typically stored on a CD. Due to such storage of data the disc's surface form a spiral track. The term "location" refers to lands or pits because their distribution is similar. The stored data is determined by the length of the location. There are nine distinct lengths available, ranging from about 900 to 3300 nm in 300 nm increments. Because of the minute imperfections in the CD writing process, small variations in the lengths of locations are introduced when data is written on a CD. For generation of unique fingerprints for each of the CDs, these minor variations in location length were used. A new method for converting generated fingerprints into 128-bit cryptographic keys was also proposed, which involves using fuzzy extractors over the Lee metric.

Photodiodes inside CD players generate electrical signals based on location lengths. However, these electrical signals are contaminated by photodiode noise. This electrical noise was also assumed to have a Gaussian distribution. The electrical noise is represented by the standard deviation of this distribution, which is assumed to be the same for all CDs and locations because it is dependent on the CD reader. This standard deviation is expected to be low in order to improve CD identification. Experiments with 100 identical CDs were carried out to validate the length and electrical noise distribution assumptions. The noise distribution was captured by reading the same CD several times, and the length distribution was captured by reading the same locations on 100 identical CDs. The results showed that the distance between histograms was large enough to identify each CD uniquely. Similarly, readings from the same location followed a Gaussian-like distribution (on hundred identical CDs). To reduce the Gaussian electrical signal noise, simple averaging was used. With a much larger dataset of 500 locations on 100 identical CDs, it was demonstrated that the length distribution is strongly Gaussian [17]. As a result, the length of each location was normalized during the results to obtain a mean value of zero.

7.3.1 Fingerprint Extraction

The data collected by the CD reader needs to be processed to remove DC offset and reading noise before generating the fingerprint. At first, each location's length was read multiple times and then averaged to remove the electrical signal noise. Each location's length was then normalized so that all location lengths would have a similar distribution. Finally, a second normalization is performed on the data set by subtracting the mean of the lengths of different locations on the same CD to remove the DC offset. Even with this averaging and normalization, the noise will not be completely removed, and there will be errors in the length data. The fuzzy extractor technique was chosen to correct these errors as it is more relevant to the CD fingerprint technique [10]. A fuzzy extractor can be defined as a technique that generates the same output string from a noiseless and noisy input. But the major challenge for the fuzzy extraction technique is the Gaussian nature of the errors. These errors will cause a shift error in the length of locations instead of a bit-flip error. So, a more suitable distance metric in this scenario is the Lee metric [22]. A threshold scheme was proposed that uses the Hamming distance and fuzzy extractor over the Lee metric. This threshold scheme permits higher noise tolerance. The proposed technique defines a threshold τ, which solves the error-correcting problem concerning the Lee distance. When the threshold is zero, the error rate will become high. But if the threshold is chosen to be very high, the Hamming distance between the fingerprints of different CDs decreases.

7.3.2   Entropy Estimation

To quantify the entropy of the variability of CD manufacturing and the noise, the Context-Tree Weighting Method (CTW) was used [36]. At first, 500 location lengths were collected for each CD, and the size of symbols was 2. The CTW algorithm resulted in 0.83 bits of entropy per bit that was extracted. When the data was averaged over more and more samples, the entropy rate dropped. This result was expected as the noise in the data set adds to the entropy, and averaging will reduce the noise level. Adjacent location on a CD has a dependency because the probability of sharp change in manufacturing variation between two locations is very low. The most significant bit of the symbol will give less entropy than the least significant bit. It was verified by applying the CTW algorithm on each of these bits separately. So, the least significant bits were used to extract the fingerprint and the key. The resultant entropy of the scheme was 0.98, and the error rate was 0.08. These results were used in the fuzzy extractor technique to generate secure keys. For more elaborate results, please refer to [17].

7.3.3   Robustness

The robustness of the CD fingerprinting technique needs to be studied with more importance as this technique is expected to be used to tie software to CDs. To protect the CD fingerprint from any physical damage, length information from different sectors of the CD can be used. Another major challenge against the robustness of this technique is aging. Viscoelastic relaxation of the disc material can cause fluctuations in bit patterns and lengths of locations. But the good news is that viscoelastic relaxation will occur mostly at the early stage and will have less effect at the latter stages of a CD lifetime. Because of the length fluctuations, the speculated lifetime of fingerprints is 10% of the CD lifetime of approximately 20 years which is good enough for the targeted use scenario. The lifetime of the fingerprint can also be increased by extracting it from a slightly aged CD as viscoelastic relaxation have an exponential dependence on time.

7.3.4   Limitations

The entropy of manufacturing variability and fingerprint robustness were thoroughly examined, but quantitative analysis of fingerprint uniqueness among different identical CDs and fingerprint reliability remains to be completed. Another significant disadvantage of this method is that the data collected by the CD reader has a high error rate. Rigid analysis is required to reduce the noise level and correct the errors. Although the on-chip data collection process is expected to reduce error rates, this has yet to be proven.

7.4   Counterfeit Deterrent Currency

To address the issue of counterfeit currency, the committees on Next-Generation Currency Design, Commission on Engineering and Technical Systems, and National Research Council collaborated to evaluate various counterfeit deterrence methodologies and make recommendations [9]. However, the recommendations are primarily aimed at the US dollar, but they are equally applicable to any currency in the world.

There were existing counterfeit deterrent mechanisms in place for US currency at the time of the committee's review, which included: special paper containing cotton and linen fibers, additional red and blue fibers, serial numbers, microprinting, security thread, fine-line engraving, and special color with a light-green tint. However, the committee discovered that the existing mechanisms could not be easily authenticated by inexperienced or untrained employees at the point of sale. As a result, trust in counterfeit detection was low, necessitating the development of newer and more innovative counterfeit deterrent features. The committee devised a number of techniques in search of novel counterfeit deterrent features. The characteristics were classified using the following criteria:

1. **Substrate-based feature:** These features include plastic substrates, watermarks, enhanced security thread, laminated substrates, enhanced fibers, tinted substrates.
2. **Color-based feature:** These features cover a wide variety of color and their properties.

3. **Ink for printing-based feature:** These features consist of color-shifting ink, photoluminescent ink, absorbing infrared ink.
4. **Design-based feature:** Some examples of design based feature are variable-sized dot patterns, barcode, aliasing line structures, see-through, latent images.
5. **Post-printed optically variable feature:** For example, diffraction-based holograms, diffraction-based pixel grams, multiple-diffraction gratings, embedded zero-order diffraction grating, thin-film interference filters.
6. **Random pattern/ Encryption:** Public-key encryption with two visible features or one visible covert feature.

Among the numerous innovative technologies, post-printed optically variable features, particularly diffraction-based holograms, stand out.

7.4.1 Diffraction-Based Hologram

Diffraction-based holograms are widely used in credit cards as a counterfeit deterrent. In addition, it has the potential to be used in currency-based counterfeit deterrents. The optical setup for this feature is shown in Fig. 7.1. Holograms are also known as holographs. A hologram is a recording of the interference pattern from an object which utilizes light diffraction to create a virtual 2D or 3D image. In the figure, the interference pattern has been developed by the picture beam and the reference beam. Both beams are coherent. In the case of counterfeit currency deterrent, this picture beam can be set to be two-dimensional, such as a presidential picture or a barcode which can be machine-readable or any other image.

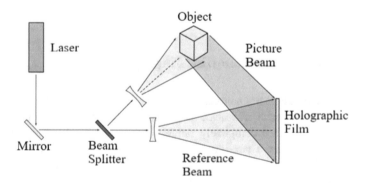

Fig. 7.1 Block diagram illustrating the optical setup for reading a hologram from a target

7.4.2 Proof of Authenticity

Whether the currency under test is authentic or not can be easily checked by looking at an appropriate viewing angle with an appropriate light source shining the hologram spot at an angle that is the same as the laser reference beam angle when the hologram was recorded. A unique feature of holography is that it can incorporate more than one picture as a fingerprint. For example, a picture of a president can be seen if the reference beam is incident at a certain angle, and a numeric value denoting the currency worth if the reference beam is incident at a different angle which adds to the obscurity of the signature. Recreation of such a signature by an attacker is next to impossible, given the extremely fine structural features of the hologram. Holograms cannot be scanned by a computer or read by a photocopier. The optical properties of the hologram make it a superior choice when it comes to providing proof of authenticity.

7.4.3 Limitations

The durability of the diffraction-based hologram feature is a significant disadvantage. This technology works well when used on a rigid substrate; however, when used on a flexible substrate, such as a currency note, the metallic film (object) wrinkles and the hologram image quickly becomes unrecognizable. To significantly increase durability, the committee recommends placing the object on a window on a flexible substrate rather than using a metallic strip. A significant challenge of this method is ensuring the durability or PUF reliability under wear and tear, which is a prime area for further investigation.

7.5 Anti-counterfeit Nanowire Fingerprint

Fingerprint generation at the nanoscale has been proposed to mitigate counterfeit risks [21]. However, such nanoscale fingerprints generated based on randomly distributed nanowires are completely random and prohibitively expensive to be duplicated by an attacker. Therefore, although only nonrepeatable patterns of silver nanowires (AgNWs) on transparent PET film were proposed, this method also applies to other types of material. Anti-counterfeit technology such as holography, laser surface authentication, and radio frequency identification has been researched extensively. Nano-level technologies, such as nano barcodes [8], quantum dot tags [31], etc., can be used for generating anti-counterfeit fingerprints. By casting a small number of nanorods upon a plate, a unique pattern can be generated. As the distribution of nanowires is naturally random, the chance of two patterns being similar is astronomically low and almost next to impossible. To incorporate more

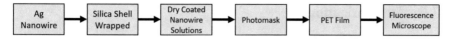

Fig. 7.2 Illustration of steps to generate unique nanowire fingerprints

randomness and increase the strength of the signature, two different colors of AgNWs were introduced in [21]. The color was introduced by adding dye to a silica coating designed to wrap around the AgNWs. Two different dyes were used, giving rise to red and green colored silica-coated AgNWs. Some silica coatings did not absorb any dye and created three distinct patterns: green, red, and dye-free nanowires. The fingerprint generation is illustrated in Fig. 7.2.

The fingerprint generation technique entails preparing green, red, and colorless AgNWs, lithographically creating an orientation marker and target marker on transparent PET. Finally, a predetermined amount of AgNW dispersion is applied to the target marker. Random writing serves as an orientation marker, and the center of the symbol "X" serves as a target marker. 1 µL of AgNW solution was found to be sufficient to prepare five or six fingerprint patterns.

7.5.1 Fingerprint Extraction

Using a fluorescence microscope fingerprinting patterns were observed. The microscope would be equipped with red and green excitation filters. The system can capture images in both bright-field and fluorescent modes. SEM and TEM were also used to observe the patterns. AgNW was chosen because it is easily visible, anisotropic in nature, and can be prepared in large quantities. Silica shell was also used because it not only acts as a dye but also increases the thickness of AgNWs, increasing visibility significantly. The NWS was easily visible and measured 10–50 um in length and 70 nm in diameter. Fingerprints were easily identified using 1000X magnified images taken from the target (center of X).

1. The shortage of AgNWs with identical length
2. Near infinite spatial degrees of freedom by multiple AgNWs
3. A varied mixing ratio of red, green, and colorless AgNWs

Combined these factors ensure that the number of potential random patterns is nearly infinite. It was determined that each unique fingerprint could be assigned a unique ID and stored in a database for authentication purposes.

7.6 Authentication

The fingerprint is positioned correctly for authentication, and both bright-field and fluorescence images are captured. It reduces the likelihood of authentication failure. This method can be automated by an algorithm that digitizes and recognizes the information (position and color) of the AgNWs. The extracted data can then be cross-checked against the database to ensure its authenticity.

7.6.1 Limitations

Along with AgNWs, there are some Ag particles in the AgNW dispersion that cannot be removed. Although these Ag particles add to the randomness of the pattern, they can make obtaining high-fidelity images during the fingerprint generation and authentication process difficult. A large number of AgNWs can form complex patterns, making counterfeiting more difficult. However, it also increases the difficulty of image processing. As a result, a compromise must be reached. Another issue is the longevity of the AgNWs' fluorescence. After two years, the fluorescence intensity of the AgNWs decreased, but the fingerprint remained distinct. This issue could be solved by using organic fluorescence wires. Although a thin transparent PET was used as a substrate, NWs can adhere to the substrate through van der Waals, hydrophobic, and hydrophilic interactions, which may not be strong enough for practical application. To get around this, to cover the entire pattern, a highly transparent polymer film can be used. Fingerprint generation is based on natural processes that are nearly impossible to duplicate. This technique has the potential to detect counterfeits and thus significantly reduce counterfeiting.

7.7 Anti-counterfeit Plasmonic Nanoparticles

The use of metallic nanoparticles (NPs) as fingerprints for anti-counterfeit applications has been proposed in literature [29]. Many researchers have previously used NPs as anti-counterfeiting measures. Complex optical barcodes are among those methods, but the manufacturing process is difficult. They also do not provide fingerprints that respond to the environment. In the work of Smith et al. [29], using drop-casting techniques, micrometer-sized, random patterns are created using Au NPs. Metallic nanoparticles have resonant wavelengths, which are also known as localized surface plasmon resonances (LSPRs). The LSPR shifts left, if the local media's RI changes. As a result, PUFs generated with metallic NPs can detect environmental exposures to pharmaceutical and electronic products. When an Ag and Au NPs mixture is deposited as fingerprints, such PUFs act as labels that provide information on tampering and age.

7.7.1 Fingerprint Generation

Fingerprint generation is done with the concept of utilizing the nano-fingerprints as optical PUFs in which the challenge is the exposure to light, and responses are the unique pattern of the diffraction-limited spots which come from the far-field scattering of the Au NPs. This scattering is used like fingerprints. This scheme is shown in Fig. 7.3.

Three simple algorithms were used in [29] to show that image registration and pattern matching techniques could be used to match NP fingerprints. When compared to other particles, the three techniques are determined by particle distance, area, and color.

An image database was created to store images of various NP patterns. The aim is to match a target image with the appropriate image database based on its corresponding pattern. The nanoparticle profiles with the most surface area are selected. These profiles are known as prominent profiles, and n prominent profiles were considered. The distance matrix calculates the distance between each particle in the database pattern and its corresponding particle in the target pattern. The distances are added up, and the matching pattern corresponds to the smallest

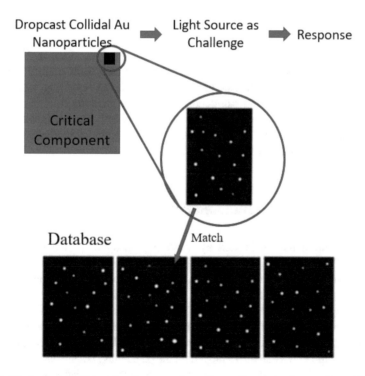

Fig. 7.3 Block diagram illustrating the generation of nano-fingerprint from unique diffraction pattern when Au NPs are exposed to light

sum. The effect of variation in illumination was also investigated by delaying measurements of the same area and moving the fingerprint in the x and x-y directions.

When test images were matched with the database for the x and x-y directions translated images or images taken consecutively, 100% accuracy was obtained in [29]. Instead of the x, y location, the color and area algorithm considers the particle profile color or area of neighboring particles. In this case, the color profile or area of the test image profile and the database image profile were compared. The color neighboring relationship was in a 3D red-green-blue (RGB) color space. Based on differences in the nearest neighbor's profile or area, an index was calculated. When profile areas were compared, 72% matching was obtained for $n = 8$ and 56% for $n = 30$; when profile colors were compared, 63% matching was obtained for $n = 8$ and 75% for $n = 30$. To see if the developed algorithm works, a cropped image was compared to the original image. The sensitivity of the Refractive Index (RI) is affected by composition, shape, and size, and it can be used as an environmental sensor for aging and tamper-evident labels. A mix of Au and Ag NPs was chosen, which scatter yellow and red wavelengths in water.

7.7.2 Limitations

It is claimed that metallic NP can detect environmental variation and aging, but strong experimental results have yet to be demonstrated to exemplify whether it works in practice. It only shows the color variation of scattering when the RI changes. However, tampering can occur without changing the RI. It has not been attempted to record the responses using various devices. The RGB values of the particles may change with different camera exposure, and the adopted threshold may not perform properly in that case.

7.8 Anti-counterfeit Random Pattern

A random pattern that is formed by scattering phosphor particles can be used as optical PUF [7]. Because the particles are randomly distributed, the physical identifier cannot be physically cloned. Robust bits can be extracted from the random pattern and used as a PUF response to encode the digital identifier. A technique for encoding and decoding the digital identifier in a random pattern is required in order to develop a user-friendly optical PUF-based anti-counterfeiting system. Many researchers are working on the development of PUF to prevent counterfeiting. Among these are PUFs that are deployed on the packaging, but are easily removable. Some PUFs are incorporated into the product, but they are costly and inaccessible to the consumer.

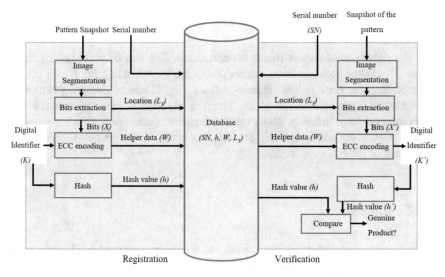

Fig. 7.4 Block diagram illustrating the registration and verification processes for the anti-counterfeiting system with a random pattern as the physical identifier

Phosphor has the property of phosphorescence, which means that it can retain its glow when exposed to energized particles such as electrons or ultraviolet light. Only UV light can reveal the pattern of phosphor particles. As a result, the product's appearance remains unchanged. There are two ways in which the phosphor particles are linked to the product:

1. Blended with the cover material of the product.
2. A tag of phosphor particle pattern is attached to the product permanently.

Since capturing variation may alter the particle pattern, an algorithm was proposed in [7] that can authenticate the product by removing variations in exposure while also extracting robust bits from the pattern snapshot. The illustration of the algorithm is shown in Fig. 7.4. The algorithm's two main components are registration and verification. During the registration process, a random and unique digital identifier is defined. To ensure confidentiality, its hash value is stored in the database. The database stores the pixel-based location of the bits or challenges, as well as the helper data generated by the ECC encoding. Durable bits are extracted from the pattern stored in the database with the same serial number during the verification process. Using the helper data, the digital identifier is decoded and the bits extracted. It is then compared to that saved in the database to ensure the product's authenticity.

Image segmentation is used to extract the pattern's location from a captured image. It is divided into two steps: separating the frame from the snapshot and separating the region from the frame. For more details on image segmentation, please refer to [7].

7.8.1 Bit Extraction and ECC Encoding/Decoding

The algorithm's next step is robust bit extraction. The size of phosphor particles varies. To capture local features, corner points are located using the Harris corner detector. Because these points are noisy and varying, K-means clustering is used to determine their centroid. A robust bit-extraction algorithm based on delta-strength was proposed for extracting distinct and robust bits from the pattern's centroids. Nonetheless, some bits escape this process; this is why ECC is used to increase the robustness of the bits even further.

7.8.2 Limitations

When it comes to product authentication, the algorithm proposed in [7] appears to be very strong. The main disadvantage is that the physical identifier is attached to the material cover. The PUFs attached to the cover are likely to be easily destroyed. When the scattering medium changes, the scattering may change. The scattering properties of air and water may differ. It is difficult to assess the reliability of the PUF without taking into account the effect of the environment on the response of the physical identifier.

7.9 Anti-counterfeit Liquid Crystal Shell

The application of Cholesteric Liquid Crystals (ChLC) as an optical PUF in object identification is an outstanding work proposed in [23]. A collection of secure ChLC shells can reflect light, creating colorful patterns that are unique, unclonable, and unpredictable. Because the fluid's viscosity varies, the movement of the shells within the medium during production gives it a distinct tag after solidification over a flat surface. The colored pattern is produced by interactions between different reflections (Bragg's reflection) of light wavelengths, which vary according to the diameter and thickness of the shells, relative distance and arrangement over the medium, angle of light incidence, and polarization. This one-of-a-kind tag, known as the Security-in-the-shell (SSh) tag, can be used to identify and authenticate any object. According to current research, SSh-tags can be used to detect counterfeit medical products and paintings.

Intrinsic uniqueness relies on physical properties instead of any artificial or mathematical properties, which is the main characteristic of a PUF, and SSh-tags have this uniqueness in terms of optical properties. Optical PUFs introduced earlier described by Pappu et al. [27] depending on scattering a laser beam, and UV-dye treated fibers [5] require specific light sources and reader. In contrast, SSh-tags work within the visible light range and can be easily read by a mobile phone or camera.

The microscope needed for analysis is very chip as well. Moreover, compared to other non-optical counterparts, SSh-tags cover a much wider range of applications which draws the community's interest in this research topic [24, 37].

7.9.1 Security Framework Model

The work in [23] uses a framework concerning the creation and processing of the tags, which is adapted from an earlier research work [3]. Figure 7.5 describes the generic proposed framework.

According to this framework, PUF is generated during the production phase by a Create() function that uses a manufacturing parameter *alpha* and some uncontrollable production variability. The number, thickness, and diameter of the shells, the production speed, and the pitch and bias of the ChLC are all creation parameters. The relative position of shells in the medium causes manufacturing variability. Furthermore, possible defects (nonuniform sphere) in the shells during production and artificial variability can contribute to production variability. Following production, the PUF is probed with the challenge x using the Eval() function, to which the Puf() function responds with a noisy output y. Both Eval() and Puf() are part of an indivisible process PF (). The extraction phase then processes the response to reduce noise using an algorithm Extract() and some helper data *beta*. (the portion of the picture to be analyzed). It removes some background color and provides a response for SSh-tag (the central, radial, and circumferential colored spot). When evaluation and extraction are combined, they yield a Physical Function System PFS (). Finally, the assessment phase compares the extracted response of a previous PUF with the currently probed PUF using an algorithm *Delta*().

A light source is used to illuminate the light bundle of ChLCs during the evaluation. The extraction phase replaces low-intensity pixels in the image with

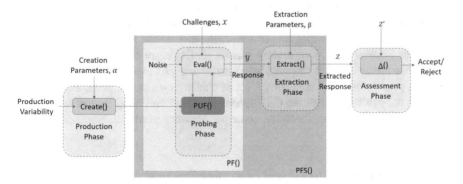

Fig. 7.5 Block diagram illustrating the generic framework of a PUF for creation and processing of tags

black pixels, removing background noise introduced during Eval(). The clustering algorithm then groups equally colored pixels while locating spheres that can be used as input to the algorithm. Following that, image alignments and calculating the distance between overlapping parts aid in determining whether two images are from the same PUF after noise removal in the assessment phase. MATLAB is used to implement all algorithms and methods. The extraction and assessment phase's security requirement is that responses from the same PUF be recognized, and responses from two different PUFs be rejected. Furthermore, if the extraction and assessment algorithms are good enough, they should contain images that are only PUF responses.

7.9.2 Security Argumentation and Limitation

The experiment is carried out on a small dataset (six PUFs with a total of 24 images), in order to meet the security requirements. It was assumed that the findings from this small dataset would be consistent with future research work with a larger dataset. Furthermore, no quantitative analysis of the PUF characteristics is performed throughout the work, only qualitative analysis. Nonetheless, the framework model for qualitative analysis presented in this research work is a noteworthy approach for future research work.

The reproduction of an SSh-tag by repositioning the shells in a new bundle is infeasible as the reproduction of the possible imperfections in the shells is not possible. So, an almost equal SSh-tag reproduced by an adversary has a different emerging colored pattern, and it can confirm its unclonability. Still, in this case, a limitation is that nothing is confirmed on whether it is possible to make a fake tag that can pass the security requirements of the extraction and assessment phase of the framework. Furthermore, whether an adversary can reproduce something like SSh-tags exploiting holograms is not investigated. Finally, unpredictability is confirmed by increasing the difficulty of a hypothetical simulation to predict a deterministic polynomial that passes the assessment requirements. The dependency of the pattern on the composition of the shells, disposition in the medium, relative distances, several reflections among various shells before capturing, and different light polarizations makes the simulation-based analysis exponentially difficult. To impede predictability by analyzing a set of challenge–response pairs, a movie can be well enough to increase the CRP space set almost unlimited instead of a single picture. However, this approach has some performance overhead.

7.9.3 Application in Object Identification and Advantages

SSh-tags are a good competitor of RFID in anti-counterfeiting applications. Having optical intrinsic physical uniqueness and unclonability, SSh-tags can be used to

Table 7.1 Advantages of SSh-tags over RFID tags

Security issue	RFID tags	SSh-tags
Tag switching	Removable	Tamper-resistant
Data integrity	Rewritable	Tamper-resistant
Imitation	Clonable	Unclonable
Data extraction	predictable	Unpredictable

prevent counterfeit medical products and painting. Although RFID tags have a wide range of applicability, their limitation and security vulnerability issues pave the way to go with the competitor SSh-tags, which exploits a technological novelty. Table 7.1 summarizes some advantages of SSh-tags over RFID in preventing various threats.

Since the local validation of SSh-tags is performed optically with low-cost instruments and no identifier is exchanged, unlike RFID, this technology shows more resilience against probing from distance attacks. However, solid research work is required to confirm the hypothesis. Furthermore, unlike RFID tags, SSh-tags can be made of different materials which are not potentially harmful to the objects to be authenticated.

7.10   FiberID: Molecular Level Identification

A novel idea of using Rayleigh Scattering in Optical Fibers as a physical unclonable function (PUF) has been proposed to provide a unique ID to each item—an IDoT (ID of Things) [6].

7.10.1   Operating Principle

There are several IDoT techniques available that are currently being used. The most common of these techniques are barcodes, QR codes, digital magnetic stripes, and RFID systems. These techniques have different limitations like large size, susceptibility to degradation and forgery, unreliability, expensive, etc. However, all these techniques suffer from a common limitation—reproducibility of the ID information. The FiberID technique addresses this common limitation by utilizing inherent irreproducibility in the molecular structure of the optical fiber. It uses a small optical fiber section to extract an irreproducible ID from its molecular imperfections in the fiber material. These imperfections in optical fiber structure are inherent to material fabrication process variation and are assumed to be irreproducible as:

1. The imperfections are random.
2. The size of the imperfections are at the molecular level.
3. The number of imperfections is large and random.
4. The size of the fiber core is very small.

Back-reflected Rayleigh scattering was proposed to extract irreproducible ID information (IDI) from molecular level imperfections. Rayleigh scattering is the elastic scattering of light by particles that have a much smaller size than the wavelength of light. When light passes through the optical fiber, it interacts with the fiber material's molecules and the imperfections in the glass matrix. As a result, a small part of the light is back-reflected, which is known as Rayleigh scattering [12]. This scattering is a unique and irreproducible IDI of a FiberID, which can be used to identify and authenticate objects.

7.10.2 FiberID System

In [6], a FiberID system has been proposed which is shown in Fig. 7.6. A section of an optical fiber can be embedded in items that we want to uniquely identify. The scanner device interrogates the items using a laser and extracts feature (IDI) using OFDR (Optical Frequency Domain Reflectometry) technology. Initially, a database is created with the IDI and PIN (Personal Identification Number) of each item. During verification, the scanner matches the extracted IDI of an item with the database stored IDIs and gives a "Yes" or "No" output. The matching process is cross-correlation which checks the similarity between two IDIs. To test the verification performance of the FiberID system, the 40 FiberIDs were used, each with a length of 1 cm. Equal Error Rate (EER) is chosen as the metric to evaluate the performance of the verification process. EER is defined as the error rate when the threshold is chosen in such a way that results in a False alarm rate equal to one minus Detection rate [19]. The experiment gave an EER of 0.06%, indicating that the FiberID technique is well-suited for anti-counterfeit applications.

Fig. 7.6 Block diagram illustrating the generation of FiberID

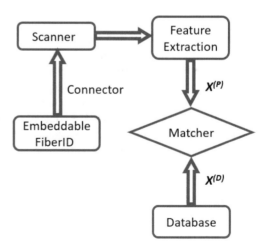

7.10.2.1 Challenges and Limitations

The effects of optical fiber length and temperature on the ROC curve of the FiberID system have been analyzed as part of the analysis. This analysis concluded that a FiberID of at least 1 cm should be used to obtain a robust verification result (EER < 1%). Also, the performance of FiberID is good between 30–55 °C as the EER was less than 0.6%. However, the performance analysis against extreme temperature and pressure is yet to be performed.

The FiberID technique is sufficiently robust for product identification and authentication. Furthermore, the cost of integrating an optical fiber with various objects is low. However, the scanner and ID verification costs are prohibitively expensive for this technique. Furthermore, FiberID is based on a contact-based scanning process, whereas many applications require a contact-less scanning approach. Nonetheless, the FiberID technique provides IDI irreproducibility for IDoT applications that no other IDoT technique currently available can.

7.11 Conclusions

From the literature review, it is evident that optical PUF provides strong unclonable signatures without the use of any electronics circuit. As a result, optical PUFs are much more difficult, if not impossible, to clone. Optical PUF provides greater security than barcodes, QR codes, and other easily replicated codes that can be copied or scanned. It was also discovered that optical PUFs could improve software licensing, deter currency counterfeiting, and provide more reliable identification than RFID. However, optical PUFs frequently suffer from durability issues, jeopardizing their long-term reliability. In addition, various authentication methods, such as a microscope, optical filters, cameras, and so on, add to the complexity of the identification process. To protect against counterfeiting, optical PUFs that are compatible with the surface of the product or package must be used in some cases, limiting the options when deciding which optical PUFs to use. Externally generated optical PUFs that are pasted on products or packages can be easily removed. Again, various signature generation mechanisms are associated with various optical PUFs. Extensive research on the quantitative analysis of optical PUF parameters for various optical signature generation mechanisms is required to enable a more accurate choice of optical PUF tailored for any specific application, which has the potential to be a very interesting research area.

References

1. (2021) Physical unclonable function. https://www.cardlogix.com/glossary/physical-unclonable-function-puf/
2. Amsaad F, Pundir N, Niamat M (2018) A dynamic area-efficient technique to enhance ROPUFs security against modeling attacks. In: Computer and network security essentials. Springer, Berlin, pp 407–425
3. Armknecht F, Maes R, Sadeghi AR, Standaert FX, Wachsmann C (2011) A formalization of the security features of physical functions. In: 2011 IEEE symposium on security and privacy, pp 397–412. https://doi.org/10.1109/SP.2011.10
4. Bauder D (1983) An anti-counterfeiting concept for currency systems. Technical Report, Technical Report PTK-11990, Sandia National Labs
5. Bulens P (2010) How to strongly link data and its medium: the paper case. IET Inf Secur 4:125–136(11). https://digital-library.theiet.org/content/journals/10.1049/iet-ifs.2009.0032
6. Chen Z, Zeng Y, Hefferman G, Sun Y, Wei T (2014) FiberID: molecular-level secret for identification of things. In: 2014 IEEE international workshop on information forensics and security (WIFS), pp 84–88. https://doi.org/10.1109/WIFS.2014.7084308
7. Chong CN, Jiang D, Zhang J, Guo L (2008) Anti-counterfeiting with a random pattern. In: 2008 second international conference on emerging security information, systems and technologies, pp 146–153. https://doi.org/10.1109/SECURWARE.2008.12
8. Demirok UK, Burdick J, Wang J (2009) Orthogonal multi-readout identification of alloy nanowire barcodes. J Amer Chem Soc 131(1):22–23. https://doi.org/10.1021/ja806396h. pMID: 19072281
9. Division on Engineering and Physical Sciences, National Research Council CoE, Technical Systems NMAB, on Next-Generation Currency Design C (1993) Counterfeit deterrent features for the next-generation currency design. The National Academies Press, Washington, DC, pp 31–80. https://www.nap.edu/read/2267/chapter/1#xii
10. Dodis Y, Ostrovsky R, Reyzin L, Smith A (2008) Fuzzy extractors: how to generate strong keys from biometrics and other noisy data. SIAM J Comput 38(1):97–139. https://doi.org/10.1137/060651380
11. Finlay R, Francis A (2019) A brief history of currency counterfeiting. RBA Bulletin, September, Viewed 20
12. Froggatt M, Moore J (1998) High-spatial-resolution distributed strain measurement in optical fiber with Rayleigh scatter. Appl Opt 37(10):1735–1740. https://doi.org/10.1364/AO.37.001735. http://www.osapublishing.org/ao/abstract.cfm?URI=ao-37-10-1735
13. Givon M, Mahajan V, Muller E (1995) Software piracy: estimation of lost sales and the impact on software diffusion. J Market 59(1):29–37
14. Guin U, Forte D, Tehranipoor M (2013) Anti-counterfeit techniques: from design to resign. In: 2013 14th international workshop on microprocessor test and verification. IEEE, Piscataway, pp 89–94
15. Guin U, DiMase D, Tehranipoor M (2014) Counterfeit integrated circuits: detection, avoidance, and the challenges ahead. J Electron Testing 30(1):9–23
16. Guin U, Huang K, DiMase D, Carulli JM, Tehranipoor M, Makris Y (2014) Counterfeit integrated circuits: a rising threat in the global semiconductor supply chain. Procee IEEE 102(8):1207–1228
17. Hammouri G, Dana A, Sunar B (2009) CDs have fingerprints too. In: Clavier C, Gaj K (eds) Cryptographic hardware and embedded systems - CHES 2009. Springer, Berlin, pp 348–362
18. Hossain MM, Vashistha N, Allen J, Allen M, Farahmandi F, Rahman F, Tehranipoor M (2022) Thwarting counterfeit electronics by blockchain. https://scholar.google.com/citations?view_op=view_citation&hl=en&user=n-I3JdAAAAAJ&citation_for_view=n-I3JdAAAAAJ:9ZlFYXVOiuMC
19. Jain A, Ross A, Prabhakar S (2004) An introduction to biometric recognition. IEEE Trans Circ Syst Video Technol 14(1):4–20. https://doi.org/10.1109/TCSVT.2003.818349

20. Jordans E (2007) Faking out the fakers. Businessweek, 2007, pp. 76–80
21. Kim J, Yun JM, Jung J, Song H, Kim JB, Ihee H (2014) Anti-counterfeit nanoscale fingerprints based on randomly distributed nanowires. Nanotechnology 25(15):155303. https://doi.org/10.1088/0957-4484/25/15/155303
22. Lee C (1958) Some properties of nonbinary error-correcting codes. IRE Trans Inf Theory 4(2):77–82. https://doi.org/10.1109/TIT.1958.1057446
23. Lenzini G, Ouchani S, Roenne P, Ryan PYA, Geng Y, Lagerwall J, Noh J (2017) Security in the shell: an optical physical unclonable function made of shells of cholesteric liquid crystals. In: 2017 IEEE workshop on information forensics and security (WIFS), pp 1–6. https://doi.org/10.1109/WIFS.2017.8267644
24. Lu H, Wilson R, Vashistha N, Asadizanjani N, Tehranipoor M, Woodard DL (2020) Knowledge-based object localization in scanning electron microscopy images for hardware assurance. In: ISTFA 2020, ASM international, pp 20–28
25. Maes R, Verbauwhede I (2010) Physically unclonable functions: A study on the state of the art and future research directions. In: Towards hardware-intrinsic security. Springer, Berlin, pp 3–37
26. Majzoobi M, Koushanfar F, Potkonjak M (2008) Lightweight secure PUFs. In: 2008 IEEE/ACM international conference on computer-aided design. IEEE, Piscataway, pp 670–673
27. Pappu R, Recht B, Taylor J, Gershenfeld N (2002) Physical one-way functions. Science 297:2026–2030
28. Pundir N, Amsaad F, Choudhury M, Niamat M (2017) Novel technique to improve strength of weak arbiter PUF. In: 2017 IEEE 60th international midwest symposium on circuits and systems (MWSCAS). IEEE, Piscataway, pp 1532–1535
29. Smith AF, Patton P, Skrabalak SE (2016) Plasmonic nanoparticles as a physically unclonable function for responsive anti-counterfeit nanofingerprints. Adv Funct Mater 26(9):1315–1321. https://doi.org/10.1002/adfm.201503989. https://onlinelibrary.wiley.com/doi/abs/10.1002/adfm.201503989. https://onlinelibrary.wiley.com/doi/pdf/10.1002/adfm.201503989
30. Spink J, Singh J, Singh SP (2011) Review of package warning labels and their effect on consumer behaviour with insights to future anticounterfeit strategy of label and communication systems. Packaging Technol Sci 24(8):469–484. https://doi.org/10.1002/pts.947. https://onlinelibrary.wiley.com/doi/abs/10.1002/pts.947. https://onlinelibrary.wiley.com/doi/pdf/10.1002/pts.947
31. Sun LW, Shi HQ, Li WN, Xiao HM, Fu SY, Cao XZ, Li ZX (2012) Lanthanum-doped ZnO quantum dots with greatly enhanced fluorescent quantum yield. J Mater Chem 22:8221–8227. https://doi.org/10.1039/C2JM00040G. http://dx.doi.org/10.1039/C2JM00040G
32. Tehranipoor M, Peng K, Chakrabarty K (2011) Introduction to VLSI testing. In: Test and diagnosis for small-delay defects. Springer, New York, pp 1–19
33. Tehranipoor M, Salmani H, Zhang X (2014) Integrated circuit authentication. Springer, Cham. https://doi.org/10.1007/978-3-319-00816-5
34. Tehranipoor MM, Guin U, Forte D (2015) Counterfeit test coverage: an assessment of current counterfeit detection methods. In: Counterfeit integrated circuits. Springer, Cham, pp 109–131
35. Vashistha N, Hossain MM, Shahriar MR, Farahmandi F, Rahman F, Tehranipoor M (2021) eChain: a blockchain-enabled ecosystem for electronic device authenticity verification. IEEE Trans Consumer Electron 68:23–37
36. Willems F, Shtarkov Y, Tjalkens T (1995) The context-tree weighting method: basic properties. IEEE Trans Inf Theory 41(3):653–664. https://doi.org/10.1109/18.382012
37. Woodard D, Tehranipoor MM, Asadi-Zanjani N, Wilson R, Lu H, Vashistha N (2022) Knowledge-based object localization in images for hardware assurance. US Patent App. 17/491150
38. Wu NC, Nystrom M, Lin TR, Yu HC (2006) Challenges to global RFID adoption. Technovation 26(12):1317–1323

39. Xiao K, Tehranipoor M (2013) Bisa: built-in self-authentication for preventing hardware Trojan insertion. In: 2013 IEEE international symposium on hardware-oriented security and trust (HOST). IEEE, Piscataway, pp 45–50
40. Xiao K, Rahman MT, Forte D, Huang Y, Su M, Tehranipoor M (2014) Bit selection algorithm suitable for high-volume production of SRAM-PUF. In: 2014 IEEE international symposium on hardware-oriented security and trust (HOST). IEEE, Piscataway, pp 101–106
41. Xiao K, Forte D, Tehranipoor MM (2015) Efficient and secure split manufacturing via obfuscated built-in self-authentication. In: 2015 IEEE international symposium on hardware oriented security and trust (HOST). IEEE, Piscataway, pp 14–19

Chapter 8
True Random Number Generators

8.1 Introduction

The rising need to implement cryptographic applications in reconfigurable hardware has led to more research in designing small, fast, and secure TRNGs with persisting entropy sources and harvesting mechanisms of electrical noise [18]. A reliable TRNG implementation with no bias in FPGA improves the overall security and performance of the system [24]. Obtaining a high bitstream of truly random bits without bias while not compromising on the design's footprint is a major challenge of TRNG implementation in FPGAs. For improving the statistical characteristics of the TRNGs, literature review shows many research techniques have been implemented. Some of these well-known methodologies have been analyzed in this chapter.

It is possible to design a true random number generator in hardware which is reconfigurable without Phase-locked loops (PLLs) [15]. Instead of PLLS, many FPGAs by Xilinx provide only delay-locked loops (DLLs). The primary distinction between DLL and PLL is that DLL lacks a voltage-controlled oscillator, whereas PLL does. On the other hand, DLL has a delay line that may self-regulate. In addition, DLL is preferable over PLL because DLL is less prone to errors [22]. This TRNG can also be implemented TRNG using only configurable logic blocks (CLBs) that use on-chip jitter [7]. Further, an advanced approach uses an open-loop structure that exploits a latch's metastability phenomenon as the source of randomness for designing a TRNG [3]. This design approach also uses post-processing techniques such as combining PRNG with TRNG or Von Neumann Corrector [13] to neutralize the bias in the generated output. Another modern design approach implements a TRNG by exploiting an RS latch's metastability [10]. This design can be implemented using LUTs by replacing RS latches with predefined placement constraints and verifying the combination of sampling interval and number of LUTs. For compact footprint, a purely digital design principle can be used for TRNG implementation [23]. This design approach can increase the random bit generator's fault tolerance against active adversaries to any degree. It exploits a mathematical model like the Urn and coupon collector problems used to calculate the final number of rings. This design's strong and short-sized code resilient function increases fault tolerance against adversarial attacks. A fast and secure TRNG can be designed with oscillator rings by sampling them at a much lower frequency to make the noise source stateless [21]. Since the post-processor is memoryless, the whole RNG becomes stateless. A simple and robust cyclic code-based resilient function improves the bias characteristics. A fast and compact TRNG can be designed by avoiding the post-processing module [27]. This TRNG design is a further advancement of the previously designed compact and pure digital TRNG architecture [23]. This design approach does not include any bias, so complicated post-processing is not needed. A flip-flop can be added after each oscillator ring to address the problem of multiple transitions during sampling. For better frequency dispersion, a three-inverter long oscillator ring can be used.

This chapter analyzes various research works on TRNG design and implementation in FPGAs, along with addressing the trade-off between parameters like area, randomness quality, and throughput of the TRNGs.

The rest of the chapter is organized as follows: In Sect. 8.2, we have briefly described the background of the problem statement and related concepts. Then, in Sect. 8.3, we have discussed the architecture of various TRNG implementations in FPGA and the limitations of each design. Finally, in Sect. 8.4, we conclude with a summary of designs.

8.2 Background

In cryptographic systems, random number generators (RNG) plays a vital role. In multiple computing applications, unpredictable bit sequences are used as cryptographic nonces, initialization vectors, block padding, challenges, and cryptographic keys. Weakness or failure in random number generators compromises the entire system's security.

There are two categories of RNG: First, the Pseudo-Random Number Generators (PRNGs), and second the True Random Number Generators (TRNGs). PRNGs are based on mathematical algorithms which are initialized with a seed to generate a random bitstream. PRNGs are fast and cost-effective, but they are not a reliable source of random bits for secure critical systems. They are unreliable because the internal state solely relies on the seed, posing a vulnerability in the cryptosystem and making it an easy target for adversaries. TRNGs, on the other hand, output random bits that are entirely based on the physical processes in the device. The quality of a TRNG design relies on three components:

1. **Entropy Source:** These are the physical variations like thermal noise, jitter, voltage variations, frequency of power supply, and radiations. A good entropy source should not exhibit any biases.

2. **Harvesting Mechanism:** It is the process of collecting entropy without disturbing the physical process. Therefore, an ideal harvesting mechanism should justify the underlying assumption about the physical process it is tapping.
3. **Post-Processing**: Most current TRNG blocks use post-processing. However, it is not mandatory to compensate for the bias and enhance the random sequences' quality, which can be quantified by metrics such as Intra-hamming distances at the basic to NIST and DIEHARD tests.

It is desirable to design TRNGs using a commonly available and inexpensive implementation. So, designing in purely digital fashion is preferred [17]. Designing a TRNG in Field Programmable Gate Array (FPGA) is more desirable than ASIC because FPGAs can be reprogrammed. It is easier to integrate with digital microprocessors, offers dynamic configuration, lower cost, and shorter time-to-market. However, this comes with challenges and threats. By introducing bias in the output one can easily attack a TRNG. Attackers tend to influence bias with several techniques to make output predictable. So, BIAS is a crucial design parameter while designing a TRNG. Challenges in designing a fast and compact TRNG with no bias come with a trade-off between the noise source's hardware cost and the post-processing's complexity. Also, it is essential to design the whole TRNG on a single chip; otherwise, an adversary can tap the output signal from that and breach into the system if an external source is used to collect the randomness.

Usually, the architecture of TRNG includes a random source of noise that is responsible for generating an analog signal that is digitized and fed into a post-processor unit. Most FPGA implementations of TRNG use ring oscillators as their noise source. In a ring configuration when odd numbers of inverters are connected, it forms a simple ring oscillator. A periodic square wave denotes the output signal of the ring oscillator ideally and the total number of inverters in the ring determines the period. Nevertheless, practically, the output signal of a ring oscillator is not in exact a perfect square wave. The ring-oscillator's period varies randomly due to local temperature and voltage variations. Designers exploit this random behavior to generate random sequences.

Another methodology of TRNG implementation uses the metastability phenomenon of a latch as the source of randomness. A latch is metastable when the output oscillates between logic levels. For example, suppose R and S inputs are activated simultaneously for the RS latch. In such a case, before eventually transitioning to one of the stable states, at the rising edge of the CLK the latch enters a metastable state. A proposed TRNG design circuit uses the delay latch chain with the option of adjustable external delay for generating randomness. However, this method suffers from unreliability and performance degradation due to environmental changes (for example, temperature and supply voltage). The metastability of an RS latch is exploited by an enhanced design which LUTs could implement in FPGAs. The LUTs are implemented as hard macro with predefined placement strategies and an already proven efficient combination of the number of LUT latches and sampling interval to obtain the output generation rate in Mbps, which affirmatively testifies the quality of randomness. The study on different types of implementation methods and the performance of each TRNG design is discussed in the next section.

8.3 TRNG Architectures

8.3.1 Technology Independent TRNG

The TRNGs are needed for various cryptographic applications; therefore, their entropy source must be highly unbiased. However, it has been seen that source of entropy could be sensitive to environmental conditions, noise, and aging of the device. As a result, attackers can manipulate temperature and voltage to create bias in the TRNGs outputs, affecting the security of cryptographic applications where they are used. Reference [18] proposed technology independent TRNG (TI-TRNG) that uses bias detection mechanisms and self-calibration mechanisms to reduce bias in output due to attack (manipulation in supply voltage) or aging. However, the proposed method specifically applies more to older technology nodes with less process variation, thus a low source of intrinsic randomness. And since government agencies are still using these higher nodes due to reliability issues, the TRNGs deployed are prone to biasing effects.

Reference [18] proposed two circuits; (1) circuit to detect the biases in the output responses due to aging/environmental changes, and (2) calibration circuit to improve the quality of PUF after bias is detected. Figure 8.1 shows the circuit used to detect bias. It includes a dedicated RO and Counter one, which counts RO's oscillations. It also includes another counter, which resets the Counter one occasionally. And finally, it includes a comparator to compare the oscillations counter which the threshold (expected count). Therefore, in case of an attack where the voltage is increased or decreased, affecting the counter value, the comparator can detect the bias.

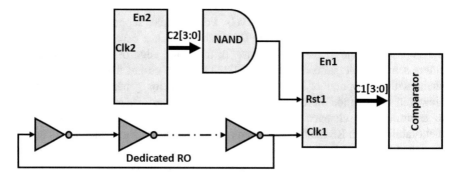

Fig. 8.1 Block diagram illustrating the circuit to detect bias in the TRNG's output

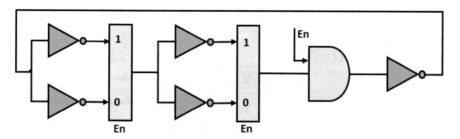

Fig. 8.2 Circuit illustrating the architecture of tunable ROs used in the TI-TRNG

When the bias is detected, the delay paths of the ROs need to be modified to avoid interlocking between ROs and a huge difference in their speed that can insert bias in the response bits. For this, [18] proposed the tunable ROs architecture. Figure 8.2 shows the architecture of a tunable RO. It consists of alternative inverters and multiplexer chains. The delay paths of the ROs are controlled by controlling the tunable select signal to the MUXes. Therefore, a large number of MUXes in the ROs provide more flexibility in calibrating the ROs; however, incurs more area overhead.

To analyze the TI-TRNG's effectiveness, [18] implemented it on older technology nodes FPGAs, i.e., 45nm, 90nm, and 130nm. The temperature was varied between 0 and 70°C, whereas voltage was varied in between 1.2 and 1.8 V. Similarly, the FPGAs were also aged by supplying excess voltage of 1.8V and temperature of 100°C for five hours using a Thermo stream. To test the quality of output response bits, [18] collected five million bits every time and used NIST statistical test suite [16] to check for randomness. The results showed that due to changes in environmental conditions, the conventional TRNG responses failed the NIST test in many cases. However, when TI-TRNG architecture was applied to detect bias and calibrate the RO's delay paths, the generated response bits passed the tests.

8.3.2 Embedded TRNG with Self-Testing Capability

This TRNG design uses FPGAs that do not contain Phase-Locked Loops (PLLs) [15]. The TRNG architecture extends the design technique in [7] that uses on-chip jitter to generate random bitstreams to PLL-less FPGAs used in cryptographic processes. Only Configurable Logic Blocks (CLBs) are used in the presented approach since they are common to all FPGAs, for implementing TRNG with good statistical characteristics (randomness). The design approach also highlights the engineering challenges that exists while extracting random bits from the proposed design's digital signals and self-assessing ability.

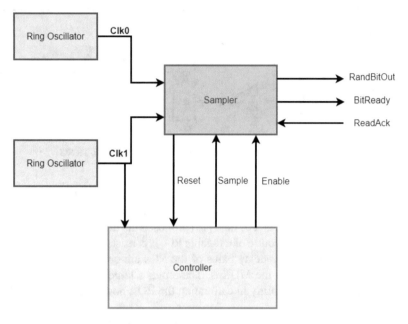

Fig. 8.3 Block level circuit design

The TRNG design uses two identical ring oscillators, a control circuit, and a sampling circuit, as shown in Fig. 8.3 [15]. The frequencies of the two oscillators are close to identical and with similar jitter amounts. Pulses generated from the ring oscillators are supplied to the sampling circuit. To obtain a sampled signal (S0), one clock pulse is used to sample another (for example, sampler uses CLK1 to sample CLK0). This is performed to obtain a sampled signal with the run of ones and a gap of zeros. To get a output bit which is random, the clock length of ones and zeros is counted modulo 2. One-bit counter is set in the sampler circuit to account for the number of cycles in CLK1 (i.e., C0). One can convert the least significant bit of sampled signal cycle length to an output bit which is random, by latching the sampled signal (S0) with the counter signal (C0). The signals obtained at the sampler circuit are shown in Fig. 8.4 [15] for reference.

The CLB ring oscillator with two transparent latches, an inverter, and a buffer has intrinsic differences in speed owing to slight physical process variations in the CLBs. The speed (through various elements propagation delays) determines the frequency of oscillators and is directly related to the TRNGs output bit rate. If the cycle difference between two signals is too small or too large, the control circuit will disable the control and output flip-flop and give no output under these conditions. The control circuit also deactivates the main circuitry of the design after random bits are latched to output flip-flop and reset the counter to eliminate the correlation in successive bits. The self-assessing mechanism is introduced from control circuitry to determine design failures.

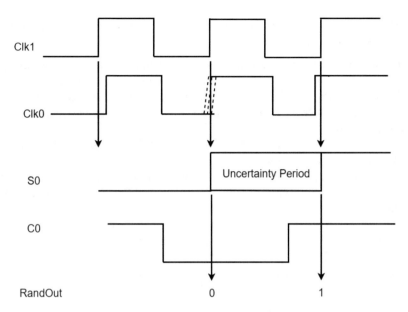

Fig. 8.4 Behavior of sampler circuit with cycle jitter

The design architecture considers the effects of placement of Ring oscillators (ROs) in Configurable logic blocks (CLB) slices and FPGA temperature on ring oscillator frequencies for the design. This design approach identifies some critical parameters such as the temperature around the CLBs for placement of ROs, the evidence for the presence of jitter (randomness), general bias reduction techniques, and bit generation speeds backed with experimental data to suggest ideal design constraints.

Xilinx FPGAs are a popular choice due to 44% share in the Programmable Logic Device market. They mostly carry delay-locked loops instead of phase-locked loops. However, DLLs fail in providing fine control of frequency synthesis required for the application mentioned in [7]. For example, the expected difference of the order 0.1% between input and output frequencies cannot be obtained with DLLs. Hence there was a need for an alternative approach to cryptographic implementations in PLL-less FPGAs. This design extends the scope for using FPGAs without PLLs to obtain random sequences for cryptographic applications.

The key advantage of this design is that it can be implemented purely on reconfigurable blocks (CLBs) without dependency on other resources in the FPGA (PLLs) or external circuitry for cryptographic systems. The design also provides the ability to examine the failure in randomness and halts the output of TRNG. In addition, the configuration approach considers a low order bit of length of the sampled signal as the output bit so the design will not fail for changes in oscillator frequencies.

Although there are some benefits of the proposed design, a design should consider the following drawbacks before implementation:

- The source of bias from metastability in the circuit was attenuated by adding a buffer to sampled signal. Still, there are other sources of bias in the implementation which is not desired for secure systems. For example, bias reduction methods significantly reduce the bit rate of output, which is not desirable (reduces output bits nearly to half).
- The output throughput of 0.5 Mbps is less than other related works on implementing TRNG in FPGAs.
- The output bit rate is proportional to the speed of ring oscillators, and identifying oscillator pairs with a more closely matched clock period (good speed ratio) is a challenge.
- The current has not implemented a bias reduction technique in hardware. Instead, it was supplied to the host computer, which can compromise the proposed design's security.
- Only the rising edge is sampled instead of both rising and falling edges of the clock signal which could have potentially doubled the output bit rate of TRNG.

The design implementation of the SLAAC-1V FPGA test system is tested using the "National Institute of Standards and Technology (NIST) Statistical Test Suite" [20]. The design had a throughput of 0.5Mbps, and a test file of 1Gbit was used for testing the design. The sequences were within the anticipated confidence levels for all the performed tests, and none of them failed. All the P-Values were in the range of 0 to 1, which is expected behavior for an ideal TRNG.

8.3.3   FPGA Vendor Independent TRNG Design

This TRNG architecture aims to design a high-speed, high-quality, compact, purely digital, FPGA-based true random number generator [21]. This design considers the shortcoming of some of the previously designed TRNGs:

1. Some FPGAs have a design that samples the jitter of the clock signal synthesized by the analog PLL. However, because certain FPGAs employ digital delay lines, this cannot be transferred to other FPGAs.
2. A design depends on an external oscillator (analog) circuit to extract jitter. Unfortunately, this design is unusable since it can easily be tampered with and not be used in reliable applications. The unreliability stems from the fact that TRNG operation can be stopped by disconnecting the external oscillator.
3. To sample each other, a design that uses two ring oscillators. Nevertheless, this design is sensitive to the oscillator's placement into CLBs since the oscillation frequency requires to be matched closely.

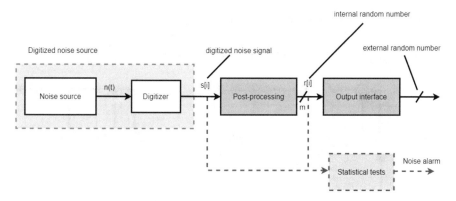

Fig. 8.5 General architecture for random number generator

This design architecture work (as shown in Fig. 8.5 [21]) has its roots from a previously developed TRNG framework, and it is based on the following deductions from the framework [23].

1. Using identical rings yields better results over pairwise relatively prime oscillator rings.
2. A combinatorial model allows us to calculate the amount of oscillators to fill with jitter events the whole spectrum of the signal n(t).
3. It is not advised to fill the entire spectrum with jitter events as that comes at the cost of using too many ring oscillators. Allow a portion of the signal to be predictable and compensate with post-processing.
4. The fill rate depends on jitter width. Larger the jitter width, the fewer the ring oscillators required to acquire the same fill rate (f).
5. It also justifies the anticipated entropy of the noise source.

This design relies mainly on the fourth deduction mentioned in the above list. The amount of jitter present in an entire period depends on the technology. To approximate the jitter and model the period of an "n" inverter long oscillator, a designer can run an experiment on an FPGA similar to Xilinx Virtex 2 Pro (XCV2VP30). From the data obtained, as shown in Table 8.1, the authors have quantified the period and concluded that a ring-oscillator's period depends linearly on the length of the ring and the single inverter's delay. It is given by:

$$T\ 0.88 * l - 0.23, \tag{8.1}$$

where T is time period and l is the length of ring oscillator (RO).

The results indicate that the jitter width increases with the number of inverters in a ring. However, the amount of jitter per period decreased with an increasing number of inverters because jitter width did not increase as fast as the period.

Table 8.1 Relationship between the RO's length and the period

Length l	1	3	5	7	9	13	19	25	31	41	57	67	83	101
Period T (ns)	2.7	3.0	5.0	6.6	7.5	10.0	15.0	20	25	38	51	58	72	90

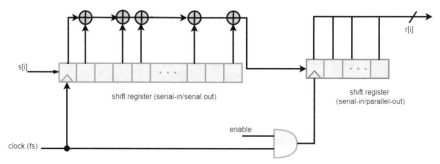

Fig. 8.6 Based on cyclic codes, implementation of a post-processing algorithm

A post-processing unit is used to compensate for the bias and to eliminate the non-random components. An error correcting-based resilient function compensates for the components that are non-random in nature in the post-processing unit. The resilient function is given by:

$$(r[i] \ldots r[i+m-1]) = (s[i] \ldots s[i+n-1]) \cdot G^T, \qquad (8.2)$$

where generator matrix for an [n,m,d] code which is linear is denoted by G, r[i] is the internal random numbers, s[i] are the digitized analog noise signals. This function can be implemented easily using a shift register and XOR gates, as shown in Fig. 8.6 [21]. The location of the taps is based on the generator matrix.

This resilient function can correct input bits m errors (d − 1) and has a compression factor m/n. This design uses rings of only three inverters and assumed, based on experimental results on Xilinx Virtex 2 Pro (XC2VP30) FPGA, that these oscillators have a period = 3ns (333MHz). It has been measured experimentally on Xilinx Virtex FPGA (XCV800) that their 13 inverters rings have a jitter with a standard deviation = 2%. Ring oscillators that are located near to each other generally tend to lock phase in practice, hence it would be safer to use more rings and also, overestimate the jitter. The noise signal is sampled at a frequency of 40MHz [23]. This makes the noise source stateless since the sampling frequency (40MHz) is smaller than the noise source bandwidth (333MHz). Also, the post-processor is memoryless because the random bits r[i] (internal) depend only on das-random bits s[i] but not on one another. So, the whole RNG is stateless, and the minimum entropy verification could be performed on post-processed random bits directly. Consequently, the same choice of [256,16,113] linear code has been used for post-processing because it helps efficiently implement cyclic codes.

8.3.4   High-Speed TRNG-Based on Open-Loop Structures

The architecture follows a new approach based on a structure of open-loop along with a simple configuration and a repetitive configuration with no ring oscillators or other distinct elements, e.g., PLL [3]. As a result, it offers relatively high data rates with respect to maximum clock frequencies.

This architecture is based on the principle of metastable memory point of CMOS, which converges to either Vss or Vdd randomly based on noise such as Gaussian thermal noise. Since the thermal noise cannot be predicted, it can be harvested as an entropy source for designing a TRNG. The basic principle behind an open-loop random generator can be explained using Fig. 8.7 [3]:

By altering the data delay d1 in relation to the clock delay d2 at the threshold of the CMOS gate's switching point around Vdd/2, the data can be sampled. The metastable state converges to a stable state within the resolving time t as a result of the noise that is present in the circuit. This programmable delay may be implemented in a digital environment by forming a chain in which each delay element of type D produces a signal di sampled by the appropriate DFF. The common clock produced by the global buffer is used by the n number of instances. To keep the stable state for one clock cycle, the outputs from all DFFs are XORed, then another DFF is used for re-synchronization. The design architecture is depicted in Fig. 8.8 [3]. The quality of a TRNG-based on the above architecture depends on the following two conditions:

1. **Seamless Condition:** The output at no time should be stuck at Vss or Vdd, which implies at the minimum one delay element should be in a state which is metastable for any temperature, supply voltage, and process.
2. **Randomness Quality:** There should not be any bias present in the generated number, and the output must pass some standard statistical tests such as NIST and DIEHARD tests.

The "ALTERA STRATIX EP1S25" FPGA was considered for performing the experiments. This testing platform works in two modes: either running the open-loop TRNG or by calibrating the initial/early delay ΔT between the data and input

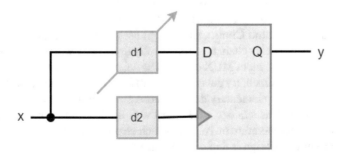

Fig. 8.7 Concept of Open-Loop TRNG

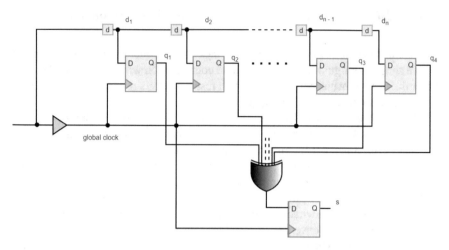

Fig. 8.8 Design of basic delay chain consisting of n-elements

clock of the delay chain for testing of the seamless condition. When the device runs in normal mode, the value of ΔT is based on a buffer which is fixed and it varies with respect to environmental parameters (temperature and voltage supply). In contrast, ΔT is set by an externally RC delay that is adjustable in the seamless test mode of operation [6]. The design principle can be tried on three different classes of implementations.

1. **DFF Chain:** The structure of this implementation is depicted in Fig. 8.8. The delay element is a carry primitive, and the DFF is internal to the logic cell LCELL. The test results show that obtaining a metastable state is nearly impossible even though the external delay is adjusted accurately.
2. **Latch:** Chain The an improved architecture with some modifications, as shown in Fig. 8.9 [3]. In this design, the DFFs are replaced by LUT latches, and a DFF resynchronizes the output with an inverted global clock. To collect metastability caused by data entry, the quantity of delay elements is increased. The latches of the same LAB are XORed, and all the outputs from LABs are XORed to obtain the TRNG output. The Seamless Condition is met when the probability of the TRNG's Vdd's output varies between 0.02 and 0.98.
3. **Auto-controlled Latch Chain:** This design uses adjustable coarse delay placed in forefront of the delay chain to automatically find ΔT in which the coarse delay is acquired from two input MUX chains. The delay chain is built either with carry chains having an elementary delay of 60ps or LCELL having an elementary delay of 150ps. Using the elementary delay value ΔT can be adjusted using the coarse delay controller. This strategy works very well if any erratic behavior is present due to varying temperature or voltage, which changes the bias point of the delay chain. The circuit design is depicted in Fig. 8.10 [3].

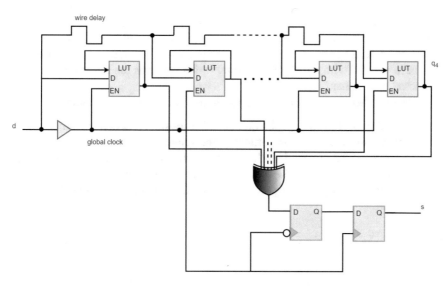

Fig. 8.9 Design of latch chain

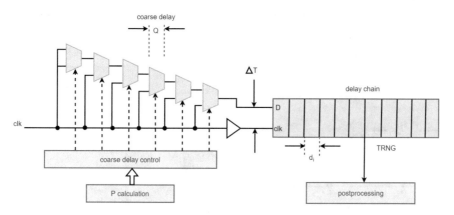

Fig. 8.10 Auto-controlled latch chain for ΔT

The open-loop structure does not allow the output probability to be exactly 1/2. A designer can try possible solutions:

1. Combine TRNG with PRNG to obtain no bias. The resulting output produces a non-unconditionally secure random flow, which is difficult to guess.
2. Von Neumann Corrector's well-known method is used to remove bias and maintain the randomness of the source by processing non-overlapping pairs of successive bits. However, this computation reduces the throughput by $P(1 - P)$.

There are some advantages of using an open-loop structure over a ring oscillator. The delay chain structure can be adjusted, giving more flexibility and robustness and the output bit rate is significantly increased compared to ring-oscillator-based TRNG. This TRNG design can be implemented in any field programmable gate array or an ASIC without needing extra components. The quality of randomness has been verified with the NIST test suite, which confirms the effectiveness of the proposed mechanism.

However, there are some shortcomings in the current design methodology:

- The quality of TRNG depends on manual placement and delicate adjustments, which would cause issues in portability and stability.
- The proposed TRNG can be affected by varying supply voltage or temperature as the wire delays are sensitive to these operational constraints.
- The usage of post-processing techniques significantly declines output bit rate.

8.3.5   FPGA-Based Compact TRNG

This TRNG design enhances the previous design [23], and it does not require post-processing to qualify for the statistical test (see Fig. 8.11 adapted from [27]). This design considers the following drawbacks of the previously designed TRNGs:

- A design that used PLLs, but this technique cannot be extended to FPGAs that do not have such analog components [7].

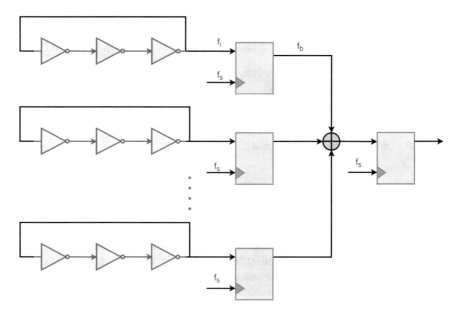

Fig. 8.11   TRNG design in FPGA without post-processing

- A design that used closely matched transparent latches poses a severe constraint [15].
- A design uses a LFSR (linear feedback shift register) as well as cellular automation shift register (CASR), whose outputs are fed to XOR to generate random signals. However, the outcome of this design went wrong, assuming TRNG to be stateless [25].
- A design that used a Galois ring oscillator (GARO) and a Fibonacci ring oscillator (FIRO) whose outputs are fed to XOR and sampled using flip-flops are used to generate random signals. However, there can be a cross-talk problem with other signals [8].

One of the primary concerns with the previous design method is the large number of transitions from oscillator rings with frequencies close to or greater than the sampling frequency at the input of the XOR tree. The setup times along with the hold times for the internal register elements and look-up tables (LUT) of an FPGA are strongly constrained by these signals.

After each oscillator ring, the proposed enhancement tackles this issue by addition of an extra D flip-flop. This addition alleviates the problem, as mentioned earlier, by making signals at the XOR tree input synchronous with the sampling clock. Moreover, added D flip-flop will not change the extraction of each rings randomness but improve the randomness as a whole of the output. A theory based on probability is proposed to observe the effects of correlation and expected values of individual oscillator rings output on BIAS of XORed oscillator rings. The expected value or bias of the XOR of two random bit sources, X and Y, with expected values μ and correlation ρ, is given by [4]:

$$E(X \oplus Y) = \frac{1}{2} - 2\left(\mu - \frac{1}{2}\right)^2 - 2\rho\mu(1-\mu)(1). \tag{8.3}$$

When $\mu = 1/2$ the above equation can be written as:

$$E(X \oplus Y) \approx \frac{1}{2}(1-\rho)(2). \tag{8.4}$$

This implies bias in the output is generated due to the correlation between the sequences. Now, the expected value of XOR of n independent random sources is given by [4]:

$$\frac{1}{2} + (-2)^{n-1}\left(\mu - \frac{1}{2}\right)^n = \frac{1}{2}[1 + (-2\varepsilon)^n](3). \tag{8.5}$$

It can be inferred from the above equation that the expected value will converge to 1/2 as n increases. So, more oscillator rings are desired to improve the bias of the output provided, and the rings are independent.

The designers have experimented with three inverters and 13 inverters long oscillators with varying numbers of FPGAs with varying numbers of rings to compare the randomness of the previous design [23] and the current design. The study of produced sequences reveals that the bias rises with the amount of rings, and the output of the prior design contains many zeroes. Nonetheless, with the suggested configuration, the bias soon converges to 1/2 and acts according to the previously established theory of XOR of independent sequences that are random.

The experiment shows that the tendency seen for the previous design is due to dependent nature or correlation in random sequences, which is caused by the difficulty with the high amount of transitions at the XOR tree's input and sampling flip-flop. Another experiment is carried out to test the prior design assumption of randomness, which states that the oscillator rings who have equal lengths would have the identical frequency. At the same time, phase drift caused by jitter causes transition areas to drift. Oscillator rings of length five and thirty one inverters are chosen to observe the frequency dispersion and distribution [27]. It is observed that five inverter oscillator rings do not follow Gaussian distribution as 31 inverter oscillator rings did. Moreover, clustering is observed in the case of 5 inverter oscillators. This prompted the researchers to undertake similar trials with different lengths. They discovered that as the number of inverters increases, so does the dispersion. The dispersion can be quantified by the variation coefficient $v(\%) = s/\mu \times 100$, where s stands for standard deviation, and μ denotes the mean. It has been discovered that oscillators with only three inverters have the greatest frequency dispersion.

To understand this behavior, researchers have analyzed the architecture of Altera Cyclone II FPGA. During the placement and routing (P&R) phase, inverters in oscillator rings are seen to be positioned at physical logic elements in distinct logic array blocks. As a result, the frequency of the oscillator ring changes with the distance between inverters. This delay is significant in the case of shorter rings because the routing delay is comparable to the period. Therefore, better dispersion, more spread of transition regions, and less interaction between oscillators are observed in the case of shorter rings. The TRNG design was implemented on Altera Cyclone II FPGA using 50 and 25 oscillator rings with three inverters without post-processing with a sampling frequency of 100MHz. Unlike the previous technique [23] of employing a combinatorial approach ("coupon collector's problem"), the amount of oscillator rings may be determined using a statistical model of TRNG by performing simulation. A total of 1000 blocks of 1Mbit random data were recorded and analyzed using "NIST (SP 800-22)" and "DIEHARD" statistical tests. Experiments revealed that passing the "NIST" and "DIEHARD" tests does not need almost 100% certainty of striking at least one transition region. The current design occupies less than 100 LEs, far less than the previous design [23], which occupies more than 1800 LEs. To investigate the randomness after start-up, an oscilloscope is often utilized to record the random result of our TRNG after it has been restarted numerous times. While the FPGA is reset, the oscillator ring results are held at zero or low levels. When the reset is switched off, the oscillator rings begin to oscillate. After triggering the oscilloscope, the output indicates that ten restart sequences are captured from the TRNG output. Because of the oscilloscope's bandwidth constraint, the measured results are not exactly square signals. This experiment illustrates how the TRNG quickly generates random output after a restart.

8.3.6 Meta-Stability-Based TRNG

This TRNG design can be implemented by exploiting the metastability characteristics of the RS latch [10]. This TRNG comprises of the digital circuits that could be merged into logic LSIs, including widely popular Field Programmable Gate Array (FPGA) technology.

The designers have considered previous research while developing their TRNG design using ring oscillators (ROs). A random bitstream by XOR'ing the outputs of several ROs that are free-running is generated by the design of TRNG [23]. Another design by [21] evaluated the proposed TRNG by [23] using 110 ROs, each consisting of three inverters, and reported the possible output bitstream of 2 Mbps. There are more instances of using ROs to achieve TRNG proposed by [8, 15, 27], which showed improved output generation rates. However, the main problem for these proposed TRNG designs using ROs is the power depletion of ROs that are free-running. Hence, the energy consumption is significant, which might be considered prohibitive for embedded systems. This energy consumption problem led the designers to exploit the metastability of RS latches to generate randomness.

The designers also considered the following metastability-based TRNG implementations while designing their TRNG, which are evaluated with full custom LSI technology:

- The design method by [1] uses the metastability of the RS latch to generate random numbers in which the circuit design is incorporated with a dedicated initialization clock to charge (precharge & discharge) internal nodes. The verification of the design is done with a two μm CMOS process.
- The R-flop is used in another method proposed by [14]. It has a differential preamplifier input which is then followed by a latch which is bistable. To manage the bias of the TRNG, a feedback loop which is negative and that fine-tunes the bias voltage of the R-flop is utilized. Verification of the design was performed by using a 0.6 μm CMOS process.
- A TRNG that uses metastability as a source of entropy to begin the switch from oscillator mode to latch mode [5]. The proposed design was verified with 0.18 μm CMOS, but the downside of this technique is the consumption of more power in oscillator mode.
- Metastability-based TRNGs designs by [11, 12] : DC-nulling type and FIR type, which passed the NIST test and verified with 0.35 μm CMOS.
- A feedback loop is utilized to maximize an RS latch resolution time using metastability. This design improved the randomness quality [26].

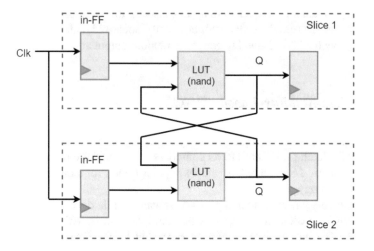

Fig. 8.12 LUT latch: an RS latch which comprises of LUTs and embedded flip-flops

However, all the above designs were implemented with custom LSI technology. In contrast, the current design is easily implementable and verifiable on an FPGA device.

As we already mentioned earlier, a similar TRNG design [3] exploits metastability caused by hold time violation of D latches implemented with LUTs in FPGA. The limitation of this design is that TRNG could be affected by the operational environment (e.g., supply voltage, temperature) because of manual placement and routing and timing adjustments of the signals with wire delays, which hamper reproducibility portability stability. This TRNG design, on the other hand, is symmetrical and resistant to changes in the operational environment.

It is feasible to design a TRNG that takes advantage of the metastability phenomenon for random number generator. If in the Clk signal there exists a skew or dispersion of elements of the circuit due to a difference in drive strengths of two NAND gates, the output may be biased, rendering metastability-based TRNGs unstable and difficult to execute in practice. Additionally, even in a balanced circuit, the output could be impacted by the prior output, which could persist after initialization as a slight voltage difference between internal nodes.

Because logic functions are performed using LUTs in an FPGA implementation, two look-up tables are used to replace the two NAND gates of an RS latch, as shown in Fig. 8.12 [10]. Quality of LUT latches is controlled and gathering entropy from a large number of LUT latches are required to harvest an adequate amount of entropy. As a result, the XOR corrector was chosen as the post-processing function for physical random generators by the designer. Figure 8.13 depicts the block layout of a TRNG architecture, which creates a single output by XOR'ing the outputs of N latches. The test was carried out on the Xilinx ML405 board.

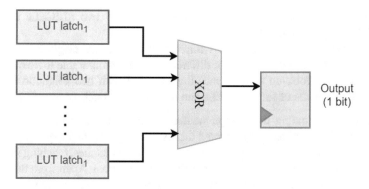

Fig. 8.13 Block diagram of LUT latch N

Each CLB in the proposed evaluation platform comprises four slices: two SLICELs located in odd columns and two SLICEMs located in even columns, considering a 2-D array structure of slices.

Each LUT latch is implemented as a hard macro according to the circuit design. LUT latches are manually created with a predetermined column layout and connections to reduce the skew of internal signals. The proposed TRNG is made up of N hard macro LUT latches (as shown in Fig. 8.13 [10]).

The in-FFs and out-FFs (as shown in Fig. 8.13 [10]) are implemented using integrated FF elements, which are used to isolate the capacitive load of the output wire from the internal node and reduce the skew from the rising edge of the peripheral clock, respectively. Based on the Diehard test, in-out-FFs are the best alternative; hence, the design may be implemented and assessed for the four combinations of in-FFs and out-FFs.

The quality and throughput of a LUT latch depend on the distance between two slices. Although longer interconnect may collect more noise and generate larger entropy, they should be circumvented because of higher consumption of interconnect resources and they incur heavier capacitive load. The entropy of a latch is inversely proportional to the skew according to the length of the wire. The designer has examined various layouts with different distances between two SLICELs of a LUT latch. Based on the Diehard test with 64 latches and a sampling interval of 320ns, it was discovered that retaining a wider distance did not increase randomness quality. Eventually, the layout dX2Y0 was adopted (dXiYj: i and j are the relative coordinates of Slice 2 in the X and Y axes) as they perform better than other layouts.

The placement of latches inside CLBs impacts the quality of the TRNG's randomness. Difficulties in routing and degradation in the randomness may be a result of implementing more latches in a pair of CLBs. Additionally, if two latches are placed in line closely to one another, then the output might follow some correlation, resulting in lower randomness. There are three possible implementations:

1. Lxxx: single look-up table latch implemented with two SLICELs from two neighboring configurable logic blocks.
2. LLxx: two look-up table latches implemented with two neighboring pairs of SLICELs residing in the same of different pairs of configurable logic blocks.
3. LLMM: four look-up table latches implemented with two neighboring CLBs, two latches made of four SLICELs and the other two made of four SLICEMs.

Based on evaluating these implementations in the Diehard test at a sampling interval of 320ns, the designer adopted the Lxxx design for their experiment.

The randomness quality improves with the number of latches, reaching saturation with 64 latches or more, beyond which the return is minimal. Apart from the number of latches, another essential property is the sampling interval, which impacts the quality of randomness. With a longer sample interval, a higher number of entropy is generated, but the throughput suffers. Therefore, different combinations of latches and sampling periods are used to find the optimal sampling interval. It is observed that LUT with 64 latches passes the Diehard test with a sampling interval greater than 120ns, while LUT with 64 latches passes the test with a sampling interval greater than 80ns.

The experiment was carried out on the Xilinx XC4VFX20 platform, containing 8544 slices. The experimental results show that TRNG with 64 LUT latches occupies around 1.7% of the total number of slices while the TRNG with 110 ROs uses around 4.2%, which is about 40% more in size than the proposed design. Furthermore, TRNG with 64 LUT latches successfully passed the NIST test (no post-processing), but TRNG with 110 ROs failed. As a result, the present TRNG design outperforms the previous TRNG design [23] with 110 ROs in terms of logic scalability and randomness quality.

The maximal generation rate of the current design generates random sequences that comply with the NIST test at various sampling intervals. For example, this TRNG with 64 latches could attain the generation rate of 3.85Mbps, while TRNG with 256 latches can generate numbers randomly at 12.5 Mbps.

This TRNG design has many advantages over other implementations, such as:

1. The design is purely based on simple digital components.
2. The design passes the standard statistical test such as the NIST test, and Diehard test without employing any post-processing.
3. The power consumption is relatively low as the logic scale is small and clock signals can be halted arbitrarily, making the design suitable for embedded applications.
4. High-quality random sequences are generated with significantly higher throughput.

However, there are some limitations of the current design that are evident considering the implementation and execution as follows:

1. Impact on performance (throughput or generation rate) in varying operating temperatures and supply voltages conditions.

2. The design parameters need to be optimized for the proposed design to perform systematically, which might not be achievable in real-world scenarios.
3. The reproducibility and reliability of the proposed design solely depend on the statistical properties, which might impact the quality of randomness.
4. The individual difference among FPGA elements needs to be examined to avoid declining throughput and entropy source.

8.3.7 TRNG Resistant to Active Attacks

This TRNG architecture design is based on the concept of sampling phase jitter in Ring Oscillators with parameters [23]. The fill rate, which is referred to as the percentage of time-domain that may be random in an analog output signal, is the key idea behind this design. In order to overcome the exponential rise in the total amount of oscillators needed to achieve a factor enhancement which is constant in the fill rate, a post-processing phase with an appropriate resilience function has been implemented. This kind of design enables the extraction of randomized samples from a signal with a fill rate that is low and few oscillators. Additionally, to withstand such attacks, fault-attack models were developed employing the characteristics of robust functions. The performance and level of resilience of the design are carefully assessed throughout the analysis using precise methods.

The quality of the good TRNG relies on three components, Entropy sources, Harvesting mechanism, post-processing, and practically. Despite that, faults and production defects are some of the essential factors given the quality of TRNG. Therefore, it is more beneficial to implement it solely using digital design techniques. Furthermore, this digital implementation permits the TRNG to implement quickly on FPGAs, which is the most popular reconfigurable platform. Unfortunately, most of the designs [2, 7, 9, 25] use both analog and digital elements faced problems at the amplification stage, thereby consuming notable amounts of power and making the analysis complex. However, purely digital designs do not require an amplification process that demands a complex harvesting outline. Indeed, all of the plotted designs have passed the statistical test, DIEHARD or NIST customarily decreasing the sampling frequency to a point until the output sequence passes the tests.

Inspired by these various designs and observations, the researchers proposed a jitter-based TRNG design that met the following requirements:

1. TRNG design should be strictly digital.
2. Simple harvesting mechanism.
3. An accurate mathematical explanation should be given for the entropy collection mechanism.
4. No correction circuits are allowed.
5. Design: Efficient and compact, high throughput in regards to per area and energy spent.

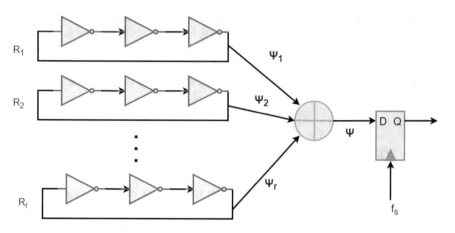

Fig. 8.14 Oscillator combining and sampling design

Oscillators possess the attributes of an output signal being a periodic square wave, and hence are a simple and potent option to build a TRNG. Coupled oscillator configuration is used in jitter harvesting. In practice, these coupled oscillators have drifting problems and exact matches in the period of two oscillators. These problems can be mitigated using a sophisticated design approach for combining and sampling oscillators.

The setup in Fig. 8.14 has r distinct oscillator rings with n inverters on each ring. The analog output signal ψ is the XOR of the individual ROs. Deterministic and transition zones will be depicted by the XORed output. The goal is to complete the full spectrum with transition zones that do not overlap. Entropy will be lost otherwise. The Urn Model and relatively prime ring lengths were used to solve this problem.

When $L < H$, the combined signal are considered to lie between H as well as L volts and time interval $I = [a, b]$ from a spectrum. The aim is to fill this interval with randomness. More precisely, for p which is the threshold, in the interval I at any point t, there exists few ring denoted as R such that $\frac{1}{4} < Prob[\psi_j > 2.5] < \frac{3}{4}$.

The general randomness of the system is based on the notion that in every open time period $(mT - \frac{T}{4}, mT + \frac{T}{4})$, there exists a distinctive point t where $(L + H)/2$ volts is crossed by the signal, and the t acts as a random variable which is distributed normally with mean mT and variance σ_j^2 and criterion. With t uniformly chosen in a random manner from the I interval, the likelihood that there exist integers m and j with $|t - mT_j| < 0.6475\sigma_j$ is at least q. Ideally, in the model, the criteria of design is inadequate to make sure the unbiased nature of the output.

The time-domain spectrum is discretized so that any transition event in any subinterval J holds the above criterion for all t times in J interval. Such subintervals are called "urns."

The following equation gives the length of the Urn:

$$J_h = \left[a + (h-1)\frac{b-a}{l}, a + h\frac{b-a}{l} \right]. \tag{8.6}$$

A J represents the urn and it is said to be full if there exists some ring R_j in the circuit who has a signal that fulfills $T_j(t) = 2.5$ some real number t in the urn J. Otherwise, the Urn is considered empty. The criterion is satisfied provided:

1. Minimum q_l of the urns are filled.
2. $l > \frac{b-a}{0.6475\sigma_j}$ for all $j = 1, 2, \ldots, k$.

If the drift in phase is ignored, any ring may probably fill out every $\pi_j = \frac{lT_j}{b-a}$ urns where π_j is called a combinatorial period. The likelihood that a randomly picked Urn would be left unfilled is provided by the equation below:

$$\left(1 - \frac{1}{\pi_1}\right)\left(1 - \frac{1}{\pi_2}\right)\ldots\left(1 - \frac{1}{\pi_r}\right). \tag{8.7}$$

By having all the ring lengths identical, this design exploits the randomized delays d_j along with the phase drift of the signals in relation to one another. With the least number of rings, the aim is to fill as many urns as possible. Coupon Collector's Problem is known as the problem of filling the long interval I with jitter. The anticipated total amount of rings required to fill all N urns is $r = N \log N$. From this equation, it is evident that design requires a very large amount of rings. Therefore, the minimum amount of rings $r = M(N, f, p)$ where N is number of urns, fill rate is denoted by $0 < f << 1$, and level of confidence is given by $0 < p < 1$. The confidence p is chosen very close to 1 and decreases the fill rate, which will be compensated with a rigorous post-processing strategy. Another assumption is that random bits act as unbiased coin flips, aware that this is a simplification which is nontrivial. So, this model uses a resilient function that reduces the bitstream to small random sequence bits to eliminate the non-random component's output. If L bits get fed into an $(L, m, L/10)$ robust function, the result must contain m truly random bits. This resilient function can be implemented in both hardware as well as software. However, there exists a trade-off between the buffer size and length of code needed to implement the resilient function.

This design considers only fault injection attacks like noninvasive and invasive attacks tolerance on each block of the design. Since the experimented design is stateless, only a finite number of bits is influenced by the noninvasive attacks like introducing the spikes in the power supply. The stronger resilient function with proper error-correcting codes with a greater minimum distance can eliminate the corrupted bits. In the Invasive attack, the adversary's goal is to damage the TRNG so that it will show bias in the outputs. To combat such invasive attacks, each part in the design must possess a level of resistance. For example, if the attacker focuses on ring oscillators and destroys one or more ROs, then the resulting output from the rings will be deterministic. So this can be easily handled by the proposed design by increasing the robustness of the resilient function [19]. The XOR networks from the invasive attack can be protected by replacing the simple XOR binary tree-like structure with an expanded XOR lattice structure. However, this structure results in multiple output lines that must be sampled in a sequential scheme. This results in using multiple flip-flops in the sampler. It is assumed that post-processing will be performed on a trusted system as it is implemented in the software; hence there is no problem of invasive attack.

The final design has been developed by selecting parameters that are realistic in nature which determine the entropy rate and also the throughput of a random bitstream which the circuit generates. The experimental design outlined various results for calculating the number of ring oscillators necessary to fill the maximum number of urns in a jitter event for various numbers of urns (N=36, 52, 100) which yielded 114 rings with 13 inverters on each ring which fills at least 60% of the urns with 99% probability. The output is fed to a resilient function that employs an extended BCH code of [256, 16, 113]. The sample output is grouped into 256 bits blocks and then fed to a resilient function which outputs 16 bits with a minimum distance of 113. In this design, the fill rate has been chosen 60%, so 40% of the bits (103 bits) will be deterministic. Since the resilience of the function is 112 in our design, the design can still tolerate up to nine more bits against fault attacks. The throughput of this model is 2.5 Mbps since the output is 16 bits per 256 bits sampled with the frequency of 40MHz (25ns). However, the downside of the whole design is that throughput got reduced because of post-processing, without which the proposed TRNG does not exhibit randomness.

8.4   Conclusions

The analysis of research work on various designs for implementing TRNG in FPGA provided insights on the trade-off between multiple parameters such as area constraint, the overall throughput, and the quality of output bits (No bias—the equal probability of ones and zeroes in the output bitstream). The design analyzed in Sect. 8.3.2, implementing TRNG in FPGAs without PLLs with self-assessing ability yielded good statistical test results. The presence of more bias in the random sequences required bias correction techniques which ultimately reduced the overall throughput of the TRNG (0.5 Mbps). For increased bit rate, Identifying the right pair of ring oscillators with a more closely matched clock period is challenging. In Sect. 8.3.3, the designers proposed a fast and secure TRNG with oscillator rings containing just three inverters. They sampled at a much lower frequency(40MHz) when compared to that oscillator ring frequency(333MHz). This makes the noise source stateless. Moreover, since the post-processor is also memoryless, the RNG becomes stateless. A simple and strong cyclic code-based resilient function is used. This function makes the design immune to attacks the bias. As a result, throughput of 2.5Mbps is achieved.

The design analyzed in Sect. 8.3.4 uses an open-loop structure that takes advantage of the metastability of a latch to generate randomness using an adjustable delay latch chain. Even though the output rate attained is much higher (around 20 Mbps), this design is less flexible because of delicate manual placement and also suffers due to operational parameters. The designers proposed a fast and compact TRNG by avoiding the post-processing module in Sect. 8.3.5. This design specification is achieved by examining the problem causing the bias in previous design [23] and alleviating it by introducing a flip-flop after each oscillator ring. A shorter three inverter long oscillator rings are chosen to obtain a better frequency dispersion. Implemented design involves 25 oscillator rings with three inverters sampled at a frequency of 100MHz. The proposed design occupies less than 100 LEs, less than 1% of the LEs present on FPGA. The final throughput of 100Mbps is achieved.

The TRNG design in Sect. 8.3.6 takes advantage of metastability characteristics of an RS latch to harvest entropy for creating randomness. The proposed design passes the NIST test without post-processing and generates bits that are random at 12.5 Mbps using 256 LUTs with five hundred and eighty slices only in Xilinx Virtex 4 FPGA. Despite that, the design is unsuitable for many practical implementations because of the rigid statistical properties and predefined placement constraints.

Finally, in Sect. 8.3.7, the designers outlined various results for calculating the number of ring oscillators necessary to fill the maximum number of urns in a jitter event for various urns ($N = 36, 52, 100$). They observed that 144 oscillator rings filled at the minimum 0.60N of the urns with the confidence probability of 99%. The ring count is decided by compromising on fill rate, keeping the confidence probability near one. Since they have to address the fill rate post-processing step is mandatory in the design, reducing the throughput of the whole design.

References

1. Bellido M, Acosta A, Valencia M, Barriga A, Huertas J (1992) Simple binary random number generator. Electronics Letters 28(7):617–618
2. Colbourn C, Dinitz J, Stinson D (1993) Applications of combinatorial designs to communications, cryptography, and networking (1999). Surveys in combinatorics, 1993, walker. London Mathematical Society Lecture Note Series 187
3. Danger JL, Guilley S, Hoogvorst P (2009) High speed true random number generator based on open loop structures in fpgas. Microelectron J 40(11):1650–1656
4. Davies RB (2002) Exclusive or (xor) and hardware random number generators. Retrieved May 31:2013
5. Epstein M, Hars L, Krasinski R, Rosner M, Zheng H (2003) Design and implementation of a true random number generator based on digital circuit artifacts. In: International workshop on cryptographic hardware and embedded systems. Springer, pp 152–165
6. Farahmandi F, Huang Y, Mishra P (2020) Automated test generation for detection of malicious functionality. In: System-on-chip security. Springer, Cham, pp 153–171
7. Fischer V, Drutarovskỳ M (2002) True random number generator embedded in reconfigurable hardware. In: International workshop on cryptographic hardware and embedded systems. Springer, pp 415–430

8. Golic JD (2006) New methods for digital generation and postprocessing of random data. IEEE Trans Comput 55(10):1217–1229
9. Gopalakrishnan K, Stinson D (1996) Applications of designs to cryptography
10. Hata H, Ichikawa S (2012) Fpga implementation of metastability-based true random number generator. IEICE Trans Inf Syst 95(2):426–436
11. Holleman J, Otis B, Bridges S, Mitros A, Diorio C (2006) A 2.92 μw hardware random number generator. In: 2006 Proceedings of the 32nd European solid-state circuits conference. IEEE, pp 134–137
12. Holleman J, Bridges S, Otis BP, Diorio C (2008) A 3μw cmos true random number generator with adaptive floating-gate offset cancellation. IEEE J Solid State Circuits 43(5):1324–1336
13. Jun B, Kocher P (1999) The intel random number generator. Cryptography Res Inc White Paper 27:1–8
14. Kinniment D, Chester E (2002) Design of an on-chip random number generator using metastability. In: Proceedings of the 28th European solid-state circuits conference. IEEE, pp 595–598
15. Kohlbrenner P, Gaj K (2004) An embedded true random number generator for fpgas. In: Proceedings of the 2004 ACM/SIGDA 12th international symposium on Field programmable gate arrays, pp 71–78
16. Marton K, Suciu A (2015) On the interpretation of results from the nist statistical test suite. Sci Technol 18(1):18–32
17. Park J, Cho S, Lim T, Bhunia S, Tehranipoor M, Wu Z, Patel H, Sachdev M, Tripunitara M, Moudallal Z, et al (2019) 1a. 1-an all-digital true random number generator based on chaotic cellular automata topology scott best, xiaolin xu 1a. 2-scr-qrng: Side-channel resistant design using quantum random number generator
18. Rahman MT, Xiao K, Forte D, Zhang X, Shi J, Tehranipoor M (2014) Ti-trng: Technology independent true random number generator. In: 2014 51st ACM/EDAC/IEEE design automation conference (DAC). IEEE, pp 1–6
19. Rahman MT, Forte D, Wang X, Tehranipoor M (2016) Enhancing noise sensitivity of embedded srams for robust true random number generation in socs. In: 2016 IEEE Asian hardware-oriented security and trust (AsianHOST). IEEE, pp 1–6
20. Rukhin A, Soto J, Nechvatal J, Smid M, Barker E (2001) A statistical test suite for random and pseudorandom number generators for cryptographic applications. Tech. rep., Booz-allen and hamilton inc mclean va
21. Schellekens D, Preneel B, Verbauwhede I (2006) Fpga vendor agnostic true random number generator. In: 2006 International conference on field programmable logic and applications. IEEE, pp 1–6
22. Sidiropoulos S, Horowitz MA (1997) A semidigital dual delay-locked loop. IEEE J Solid State Circuits 32(11):1683–1692
23. Sunar B, Martin WJ, Stinson DR (2006) A provably secure true random number generator with built-in tolerance to active attacks. IEEE Trans comput 56(1):109–119
24. Tehranipoor M (2012) Physical unclonable functions and true random number generator
25. Tkacik TE (2002) A hardware random number generator. In: International workshop on cryptographic hardware and embedded systems. Springer, pp 450–453
26. Tokunaga C, Blaauw D, Mudge T (2008) True random number generator with a metastability-based quality control. IEEE J Solid State Circuits 43(1):78–85
27. Wold K, Tan CH (2008) Analysis and enhancement of random number generator in fpga based on oscillator rings. In: 2008 International conference on reconfigurable computing and FPGAs. IEEE, pp 385–390

Chapter 9
Hardware Security Primitives Based on Emerging Technologies

9.1 Introduction

Due to globalization of the semiconductor industry, outsourcing, and the involvement of multiple vendors over the years, security of integrated circuits is a major concern [18, 65]. Due to supply chain globalization, several vulnerabilities and threats could be introduced in hardware [16, 54, 60, 63]. Various hardware security primitives have been designed over the years to provide countermeasures and defensive mechanisms against adversarial attacks. Physical unclonable functions (PUFs) and true random number generators (TRNGs) are two examples of hardware security primitives, as there are numerous applications developed to help with various aspects of hardware security [34, 36, 54, 67]. However, scaling the pre-existing security primitives with technological advancements has proven difficult. The existing security primitives are based on CMOS technology, which is nearing saturation [37] or comes with high overheads [6]. Furthermore, current primitives or countermeasures based on CMOS technology are insufficient for dealing with emerging attack models, threats, and vulnerabilities [37]. Exploring new alternatives for developing hardware security primitives is therefore critical.

Recent applications show that emerging devices and technologies demonstrate unique features. Examples include atomic switches, graphene transistors, Mott FET, memristors, spin FET, nanomagnetic, orthogonal spin-transfer torque (OST) RAM, all-spin logic, magnetoresistive random access memory (MRAM), and spintronic devices [13, 44, 53]. These emerging devices' unique features can improve electronic circuits' performance and device security by simplifying circuits. However, there are three fundamental concerns regarding their usage in hardware security. First, can emerging devices and technologies (instead of CMOS technologies) offer more efficient infrastructure against emerging attack models, threats, and vulnerabilities? Second, how can emerging technology-based structures better support software security schemes? Third, what should be the different characteristics of different hardware based on emerging technologies? [6]. Finally, establishing security versus performance trade-off accurately for emerging devices, as speed, power, and reliability (performance-based features) are given high priority compared to security-based features.

There is a well-established link between new devices with specific properties and a broader range of hardware security applications [37]. It has been established, for example, how the specific properties of emerging devices such as phase change memory (PCM), carbon nanotube (CNT), and graphene can give solutions to various threats and vulnerabilities beyond standard security solutions. A collection of design concepts, on the other hand, can use the distinctive features of emerging technologies, such as silicon nanowire (SiNW) FETs, NCFETs, TFETs, and graphene SymFETs, to guard against IP infringement and hardware attacks [6].

Static random access memory (SRAM), dynamic random access memory (DRAM), and flash-memory-based hardware security primitives are much more attractive for various security primitive applications [22, 33, 35, 52, 68]. This is because they are cost-effective compared to other conventional CMOS technologies. Furthermore, DRAM-based security primitives are much more advantageous than other security primitives because they are omnipresent in modern computer systems, provide extensive storage capability, and can be accessed very quickly, even in runtime [3]. This large storage size guarantees an adequate entropy for primitive implementations, and a dedicated part can be used for this.

Many CMOS-based emerging technologies are not good enough because of design overheads while implementing security applications. Security threats such as reverse engineering (RE) ICs, IP piracy, insertion of hardware Trojan, and various attacks need to be addressed [11, 46, 60, 61, 64]. We discuss an advanced method based on a giant spin Hall effect (GSHE) switch, which has been used to construct polymorphic gates for advanced security and analyzed with probabilistic computing [40]. A polymorphic gate can efficiently execute several Boolean functions that support logic locking and camouflaging [21]. Unlike CMOS devices, it is possible to interchange between locking and camouflaging because of the polymorphic nature of the primitive [70]. Therefore, the polymorphic, GSHE-based device provides highly effective camouflaging capabilities.

This chapter will explore the pros and cons of choosing emerging technologies as an alternative to existing CMOS-based technology for developing hardware security primitives and point out the research directions that require establishing a bridge between the device industry and the hardware security community [6, 37].

9.2   Background

New emerging devices present several challenges for their implementation in high-speed logic and memory implementations. However, the same erratic behavior of these devices can be leveraged to implement hardware security primitives such as PUFs. Some of the security applications in which these emerging technologies can be used are:

- Identification and Authentication: Hardware security primitives can be used to identify IP to protect against reverse engineering thefts and for authentication of the chips.
- Camouflaging: Security primitives can be used to obfuscate the chips to prevent reverse engineering, making deciphering the structure of circuits complex and tedious.
- Counterfeit Detection: Can be used to identify counterfeit chips, which pose a severe threat to the security and reliability of the electronic circuits.
- Protection Against Side-Channel and Fault Injection Attacks: By using side-channel attacks such as laser, timing, power, EM, etc., adversaries can leak internal secrets. Security primitives based on emerging technologies can pave the way to protect against such threats [62].

In this section, we provide a brief overview of various such emerging devices.

9.2.1 Emerging Devices

Carbon nanotubes (CNT), transition metal dichalcogenides (TMDs), graphene, and silicon nanowires (SiNWs) have the property of configurable polarity, which changes carrier type in the channel [12]. Due to different doping concentrations, tunnel field-effect transistors (TFETs) have ambipolarity properties. As a result, TFET can serve as either an n-type or p-type device [59]. Devices with tunable polarity can be employed to protect IP and prevent counterfeiting. Emerging transistor technologies, Symmetric graphene FETs (SymFETs) and ThinTFETs, demonstrate bell-shaped I–V curves, exhibiting a robust, negative differential resistance (NDR) region. Negative capacitance field-effect transistors (NCFETs) have tunable hysteresis, which allows for non-linear capacitance by varying the thickness of the FE material. Emerging nanoscale CNT-based electronics and graphene have been considered as alternatives [8]. Due to process variation, their energy-band diagram variations determine their unique properties. CNTFET as a channel can be employed to build robust nano-CMOS architecture in terms of ultrahigh-speed, RF applications, and ultralow-power logic due to its low electric field. Emerging CNT ensures unique properties can be leveraged for hardware security primitive applications such as supply chain security, prevention of physical tampering, and reverse engineering attacks.

Phase change memory (PCM) can be a good substitute for non-volatile storage in security applications as they have very fast read/write operations and very high storage density. Chalcogenide material-based PCM such as $Ge_2Sb_2Te_5$ (GST) works on the principle of fast reversible phase transition between an amorphous high-resistance and crystalline low-resistance phase. The resistance difference between these two states is relatively large. Unique features of PCM cells that make them suitable for primitive applications are as follows: programming variability,

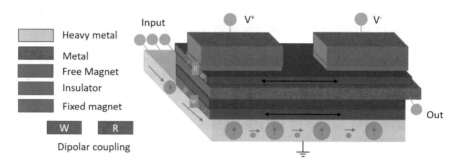

Fig. 9.1 Illustration of GHSE switch structure

resistance drift, random telegraph noise (RTN) and 1/f noise, multi-bit storage per cell and variability, and initial forming step.

All modern computer systems these days comprise DRAM volatile memories. These memories can only store values under a continuous power supply. DRAMs are similar to SRAMs, as both are volatile. However, in contrast to SRAMs, DRAMs require periodic refresh of charge to hold the stored value properly. A typical SRAM cell comprises six transistors, but a DRAM cell contains only an access transistor and a capacitor, where a capacitor is used to store charge. For this reason, DRAMs have relatively higher memory density than SRAMs, so they are widely available in a range of gigabytes (GBs). In contrast, SRAMs are available in kilobytes (KBs) only. The inherent properties of DRAMs make them very much suitable for PUF and TRNG implementations. Primarily, they are preferable for resource-constrained IoT applications as DRAM-based primitives do not require extra hardware to ensure security.

The authors [1] make use of all-spin logic (ASL), while the authors [66] make use of LUT based on spin-transfer torque (STT). They are, however, restricted to a low-cost, selective use with significant security concerns. The GSHE switch is the main core component for new proposed primitives (NM). As shown in Fig. 9.1, the GSHE switch's write (W) and read (R) nanomagnets are heavy metal spin Hall layers (purple) (NM). Mutual negative dipolar coupling is seen in these nanomagnets (W-NM and R-NM). R-NM is composed of two fixed ferromagnetic layers with opposite magnetization directions. It conducts spin-polarized current, which introduces spin-transfer torque (STT) to W-NM [48], depending on the applied charge current in the bottom layer. The top fixed ferromagnets will then be parallel to one magnetization direction and anti-parallel to the other.

Finally, depending on polarity, the output current might be inward or outward—this presents the GSHE switch operation, which is represented by Fig. 9.2 and, by nanomagnetic dynamics, which is simulated that describes the GSHE switch performance here represented by Fig. 9.2. The distributions are obtained from 100,000 simulations.

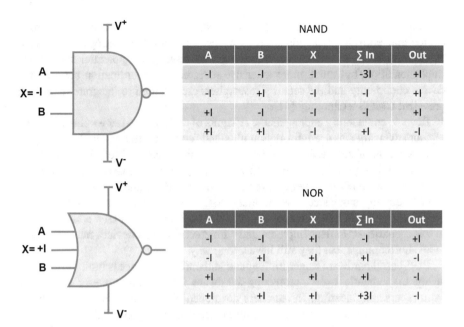

Fig. 9.2 Truth tables representation for the NAND and NOR gate functionalities for GHSE-based primitives

9.2.2 Security Applications of Emerging Devices

PUF is a one-way function that uses the device's inherent process variation to provide a unique and unclonable output response to a specific challenge given as an input. There are numerous sorts of PUF implementations, which can be classified as strong or weak depending on the number of challenge–response pairs (CRPs). It is challenging to say what number of CRPs distinguishes the PUFs in these two types since there exist no such margins. PUFs are widely used for identification, authentication, and hardware metering. Uniqueness, uniformity, and reliability are three significant quality indicators for evaluating PUF performance. The inter-chip Hamming distance (inter-HD) is used to determine the uniqueness of a PUF, with 50% being ideal. Uniformity stands for randomness in the output responses provided by PUF. A PUF's reliability is measured by its capacity to create the same CRPs under different environmental conditions and over time, which should be 100% ideally. PUFs often require error-correcting codes as they suffer from reliability issues over time [5].

TRNGs are also commonly used in a variety of security applications to create one-time passwords (OTPs), session keys, random seeds, nonces, and PUF challenges. Random numbers can be generated in two ways: by using software algorithms to generate pseudo-random numbers based on seeds that appear to be random or by relying on inherent process variation of hardware to generate truly random numbers. The effectiveness of TRNGs largely depends on the entropy source and entropy extraction unit, as a very high level of entropy is expected at the output of a TRNG. Emerging devices have great potential for TRNG implementation, but much more extensive research and coordination between the device community and the security researchers are required to integrate them with prevalent CMOS technology [41].

Reverse engineering endangers the IP rights of designers. After reverse engineering, an adversary can use information obtained for counterfeiting and IP piracy. At various levels of abstraction, several solutions have been offered: RTL, gate and layout, and finally, at PCB level. Hardware fingerprinting and IP authentication can protect hardware IPs from being reverse-engineered. For authentication, PUFs have been frequently suggested. Nevertheless, ML-based modeling attacks have been developed to predict PUF responses, and the application of PUFs adds no value in this scenario. Primitives implemented with emerging nanodevices may overcome these shortcomings, but they still have a long way.

PUF-based hardware fingerprinting and authentication process limit the device's utility to authentic users' protecting hardware intellectual property (IP) cores from reverse engineering. Researchers developed attack methods to diminish the authentication security level anticipating the challenge–response pair [42]. Device-level obfuscation or logic encryption makes it difficult for attackers, although there is an overhead of CMOS camouflaging gates [38, 39, 58]. Counterfeit ICs, whether recycled, remarked, cloned, overproduced, or defective, threaten the worldwide semiconductor supply chain [55]. PUF-based sensors have been employed for detecting counterfeit products. For example, the CDIR sensor is designed to combat recycled ICs, and metering techniques are employed to control overproduction [2, 19]. However, current MOSFET-based countermeasures incur overhead. Anti-tamper design safeguards intellectual property (IP), cryptographic keys, and sensitive data from unwanted access, denial-of-service attacks, and cloning. An adversary can tamper from both the back and front sides of the chip to access the critical nets and bus. The protection against tampering can be classified into tamper-evident and a tamper-resistant design, which erases the secret key due to unauthorized physical access [47].

By the implementation of 16 Boolean logic functions with the application of clock signal to the defined ferromagnets, terminals by the proposed primitive in Fig. 9.3 can be able to elaborate the primitive to flip-flops and clock latches. The GSHE switch alone can represent 16 Boolean functions, and from the convention layout of GSHE in Fig. 9.4, it can be seen that only three wires are used for all Boolean gates. Dummy wire can be used to protect the model, depending on the chip-level model and concept to implementation [32]. This type of physical layout protects the ICs from optical-imaging-based RE and side-channel attacks.

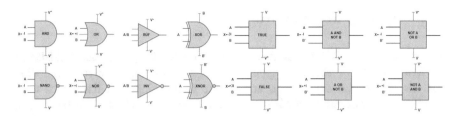

Fig. 9.3 Possible Boolean logic gates for the two inputs using the primitive

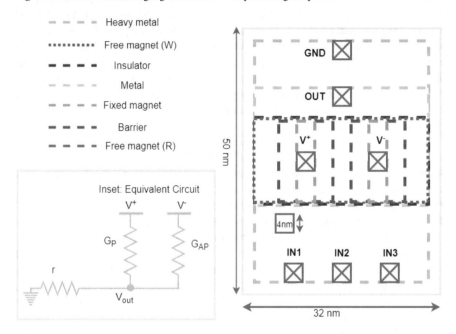

Fig. 9.4 Illustration of layout and circuit of GHSE switch

9.3 Emerging Technologies in Hardware Security

Emerging nanoscale technologies have many inherent hardware security features ranging from PUF/TRNG and in the supply chain to tampering detection. Outstanding features of emerging technologies have introduced security and trust applications. Emerging transistor technologies facilitate providing a robust system in the security primitives domain.

9.3.1 Security Beyond CMOS

Reference [37] looked for unique features of CNT/Graphene and PCM-based devices that can be useful in applications of hardware security such as PUFs, TRNGs, design for anti-tamper, anti-reverse engineering, and design for anti-counterfeit. For example, PCMs show features such as stochastic resistance variability due to process variation and MLC (multilevel cell) operation with a non-deterministic resistance window, which can be used to create PCM-based PUFs. Furthermore, a crossbar architecture is used where the fluctuations between PCM cells in the array can be used to generate challenge–response pairs. Moreover, PCM displays 1/f noise and random telegraph noise (RTN)—these are excellent variability sources and can be used for TRNGs. Furthermore, PCMs have a feature called the initial forming step, which could be used for tamper detection. Another feature called "resistance drift" could be employed to develop passive aging sensors in PCM cells to prevent counterfeiting by detecting the chip for how long it has been out from the supply chain [37].

Another emerging nanodevice that has great potential in creating PUFs is a carbon nanotube (CNT). A CNT-based PUF utilizes the characteristic fluctuation between the semiconducting and metallic characteristics of CNTs. The manufacturing and placement uncontrollability of CNT can be used to formulate a PUF's lightweight implementation. The stochastic physical and electrical variability of CNTs and graphene-based nanodevices has made them potential candidates for TRNGs. In combating invasive and semi-invasive attacks, CNT- and graphene-based optical, mechanical, and chemical sensors can be used. Another application of CNT- and graphene-based nanodevices is counterfeit detection. Over conventional, the main advantage of graphene-based printing electronics, as mentioned in [37], is that the IP owner can generate digital fingerprinting using them and ensure the product's security in the supply chain. Though graphene- and CNT-based designs promise hardware security, integrating them into a CMOS platform comes with its own set of hurdles in terms of yield, cost, and large scale manufacturing. A comparative analysis has been provided between ideal device intrinsic features and potential nanodevice features for hardware security applications. The challenges for developing security primitives of nanodevice-based have been pointed out in Table 9.1.

Their approach to addressing the challenges of using nanoscale devices for designing hardware security primitives is unique. They indicated in which direction researchers should focus while considering the security perspectives of the distinct features of nanodevices—developing necessary designs, experimental demonstrations, and vulnerability analysis. The key takeaways from this research are the following aspects that should be considered as high-level recommendations for possible research directions [37]:

1. Security Evaluation: Necessary metrics should be developed for the security-oriented evaluation of the emerging nanodevices to quantify the device's intrinsic

Table 9.1 Challenges for developing security primitives and countermeasures of emerging nanodevice-based hardware

Challenges	Challenges	Device and security applications					
		PUF/TRNG		Anti-tampering		Anti-counterfeit	
		PCM	CNTs/Graphene	PCM	CNTs/Graphene	PCM	CNTs/Graphene
Device modeling and metrics	Statistical model and metrics for variability	x	x	x	x	x	x
	Process variation	x	x	x	x	x	x
	Environmental variations & Aging	x	x				x
	Measurable entropy	x	x				
	Resistance drift	x				x	
Circuit and design optimization	Cycle-to-cycle stochasticity	x		x		x	
	Entropy extraction	x	x				
	Efficient post-processing	x	x		x		x
	Device forming step			x			
	Data retention loss					x	
Physical demonstration	Platform compatibility		x		x	x	x
	Integrate analog sensors into package/die				x		x
	Sensor printability				x		x
	Noise amplification	x	x				
	Initial device resistance	x				x	
Resistance to attacks (invasive/non-invasive)	Metrics to assess attack vulnerability	x	x	x	x	x	x
	Passive sensor operation				x		x
	Can sensor be bypassed?				x		x
	Side-channel leakage via remanence	x		x			

and extrinsic parameters, such as tampering susceptibility, variability, entropy, and sensor gain at the device level, etc.

2. Design and Modeling: Appropriate device-level models and security-specific designs are required for emerging nanodevices to withstand different hardware attacks and vulnerabilities. Moreover, efficient statistical models should be highly sought after so that various devices' security attributes, such as process variation of sources, aging impact, entropy, variation in environmental conditions, and changes due to tampering, can be evaluated.

3. Integration and Demonstration: Due to the lack of technological maturity of many emerging devices, e.g., CNTFETs, it is challenging to integrate them into CMOS platforms. Demonstrating the proposed applications of the emerging nanodevices is of utmost importance since it is necessary to ensure that these nanodevice-based designs are protective against emerging hardware attacks (such as FIB-based attacks or side-channel attacks).

9.3.2 DRAM-Based Security Primitives

The rapid research pace, along with a large amount of publications in the area of DRAM-based security primitives, makes it challenging to track advancements in the field. As a result, it retains a very high level of obscurity. Moreover, it is a general notion that hardware is rather challenging to attack compared to software [23].

Hardware-based security primitives exploit inherent manufacturing process variations and noises to extract a high level of entropy. There always exists some imperfections in every manufactured hardware, and these imperfections should not be significant enough to impact the hardware's correct operation. However, they can be harnessed to implement security primitives without any additional cost in terms of area and performance. The outputs obtained from these security primitives can be either stable or very unstable [24]. If the outputs are very stable, the primitive can be used for authentication and identification; otherwise, it can generate a cryptographic nonce, one-time pads, and true random keys. Inherent characteristics of DRAM make it very much suitable for both PUF and TRNG applications. Various aspects of DRAM have been discussed, and their potential for efficient primitive implementations, especially for resource-constrained applications, will be presented in upcoming paragraphs.

First of all, DRAM cells may assume different values at power on, and this scenario is very similar to SRAM cells. At start-up, the charge of the storage capacitor of each DRAM cell may be slightly over (VDD/2) threshold or slightly below this, which can be treated as logic 1 or 0, respectively. Second, DRAM cell capacitors leak charges and therefore are periodically refreshed. This charge decay characteristic for all DRAM cells is not the same. As a result, the DRAM cells exhibit different times of charge remanence. These properties of DRAM cells can be significantly exploited to extract a high degree of entropy. DRAM cells always require frequent refreshment. Otherwise, the stored bit can be flipped, resulting in a

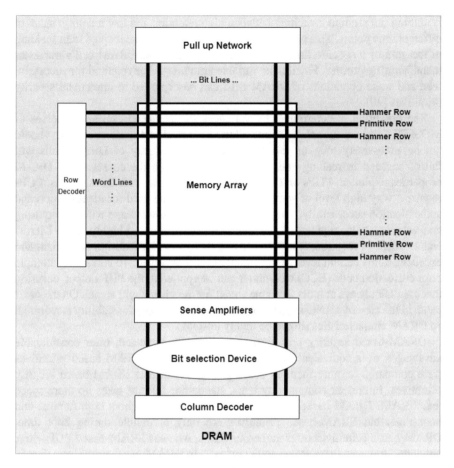

Fig. 9.5 Order of row hammering for security primitive implementation

high degree of instability. Researchers have done a couple of things to achieve like stopping refresh operation, which exploits retention of different cells, cutting power of the entire chip, which takes advantage of data remanence of the cells [57] and forcing different low voltages in word lines in order to deepen the decay effect [15]. Moreover, to enhance the number of bit flips significantly over a period of time, the row hammer process has been utilized [43]. In this method, rows are written continuously at a stretch with 0s and 1s in an alternate fashion, and due to excessive leakage, values of the SRAM cells flip in other rows. Alternate ordering helps to maximize the effect as shown in Fig. 9.5. As a result, it is clear that DRAM retention is significantly dependent on capacitor leakage.

One DRAM cell can be affected by leakage of nearby cells lying in the same or other rows. Also, the variable retention time (VRT) phenomenon of the DRAM cells can add more instability to the retention characteristic of the cells [28]. VRT is nothing but random switching between high retention and low retention states at different time points. This particular behavior devolves on unoccupied traps residing in the gate of a specific time and is solely related to the DRAM cell's transistor manufacturing process. Finally, the variable amount of time required for successful read and write operations of DRAM cells can be exploited to implement security primitives [20].

The condition is determined by variances in the manufacturing processes of MOSFETs and capacitors. The time allotted for read and write operations should not be excessively long or short; otherwise, the majority of DRAM cells will fail or succeed in reading and writing. Depending on the constancy of DRAM properties exploited, PUFs or TRNGs can be carried out with DRAM cells. PUFs require a very high level of stability so that they can be used for identification and authentication successfully. They may act as secure key storage without requiring non-volatile key storage techniques. For these applications, a highly stable DRAM characteristic is required. It is possible that though the PUF design is efficient and excellent, some certain level of uncertainty is still observed. To solve this, multiple error correction code (ECC) techniques can be applied to the PUF output, ensuring that the PUF always outputs the same signal for the given input signal. On the other hand, in the case of TRNG applications, a very high degree of instability is required, so DRAM characteristics should be highly unstable.

DRAM-based security primitives, as previously indicated, offer considerable advantages over equivalent hardware-based primitives. DRAM-based primitives were compared to other memory-based primitives, such as SRAM-based security primitives. In current computer systems, standalone DRAM takes up more space than SRAM. DRAM-based primitives are accessible throughout both runtime and boot time, but SRAM-based primitives are only accessible during boot time. DRAM-based primitives offer numerous CRPs, whereas SRAM-based PUFs offer only one. However, there are several drawbacks to DRAM-based primitives. First of all, when primitives are based on the retention times of DRAM cells, they require a long generation time. Second, DRAM modules are typically removable, so there is always a risk of the DRAM chips being stolen or replaced. However, this problem can be solved by using onboard DRAM modules. Last but not least, DRAM-based primitives are so much susceptible to temperature changes. The vast expansion of IoT devices has dramatically increased the demand for enough security resolutions. IoT devices are too resource-constrained, and DRAM-based primitives are much more attractive solutions.

Finally, the authors of [3] assessed the security of DRAM-based primitives. It is critical to look into the actual amount of security given by DRAM-based primitives and ways to assess them properly. They meticulously examined various factors for security. The examined factors included the common attack vectors, the time needed for primitives generation, the number of unique outputs provided, whether can DRAM module be removed, can primitives generate runtime outputs, and if the variations in environmental constraints impact the output responses. Several attacks and countermeasures have been investigated regarding DRAM-based primitives. Invasive attacks [50, 51] can be easily prevented if DRAM modules are designed to be tamper-proof or tamper-evident. In case of attacks, the DRAM module stops working, or its behavior is altered severely. The DRAM module can only be accessed by a dedicated security module or a privileged code, where access is constrained. DRAM-based primitives are also immune against non-invasive attacks such as memory readout attacks [68]. Machine-learning-based and row hammer attacks are very similar to memory readout attacks, and they can be prevented by DRAM-based primitives providing a large number of CRPs. Moreover, the entropy of the DRAM-based primitives' output is very high since outputs are highly unpredictable and contain almost equal numbers of 0s and 1s. Furthermore, if the size of the output is enormous and the outputs have no correlation, DRAM-based primitives are highly effective against brute force attacks, dictionary, and guessing attacks [56]. Replacement or stealing attacks can be prevented if onboard DRAM modules are used, and detectors can constantly monitor their presence. The use of temperature and voltage sensors can help to prevent environmental threats. Various forms of protocol assaults can be mitigated by employing cryptographic nonce and authentication channels, and in these circumstances, a high degree of redundancy is advantageous. Another method of defending against protocol attacks is out-of-band (OOB) authentication. Collision and spoofing attacks [25, 27] work in a way similar to protocol attacks. Onboard DRAMs with high entropy can help to fight against these attacks. Aging-based attacks can be adequately mitigated by using anti-aging techniques and using DRAMs resilient to aging.

9.3.3 Security Beyond PUF Using Emerging Technologies

Hardware security based on tunable polarity: SiNW FETs and TFETs provide logic and layout obfuscation that serves as a robust system for IP protection, IP prevention, and countermeasure against hardware Trojans to serve as novel security primitives. Previously polymorphic logic gates have been employed rarely in CMOS technology because of design complexity. However, attackers cannot recognize the polymorphic logic encryption-based circuit functionality. In this chapter, SiNW FET as polymorphic logic gates has been employed to prevent IP piracy. NAND based on SiNW FET can be turned into NOR gate by changing power and ground, whereas CMOS is unable to behave that way. Polymorphic gates based on SiNW FET can easily alter the circuit functionality by performing the control gate (CG) as standard input, and the polarity gate (PG) is acted as the polymorphic control input [4]. TFET-based polymorphic logic circuits can be operated as n-type or p-type transistors depending on biasing gate. By connecting the n-doped regions of the two parallel TFETs to VDD and the p-doped regions to GND, the circuit acts as a NAND gate. The circuit also acts like a NOR gate by simply changing the VDD and GND position.

MUXes have been employed to select two types of connections to make it function as a polymorphic gate. SiNW FETs and TFETs can be used to build low-cost polymorphic logic gates to perform desired computation. In an ASIC design,

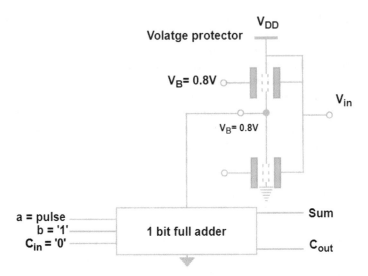

Fig. 9.6 Illustration of SymFET-based protector against fault injection

the chip is encrypted using polymorphic gates with SiNW FETs and TFETs to provide locking functionality from invalid users and attackers. Hence, SiNW FET- and TFET-based polymorphic gates can be employed to prevent IP cloning and IP piracy.

Hardware security based on switching behaviors can be introduced by some devices that exhibit the property of bell-shaped I–V curves and tunable hysteresis for designing hardware security primitives. Those characteristics of devices enable lower operating voltage and power.

Attackers can access the internal circuit signals using side-channel analysis without destroying the chips. Researchers proposed countermeasures to monitor delay and power consumption during encryption/decryption performance [29]. Graphene SymFETs have properties of voltage control and unique peak current. Graphene SymFET-based power supply protectors can monitor supply voltage within a predefined range, ensuring low-cost, high-sensitivity circuit protectors [4].

In the case of fault injection, we can protect the circuit from fault injection attacks by monitoring power fall due to decreasing supply voltage. As shown in Fig. 9.6, Vout has been used as a power supply for SymFET-based protector. SymFET-based protectors can lower VDD because of the power supply's bell-shaped I–V characteristic of the SymFET. Therefore, current/voltage can be monitored and protected in the circuit design using a SymFET-based protector. In comparison, a CMOS-based protector requires op-amps for voltage comparison.

NCFETs and ionic FETs act as a non-volatile storage element or a switch due to their tunable hysteresis property. As a result, the devices can design power-efficient logic-in-memory (LiM) cells that decrease memory attack vulnerability, reducing communication between CPU and memory [14]. In Figs. 9.7 and 9.8, different LiM

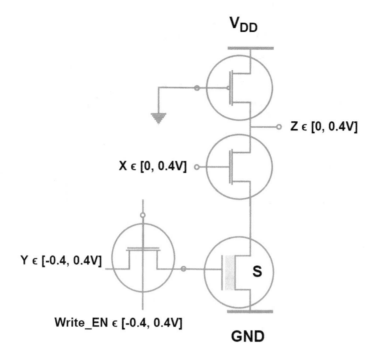

Fig. 9.7 Illustration of pseudo-NMOS logic-based NAND-LiM

cells using NCFETs have been developed based on the pseudo-NMOS logic style. Both LiM cells can be operated in update and hold modes. In the hold mode, when Write_EN= 0, the output Z is given as Z=X.S, where S is the fixed bit value. In the update mode (Write_EN= 1), the output is given as Z=X.Y.

Tunable hysteresis property-based devices can be employed as a switch from a non-volatile storage element that can function as logic obfuscation. In order to have tamper-resistant circuitry, such a non-volatile storage element device can be adjusted based on retention time. However, for MOSFETs without tunable hysteresis property, it is not easy to achieve that functionality.

9.3.4   Emerging Transistor Technologies for Hardware Security

The authors of [5] discuss briefly how emerging transistor technologies can help improve hardware security. They have proposed the usage of I–V characteristics of emerging transistors such as SiNW FETs, TFETs, and NCFETs to enhance security and have provided a great emphasis on moving beyond direct implementations of PUFs and TRNGs. These emerging transistor technologies can be applied effectively for sophisticated logic obfuscation for IP protection, prevention of fault

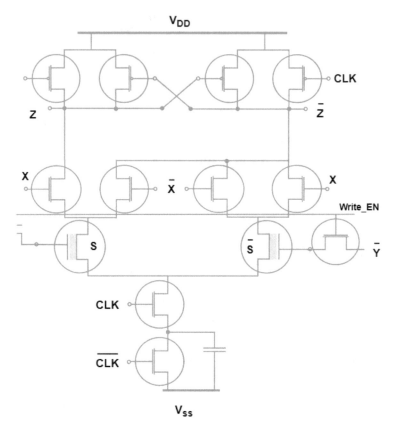

Fig. 9.8 Illustration of DyCML-based AND/NAND-LiM

injection attacks, and differential power analysis (DPA) attacks for lightweight cryptosystems. The uniqueness of this paper [5] lies in the fact that the authors have discussed the potential application of emerging transistor technologies beyond designing PUFs and TRNGs with emerging technologies. While, emerging technologies being studied in the existing literature are targeted only to PUF and TRNG applications that exploit device-to-device process variations. Nevertheless, let us look through the lens of reliable logic or memory. These variations will often represent shortcomings, and for this reason, the authors have discussed security-related applications other than PUFs and TRNGs. However, they do not rely on device variances as a means to an end. The main focus of the authors was on how developing technologies may have an impact on encryption engines, potentially leading to more sophisticated and robust encryption ciphers, while keeping resource limitations in mind how new gadgets could make ciphers more resistant to DPA attacks. Going beyond standard CMOS transistors' unique I-V properties might enable novel hardware primitives and help protect the IC supply chain by preventing side-channel attacks.

In nanoscale FETs in normal bias conditions, there is an observable superposition of n-type and p-type carriers. This creates ambipolarity that exists in various materials discussed in [9, 30, 31]. After deployment, the polarity of the device can be adjusted by adjusting ambipolarity. CNT, graphene, silicon nanowires (SiNWs), and TMD transistors all have programmable polarity. The operation of a SiNW FET is enabled by the management of Schottky barriers at the S/D junctions. A control gate (CG) is a type of MOS transistor that uses gate voltage to switch a device on or off. After manufacturing, a polarity gate (PG) near S/D Schottky junctions can switch the device polarity between n-type and p-type. Because their i/o voltage levels are comparable, logic gates can be readily cascaded [12]. For TFETs, ambipolarity is also an option. Different levels of doping profiles are applied in this situation for the drain and source. To activate the TFETs as n-type and p-type devices, proper biasing of the n-doped, p-doped, and gate areas is necessary, but no polarity gate is required. Tunable hysteresis is another important aspect of these new technologies. In a device's I–V curve, a hysteresis loop may occur, which can be shifted to different points along with the applied voltage levels or caused to disappear entirely. A device with this feature is a negative capacitance FET (NCFET). The ferroelectric material, which has high polarizability, is introduced to the gate stack. As a result, for certain electric field levels, CFETs display non-linear capacitance and turn negative, enabling step-up voltage conversion of applied gate bias to the surface potential. The composition of the gate stack material can be changed, and modifying the thickness of the FE material allows for the fabrication of an NCFET with or without hysteresis.

Current mode logic is a type of differential logic that includes a current-steering core, a current source, and a differential load [7, 69]. A CML circuit's power consumption is stable, providing good resiliency to DPA attacks. Attacks focusing on power consumption during circuit transitions are known as DPA attacks. Power is consumed in static CMOS circuits when the output flips from 0 to 1 or 1 to 0. As a result, any cryptographic algorithm is always subject to DPA attacks. Conversely, a CML is naturally resistant to DPA attacks since the transitions use a relatively consistent amount of power. Power traces are analyzed using both static XOR gates and differential XOR gates based on TFETs.

A significant amount of power overshoot is observed in the power trace of the TFET-based static XOR gate. Nevertheless, the power dissipated by the CML XOR gate is virtually constant. So, an attacker could utilize the information leaked by a TFET-based static XOR gate to characterize the internal activity of a cryptographic system. However, it is not possible for CML gates. TFET-based CML can improve DPA resistance while still preserving low-power consumption. Research [8] has shown that a TFET CML circuit consumes only 20% of the power consumed by a CMOS-based CML circuit, but both provide similar DPA resiliency levels.

9.3.5   SiNW and Graphene SymFET-Based Hardware Security Primitive

Reference [6] introduced hardware attack prevention and IP protection using two emerging technologies, i.e., graphene SymFETs and silicon nanowire (SiNW) FETs. They used polarity-controllable SiNW FETs to demonstrate a novel way of building polymorphic gates. They showed that SiNW FET-based camouflaging layout and polymorphic gates had been used to ensure the protection of the IPs. They further proposed SymFET circuit protectors as a countermeasure against fault injection attacks. Finally, they presented a lightweight SymFET-based XOR for cryptographic function implementation.

Camouflaging is a popular method for preventing hardware attackers from gaining access to the circuit schematic through reverse engineering. Because PMOS and NMOS have fixed polarities, CMOS camouflaging gates have a substantially wider area than conventional gates. Because of their unique properties, polarity-controlled SiNW FETs can aid in the design of camouflaging gates without the use of additional FETs. Figure 9.9 illustrates a more complicated camouflaging gate that can behave as a NAND, NOR, XOR, or XNOR depending on which dummy connections are used. In comparison to the CMOS-based camouflaging gate, which requires 12 transistors for a NAND–NOR–XOR gate, the proposed architecture in Fig. 9.9 can reduce the transistor count to 4. Table 9.2 depicts multiple contact connections of the camouflaged arrangement that might result in four distinct processes for the identical input signals [6].

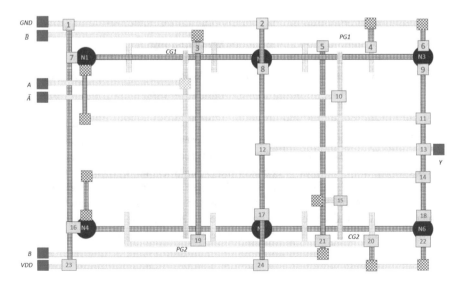

Fig. 9.9   Illustration of camouflaging layout that can be used as NAND, NOR, XOR, or XNOR

Table 9.2 Different connection combinations of the camouflaged layout to behave as different gates

Function	Contacts	
	True	Dummy
NAND	1,4,8,9,11,13,15,16,18,20,24	2,3,5,6,7,10,12,14,17,19,21,22,23
NOR	2,4,7,9,13,14,15,17,18,20,23	1,3,5,6,8,10,11,12,16,19,21,22,24
XOR	1,3,6,8,10,11,12,16,17,18,21,22	2,4,5,7,9,13,14,15,19,20,23,24
XNOR	1,5,6,8,10,11,12,16,17,18,19,22	2,3,4,7,9,13,14,15,20,21,23,24

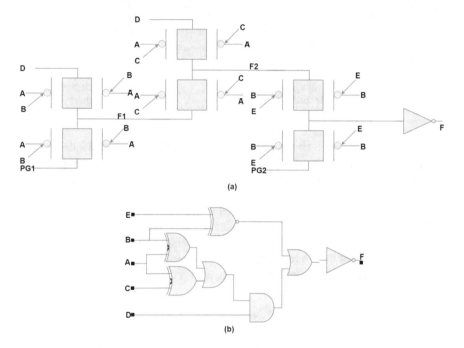

Fig. 9.10 Illustration of a SiNW FET complex gate. (**a**) Schematic diagram of a transistor. (**b**) Schematic diagram of a gate

As explained in Fig. 9.10, the unique technique is to create polymorphic gates utilizing polarity-controllable SiNW FETs. Table 9.3 depicts the simulation results of the SiNW FET and CMOS five-input polymorphic function [22]. Because the critical path has the same number of cells, the SiNW FET approach reduces the total dynamic power by using fewer cells with a longer delay. Furthermore, despite having lower leakage power than CMOS, SiNW FET polymorphic logic circuits surpass CMOS circuits in terms of power and time while providing comparable levels of safety.

Reference [6] proposed two SymFET-based circuit protectors that safeguard circuits from voltage glitch attacks by utilizing the specific I–V properties of SymFETs. The circuit protectors can be categorized as current-based circuit pro-

Table 9.3 Comparison between FinFET and SiNW FET technologies

Technology	Static power (nW)	Switching average power (uW)	Average delay
FinFET 22nm LSTP	0.755	4.04	80
FinFET 22nm HP	491	5.4	60
SiNW 20nm	0.01	2.5	100

Table 9.4 Overview of security applications using SiNW FET and SymFET

Key topics	SiNW FETs	Graphene SymFETs
Benefits over CMOS	Polarity configurable, low static power, fewer transistors for applications	Lower power, built-in negative differential resistance
Challenges	Larger area per-transistor, large dynamic power	Current-based designs, non-Boolean computation
Opportunities	IP protection, logic encryption, other security applications	Side-channel attack prevention, Cryptographic circuits

tectors and voltage-based circuit protectors. Because CMOS-based standard power regulators used for circuit protection have a huge size and power consumption, the suggested architecture, which uses only three SymFETs, is an effective alternative. Table 9.4 provides a summary of the applications of SiNW FET and SymFET in hardware security applications, as discussed in the work of [6].

The critical contribution of [6] is in evaluating the unique properties of emerging nanodevices in protecting circuit designs and preventing IP piracy. Another important takeaway is that they demonstrated how the unique properties of these devices could provide high-level circuit protection with a significantly low-performance overhead if used effectively.

9.3.6 Polymorphic and Stochastic Spin Hall Effect Devices

References [6, 71] proposed security primitives that contrast camouflaging and logic locking as interchangeable motions without loss. Additionally, [70] concentrated on CMOS-centric GSHE-based techniques. All techniques indicate that the proposed primitive is powerful against SAT attacks [49]. The results highlighted that the proposed primitives were more effective in camouflaging across different benchmarks. However, some computational failures impose practical limitations on the SAT attacks. Not only against traditional SAT attacks, but this method can also show resiliency against Double DIP attack [45]. The attack rules out at minimum two incorrect keys in each iteration.

As the delay of the GSHE switch is higher than CMOS, some practical study shows that some paths have a shorter delay, which replaces CMOS gates with the GSHE-based primitive, which can conduct camouflage 5–15%. This proposed primitive is way better to protect ICs and IPs with no excessive layout (PPA) overheads. Furthermore, this GSHE switch supports tunable probabilistic computation. So, by tuning this GSHE, 95% accuracy can be achieved, which implies that SAT attack will fail in such a case. As GSHE switches experience stochasticity (thermally induced) and can be tuned and superposed with each other, they conduct stochastically correlated behavior at the primary outputs that effectively protect the circuit against SAT attack.

As their conventional layout is uniform, this proposed primitive is very effective against optical-imaging-based RE and electron microscopy (EM) [10]. This is because GSHE is significantly smaller than CMOS devices and the primitive is polymorphic, with the ability to modify behavior at runtime. So implementing runtime polymorphism at the chip level can make the ICs independent of reverse engineering threats. Reference [26] proposed an idea that emphasizes this, enabling dynamic protection to the entire circuit to render runtime-intensive attacks incapable (SAT attacks in particular). Reference [17] have outlined that spintronics devices can be attacked magnetically and changing temperature. This GSHE design, on the other hand, can assure a strong coupling between R and W nanomagnets. Because GSHE is a tiny form of switch, the states of their W and R magnetizations, the fixed magnet's orientation, and voltage polarities are difficult to ensure. While temperature-driven attacks occur, inherent thermal noise in nanomagnets conducts stochastic behavior.

9.4 Future Research Directions

Emerging nanodevices such as SiNW FETs, TFETs, NCFETs, PCM, and CNT, are lucrative solutions for implementing the existing and new security primitives, but they still have a long way to go. Integrating those emerging nanodevices with prevalent CMOS technology has been challenging as these emerging technologies are still immature. Moreover, proper security evaluation of these emerging devices is yet to be done since most device-level metrics developed so far target performance issues only. Additionally, emerging nanodevices still lack proper statistical models capturing security features of the device since these features highly depend on device intrinsic process variation, which is very hard to model correctly. Moreover, security primitives implemented by emerging nanodevices create area and power overhead issues in the design to be protected. Hence, there is always a need for extensive research in these areas as so many challenges remain to be adequately addressed to incorporate security primitives implemented with emerging nanodevices. There is always a dire need for interdisciplinary collaboration between researchers from both hardware security and device domains. Researchers in hardware security can concentrate on generating effective security metrics, while device researchers can utilize these metrics to influence the design and modeling of upcoming devices. We hope that future extensive research will address the existing issues adequately and open a new era of successful implementation and integration of security primitives with emerging nanotechnology.

9.5 Conclusions

This chapter identifies the pros and cons of using emerging nanodevices in developing hardware security primitives and explores their potential to establish protection against different security vulnerabilities. A precise and brief analysis of state-of-the-art applications of emerging nanotechnologies based on the existing literature has been provided. The collaboration of the security and device community can be urged to look for further improvement of these devices with appropriate evaluation metrics and statistical models. It can be concluded that this discussion will provide a better perspective on leveraging emerging technologies in security applications instead of using conventional CMOS technology.

References

1. Alasad Q, Yuan J, Fan D (2017) Leveraging all-spin logic to improve hardware security. In: Proceedings of the on Great Lakes symposium on VLSI 2017, pp 491–494
2. Alkabani Y, Koushanfar F (2007) Active hardware metering for intellectual property protection and security. In: USENIX security symposium, pp 291–306
3. Anagnostopoulos NA, Katzenbeisser S, Chandy J, Tehranipoor F (2018) An overview of DRAM-based security primitives. Cryptography 2(2):7
4. Bi Y, Gaillardon PE, Hu XS, Niemier M, Yuan JS, Jin Y (2014) Leveraging emerging technology for hardware security-case study on silicon nanowire FETs and graphene SymFETs. In: 2014 IEEE 23rd Asian test symposium. IEEE, pp 342–347
5. Bi Y, Hu XS, Jin Y, Niemier M, Shamsi K, Yin X (2016) Enhancing hardware security with emerging transistor technologies. In: Proceedings of the 26th edition on Great Lakes symposium on VLSI, pp 305–310
6. Bi Y, Shamsi K, Yuan JS, Gaillardon PE, Micheli GD, Yin X, Hu XS, Niemier M, Jin Y (2016) Emerging technology-based design of primitives for hardware security. ACM J Emerg Technol Comput Syst (JETC) 13(1):1–19
7. Cevrero A, Regazzoni F, Schwander M, Badel S, Ienne P, Leblebici Y (2011) Power-gated MOS current mode logic (PG-MCML): A power aware DPA-resistant standard cell library. In: 2011 48th ACM/EDAC/IEEE design automation conference (DAC). IEEE, pp 1014–1019
8. Chen A, Hutchby J, Zhirnov V, Bourianoff G (2014) Emerging Nanoelectronic Devices. Wiley
9. Colli A, Pisana S, Fasoli A, Robertson J, Ferrari A (2007) Electronic transport in ambipolar silicon nanowires. Physica Status Solidi (b) 244(11):4161–4164
10. Courbon F, Skorobogatov S, Woods C (2016) Direct charge measurement in floating gate transistors of flash EEPROM using scanning electron microscopy
11. Cruz J, Farahmandi F, Ahmed A, Mishra P (2018) Hardware trojan detection using ATPG and model checking. In: 2018 31st international conference on VLSI design and 2018 17th international conference on embedded systems (VLSID). IEEE, pp 91–96
12. De Marchi M, Sacchetto D, Frache S, Zhang J, Gaillardon PE, Leblebici Y, De Micheli G (2012) Polarity control in double-gate, gate-all-around vertically stacked silicon nanowire FETs. In: 2012 International electron devices meeting. IEEE, pp 8–4
13. Di Ventra M, Pershin YV, Chua LO (2009) Circuit elements with memory: memristors, memcapacitors, and meminductors. Proc IEEE 97(10):1717–1724
14. Elliott DG, Stumm M, Snelgrove WM, Cojocaru C, McKenzie R (1999) Computational RAM: Implementing processors in memory. IEEE Des Test Comput 16(1):32–41

15. Fainstein D, Rosenblatt S, Cestero A, Robson N, Kirihata T, Iyer SS (2012) Dynamic intrinsic chip ID using 32nm high-K/metal gate SOI embedded DRAM. In: 2012 symposium on VLSI circuits (VLSIC). IEEE, pp 146–147
16. Farahmandi F, Huang Y, Mishra P (2017) Trojan localization using symbolic algebra. In: 2017 22nd Asia and South Pacific design automation conference (ASP-DAC). IEEE, pp 591–597
17. Ghosh S (2016) Spintronics and security: Prospects, vulnerabilities, attack models, and preventions. Proc IEEE 104(10):1864–1893
18. Guin U, Huang K, DiMase D, Carulli JM, Tehranipoor M, Makris Y (2014) Counterfeit integrated circuits: A rising threat in the global semiconductor supply chain. Proc IEEE 102(8):1207–1228
19. Guin U, Zhang X, Forte D, Tehranipoor M (2014) Low-cost on-chip structures for combating die and IC recycling. In: 2014 51st ACM/EDAC/IEEE design automation conference (DAC). IEEE, pp 1–6
20. Hashemian MS, Singh B, Wolff F, Weyer D, Clay S, Papachristou C (2015) A robust authentication methodology using physically unclonable functions in DRAM arrays. In: 2015 Design, automation & test in Europe conference & exhibition (DATE). IEEE, pp 647–652
21. Kamali HM, Azar KZ, Farahmandi F, Tehranipoor M (2022) Advances in logic locking: Past, present, and prospects. Cryptology ePrint Archive
22. Katzenbeisser S, Kocabaş Ü, Rožić V, Sadeghi AR, Verbauwhede I, Wachsmann C (2012) PUFs: Myth, fact or busted? A security evaluation of physically unclonable functions (PUFs) cast in silicon
23. Keller C, Felber N, Gürkaynak F, Kaeslin H, Junod P (2010) Physically Unclonable Functions for Secure Hardware. Swiss National Science Foundation (SNSF), Nano-Tera CH, RTD
24. Keller C, Gürkaynak F, Kaeslin H, Felber N (2014) Dynamic memory-based physically unclonable function for the generation of unique identifiers and true random numbers. In: 2014 IEEE international symposium on circuits and systems (ISCAS). IEEE, pp 2740–2743
25. Kirihata T, Rosenblatt S (2017) Dynamic Intrinsic Chip ID for Hardware Security. In: VLSI: Circuits for emerging applications. CRC Press, pp 357–388
26. Koteshwara S, Kim CH, Parhi KK (2017) Key-based dynamic functional obfuscation of integrated circuits using sequentially triggered mode-based design. IEEE Trans Inf Foren Secur 13(1):79–93
27. Kumar R, Xu X, Burleson W, Rosenblatt S, Kirihata T (2016) Physically unclonable functions: A window into CMOS process variations. In: Circuits and systems for security and privacy. CRC Press, pp 183–244
28. Liu J, Jaiyen B, Kim Y, Wilkerson C, Mutlu O (2013) An experimental study of data retention behavior in modern DRAM devices: Implications for retention time profiling mechanisms. ACM SIGARCH Comput Architect News 41(3):60–71
29. Mamiya H, Miyaji A, Morimoto H (2004) Efficient countermeasures against RPA, DPA, and SPA. In: International workshop on cryptographic hardware and embedded systems. Springer, pp 343–356
30. Martel R, Derycke V, Lavoie C, Appenzeller J, Chan K, Tersoff J, Avouris P (2001) Ambipolar electrical transport in semiconducting single-wall carbon nanotubes. Phys Rev Lett 87(25):256,805
31. Mishchenko A, Tu J, Cao Y, Gorbachev RV, Wallbank J, Greenaway M, Morozov V, Morozov S, Zhu M, Wong S, et al (2014) Twist-controlled resonant tunnelling in graphene/boron nitride/graphene heterostructures. Nature Nanotechnology 9(10):808–813
32. Patnaik S, Ashraf M, Sinanoglu O, Knechtel J (2020) Obfuscating the interconnects: Low-cost and resilient full-chip layout camouflaging. IEEE Trans Comput Aided Des Integr Circuits Syst 39(12):4466–4481
33. Prabhu P, Akel A, Grupp LM, Wing-Kei SY, Suh GE, Kan E, Swanson S (2011) Extracting device fingerprints from flash memory by exploiting physical variations. In: International conference on trust and trustworthy computing. Springer, pp 188–201

34. Pundir N, Amsaad F, Choudhury M, Niamat M (2017) Novel technique to improve strength of weak arbiter PUF. In: 2017 IEEE 60th international midwest symposium on circuits and systems (MWSCAS). IEEE, pp 1532–1535
35. Rahman MT, Forte D, Tehranipoor M (2015) Robust SRAM-PUF: Cell stability analysis and novel bit selection algorithm. TECHCON
36. Rahman MT, Rahman F, Forte D, Tehranipoor M (2015) An aging-resistant RO-PUF for reliable key generation. IEEE Trans Emerg Top Comput 4(3):335–348
37. Rahman F, Shakya B, Xu X, Forte D, Tehranipoor M (2017) Security beyond CMOS: fundamentals, applications, and roadmap. IEEE Trans Very Large Scale Integr (VLSI) Syst 25(12):3420–3433
38. Rajendran J, Sam M, Sinanoglu O, Karri R (2013) Security analysis of integrated circuit camouflaging. In: Proceedings of the 2013 ACM SIGSAC conference on computer & communications security, pp 709–720
39. Rajendran J, Zhang H, Zhang C, Rose GS, Pino Y, Sinanoglu O, Karri R (2013) Fault analysis-based logic encryption. IEEE Trans Comput 64(2):410–424
40. Rangarajan N, Parthasarathy A, Kani N, Rakheja S (2017) Energy-efficient computing with probabilistic magnetic bits–performance modeling and comparison against probabilistic CMOS logic. IEEE Trans Magnet 53(11):1–10
41. Rostami M, Majzoobi M, Koushanfar F, Wallach DS, Devadas S (2014) Robust and reverse-engineering resilient PUF authentication and key-exchange by substring matching. IEEE Trans Emerg Top Comput 2(1):37–49
42. Rührmair U, Sehnke F, Sölter J, Dror G, Devadas S, Schmidhuber J (2010) Modeling attacks on physical unclonable functions. In: Proceedings of the 17th ACM conference on computer and communications security, pp 237–249
43. Schaller A, Xiong W, Anagnostopoulos NA, Saleem MU, Gabmeyer S, Katzenbeisser S, Szefer J (2017) Intrinsic Rowhammer PUFs: Leveraging the Rowhammer effect for improved security. In: 2017 IEEE international symposium on hardware oriented security and trust (HOST), IEEE, pp 1–7
44. Schwierz F (2010) Graphene transistors. Nature Nanotechnology 5(7):487–496
45. Shen Y, Zhou H (2017) Double DIP: Re-evaluating security of logic encryption algorithms. In: Proceedings of the on Great Lakes symposium on VLSI 2017, pp 179–184
46. Shi Q, Vashistha N, Lu H, Shen H, Tehranipoor B, Woodard DL, Asadizanjani N (2019) Golden gates: A new hybrid approach for rapid hardware trojan detection using testing and imaging. In: 2019 IEEE international symposium on hardware oriented security and trust (HOST). IEEE, pp 61–71
47. Skorobogatov SP (2005) Semi-invasive attacks: a new approach to hardware security analysis
48. Slonczewski JC (1996) Current-driven excitation of magnetic multilayers. J Magnet Magnet Mater 159(1-2):L1–L7
49. Subramanyan P, Ray S, Malik S (2015) Evaluating the security of logic encryption algorithms. In: 2015 IEEE International symposium on hardware oriented security and trust (HOST). IEEE, pp 137–143
50. Sutar S, Raha A, Raghunathan V (2016) D-PUF: An intrinsically reconfigurable DRAM PUF for device authentication in embedded systems. In: 2016 International conference on compliers, architectures, and synthesis of embedded systems (CASES). IEEE, pp 1–10
51. Sutar S, Raha A, Raghunathan V (2018) Memory-based combination PUFs for device authentication in embedded systems. IEEE Trans Multi-Scale Comput Syst 4(4):793–810
52. Talukder B, Ray B, Tehranipoor M, Forte D, Rahman MT (2018) LDPUF: exploiting DRAM latency variations to generate robust device signatures. Preprint arXiv:180802584
53. Tehrani S, Slaughter J, Chen E, Durlam M, Shi J, DeHerren M (1999) Progress and outlook for MRAM technology. IEEE Trans Magnet 35(5):2814–2819
54. Tehranipoor M, Wang C (2011) Introduction to Hardware Security and Trust. Springer Science & Business Media
55. Tehranipoor MM, Guin U, Forte D (2015) Counterfeit integrated circuits. In: Counterfeit integrated circuits. Springer, pp 15–36

56. Tehranipoor F, Karimian N, Yan W, Chandy JA (2016) DRAM-based intrinsic physically unclonable functions for system-level security and authentication. IEEE Trans Very Large Scale Integr (VLSI) Syst 25(3):1085–1097

57. Tehranipoor F, Yan W, Chandy JA (2016) Robust hardware true random number generators using DRAM remanence effects. In: 2016 IEEE international symposium on hardware oriented security and trust (HOST). IEEE, pp 79–84

58. Tehranipoor MM, Forte DJ, Farahmandi F, Nahiyan A, Rahman F, Rahman MS (2022) Protecting obfuscated circuits against attacks that utilize test infrastructures. US Patent 11,222,098

59. Vasen TJ (2014) Investigation of III-V Tunneling Field-Effect Transistors. University of Notre Dame

60. Vashistha N, Lu H, Shi Q, Rahman MT, Shen H, Woodard DL, Asadizanjani N, Tehranipoor M (2018) Trojan scanner: Detecting hardware trojans with rapid SEM imaging combined with image processing and machine learning. In: ISTFA 2018: Proceedings from the 44th international symposium for testing and failure analysis. ASM International, p 256

61. Vashistha N, Rahman MT, Shen H, Woodard DL, Asadizanjani N, Tehranipoor M (2018) Detecting hardware trojans inserted by untrusted foundry using physical inspection and advanced image processing. J Hardware Syst Secur 2(4):333–344

62. Vashistha N, Rahman MT, Paradis OP, Asadizanjani N (2019) Is backside the new backdoor in modern SoCs? In: 2019 IEEE international test conference (ITC). IEEE, pp 1–10

63. Vashistha N, Hossain MM, Shahriar MR, Farahmandi F, Rahman F, Tehranipoor M (2021) eChain: A blockchain-enabled ecosystem for electronic device authenticity verification. IEEE Trans Consumer Electron

64. Vashistha N, Lu H, Shi Q, Woodard DL, Asadizanjani N, Tehranipoor M (2021) Detecting hardware trojans using combined self-testing and imaging. IEEE Trans Comput Aided Des Integr Circuits Syst

65. Wang X, Tehranipoor M, Plusquellic J (2008) Detecting malicious inclusions in secure hardware: Challenges and solutions. In: 2008 IEEE international workshop on hardware-oriented security and trust. IEEE, pp 15–19

66. Winograd T, Salmani H, Mahmoodi H, Gaj K, Homayoun H (2016) Hybrid STT-CMOS designs for reverse-engineering prevention. In: 2016 53rd ACM/EDAC/IEEE design automation conference (DAC). IEEE, pp 1–6

67. Xiao K, Rahman MT, Forte D, Huang Y, Su M, Tehranipoor M (2014) Bit selection algorithm suitable for high-volume production of SRAM-PUF. In: 2014 IEEE international symposium on hardware-oriented security and trust (HOST). IEEE, pp 101–106

68. Xiong W, Schaller A, Anagnostopoulos NA, Saleem MU, Gabmeyer S, Katzenbeisser S, Szefer J (2016) Run-time accessible DRAM PUFs in commodity devices. In: International conference on cryptographic hardware and embedded systems. Springer, pp 432–453

69. Yamashina M, Yamada H (1995) An MOS current mode logic (MCML) circuit for low-power GHz processors. NEC Res Dev 36(1):54–63

70. Yasin M, Mazumdar B, Rajendran JJ, Sinanoglu O (2016) SARLock: SAT attack resistant logic locking. In: 2016 IEEE international symposium on hardware oriented security and trust (HOST). IEEE, pp 236–241

71. Zhang Y, Yan B, Wu W, Li H, Chen Y (2015) Giant spin Hall effect (GSHE) logic design for low power application. In: 2015 Design, automation & test in Europe conference & exhibition (DATE). IEEE, pp 1000–1005

Chapter 10
Hardware Camouflaging in Integrated Circuits

10.1 Introduction

As attacks on integrated circuits become more common, reverse engineering and intellectual property theft/counterfeiting are becoming a growing concern in the industry [3, 4, 8]. As a result, intellectual property owners and integrated circuit designers must keep up with attackers by developing increasingly resilient countermeasures to reverse engineer attempts [1]. Because reverse engineering of ICs is becoming more robust, the techniques and methods for protecting ICs must evolve. Every year, the semiconductor industry loses billions of dollars due to intellectual property (IP) infringement [17]. Designers typically use layout-level countermeasures to prevent reverse engineers from obtaining the gate-level netlist and thus protecting the IP. If the attacker cannot obtain an accurate gate-level netlist, he will be unable to determine the true structure of the design [6]. The designer's goal is to conceal the gates so that their true logic function is ambiguous, implying that their logic functions are unknown [9].

In this chapter, various defense techniques are evaluated for their ability to camouflage an IP. These techniques entail obscuring physical logic cells and creating new types of cells that resemble existing ones. It is critical that the camouflaged cells do not add extra overhead and degrade the IC's performance. These methods must also be affordable. When these conditions are met, they can be easily deployed in industries where time-to-market and price are critical.

Typically, a reverse engineering attack involves the following five phases:

1. Decapsulation: To remove the IC packaging, a reverse engineer must use corrosive chemicals or some form of etching [14, 15].
2. Delayering: Because an integrated circuit is made up of multiple layers, the engineer must use corrosive or etching to remove each individual layer [13, 15].
3. Imaging: Each layer must be photographed and saved. A scanning electron microscope (SEM) is commonly used for this [7, 15, 16].

M. Tehranipoor et al., *Hardware Security Primitives*,
https://doi.org/10.1007/978-3-031-19185-5_10

4. Annotation: Once each individual layer has been captured, it must be stitched together to reconstruct the entire IC digitally.
5. Extraction of Gate-Level Netlist: The standard cells can be identified using the annotated images, and the gate-level netlist can be extracted.

Obfuscation of circuit logic, specifically the camouflaging of key logic gates specially selected by the designer, is an effective countermeasure that hides the type of logic gate and thus its functionality from potential attackers looking to enter the chip and capture images of its inner workings [11]. This works by effectively adding an extra layer atop certain gates, such as XNOR and XOR, that conceals the location of the logic gate's true connections (which is the only distinguishing factor between certain gates). If the attacker chooses to delay the IC in order to physically examine the layout to determine which cell is implemented, they will be perplexed because the cell appears to implement multiple gates.

However, due to the added overhead from the camouflaging methods, there is a direct relationship between the level of logic obfuscation from gate camouflaging and degradation to speed, cost, and space inside the IC. It is a common theme that with increased circuit protection comes increased unwanted overhead, and it is critical that the designer strikes a balance between anti-piracy resilience and extra cost, loss of space, and loss of speed/efficiency. Various existing research into the implementation and different methodologies of IC camouflaging and methods of attack against obfuscated circuits have been explored and summarized in this chapter to resolve the functionality of camouflaged gates.

10.2 Background

It is reasonable to assume that an attacker has reverse engineering tools, allowing him to distinguish between regular and camouflaged gates and, as a result, know the functions implemented on the circuit. An attacker will buy two copies of the same integrated circuit (IC), extract the netlist from the first chip, compute input patterns, apply the computed patterns to the second chip, and resolve the functionality of the camouflaged gates [2].

Attackers will use justification and sensitization to successfully reverse engineer a chip. Sensitization refers to setting each gate's side inputs between the gate's non-controlling value, whereas justification refers to fixing the output of a gate by setting the input lines accordingly. The attackers can use these two techniques to determine the functionality of the camouflaged gates. When the paths of two camouflaged gates intersect, or when camouflaged gates interfere, reverse engineering becomes more difficult. In some cases, the functionality of the gates can be easily determined (fully resolvable), while in other cases, it can only be determined through brute force methods (non-resolvable) [2].

To reverse engineer a camouflaged IC, the attacker first applies input patterns to the isolated gates. Following that, the fully resolvable gates are targeted using the previously mentioned justification and sensitization. Finally, partially resolvable and non-resolvable gates are targeted using brute force methods. The majority of the time spent determining the identities of the gates is spent performing the necessary brute force [2]. IC camouflaging could be used to defend against reverse engineering attacks. IC camouflaging allows a camouflaged cell to resemble another standard cell, keeping its functionality hidden from reverse engineering attacks. To be successful with IC camouflaging, the functionalities of camouflaged gates must be hidden from attackers, and the outputs of the deceiving netlist must be corrupted, or different from the original [2].

10.3 CamoPerturb

The most common camouflaging technique is to obscure the physical layout of the logic that an IC designer wishes to hide. CamoPerturb is a technique that goes a step further by attempting to obfuscate the logic by modifying the circuit's minterms [17]. The motivation for this is that using a technique known as the DeCamo attack makes it easier to distinguish the actual gate from a camouflaged gate. The DeCamo attack generates a list of discriminating inputs (DI) (incorrect inputs) that will result in incorrect outputs. The attacker can quickly narrow down the possible incorrect values with well-thought-out ordered sets of DIs. As a result, the attacker will be able to ascertain the truth relatively quickly.

CamoPerturb logic comprises two logic blocks: *Cpert*, the perturbed logic circuit block, and the CamoFix block. Figure 10.1 shows the various blocks in CamoPerturn method. *Cpert* is the circuit that needs to be protected from reverse engineering and that has had one minterm changed. This perturbed minterm is one

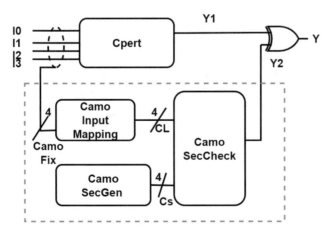

Fig. 10.1 Illustration of the logic circuit that is camouflaged using the CamoPerturb method that shows the internals of the CamoFix block

that was either added to the original circuit or removed from the original circuit. The CamoFix block contains the logic necessary to decrypt and restore the *Cpert* output to the proper minterms. The original minterms are restored by XORing the outputs of the *Cpert* and CamoFix blocks. CamoFix is also made up of three subcomponent blocks: Camo Input Mapping, CamoSecGen, and CamoSecCheck. The CamoSecGen block contains the hard-coded secret in the form of camouflaged gates that could be inverters or buffers. There are as many of these gates as inputs required to encrypt and decrypt, and by connecting them to VDD, they generate a vector of 1s and 0s that comprise the secret. To prevent the secret from being decrypted, the Camo Input Mapping block obfuscates the CamoFix inputs by using the same camouflaged gates as in the SecGen block to mask both CamoSecCheck inputs. The CamoSecCheck uses XNOR gates to determine whether the masked input from the Camo Input Mapping block matches the secret generated by the CamoSecGen block. The CamoSecCheck block will set the output of the CamoFix block to high if the two values match. The outputs of the *Cpert* and CamoFix blocks are finally XORed together to decrypt the perturbed minterm and restore the original logic. Figure 10.2 illustrates the general functioning of CamoSecGen and Camo Input Mapping block. Finally, Fig. 10.3 illustrates how CamoSecCheck output is generated from CamoSecGen and Camo Input Mapping block inputs.

The CamoPerturb method is significant because the attacker will not be able to distinguish multiple incorrect inputs from a single DI. This is due to the CamoPerturb block's output going high only when the camouflaged input matches the camouflaged secret. Furthermore, because one DI can only reveal one unique set of incorrect inputs, the number of DIs required cannot be reduced. The number of DIs possibly needed is the same as the number of possible inputs, which is $2^k - 1$, where k is the number of inputs. This complexity is desirable in a camouflaging technique and distinguishes it from other currently used techniques.

CamoPerturb's effectiveness is demonstrated in [17] and compared to more traditional camouflaging techniques such as Clique-Based Selection (CBS). When both CamoPerturb and CBS are applied to the same circuit, CamoPerturb requires $2k - 1$ DIs to decrypt, whereas CBS can be broken relatively easily with $DI \leq 10$. CBS camouflaging usually takes less than a second to solve due to the low number of DIs required. On the contrary, as the number of CamoPerturb possible DIs grows exponentially, so does the time it takes to execute the attack. In addition to the improved complexity that CamoPerturb provides, it has lower delay than other techniques. Since it needs to create two sets of camouflaged blocks in CamoPerturb, CamoFix, and *Cpert*, it has a higher area requirement. Table 10.1 summarizes the metrics of multiple camouflaging techniques.

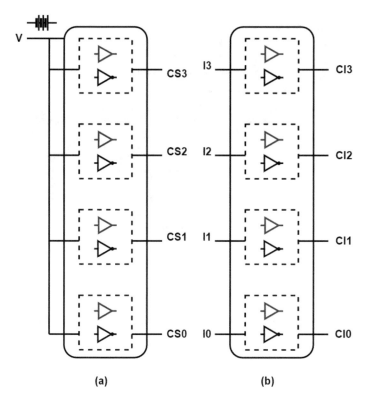

Fig. 10.2 (**a**) The illustration of CamoSecGen block's layout with four camouflaged inverter/buffers that generate the secret code. (**b**) Illustration of Camo Input Mapping block consisting of four camouflaged inverters/buffers to hide the transformations. The red gates denote the correct assignment

10.4 Covert Gates—IC Protection Method Using Undetectable Camouflaging

The existing countermeasures to IC reverse engineering involve creating cells with ambiguous true logic functions. These "special cells" can be distinguished from the standard logic cells that constitute digital ICs [11]. Cells are typically larger and have a greater number of contacts. This highlights which cells are camouflaged and work in the reverse engineer's favor because they know which gates to attack.

An adversary can use both invasive and non-invasive attack methods. An invasive method would be using a focused ion beam (FIB) to mill metal contacts and determine which contacts are dummy contacts. Non-invasive methods include SAT and test-based attacks. Both types of attack methods have been shown to easily determine the identity of camouflaged cells [11].

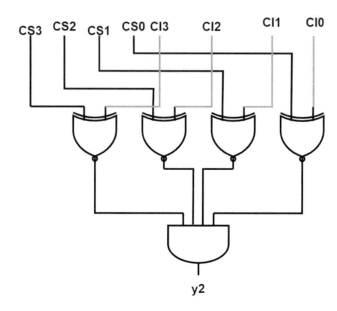

Fig. 10.3 The illustration of how the CamoSecCheck generates its output from the CamoSecGen-block and Camo Input Mapping block

Table 10.1 The table compares multiple camouflaging techniques across different key metrics

Metric	RS	OCS	CBS	CBS+OCS	CamoPerturb
$|SDI|$	26.0	16.0	15.0	27.0	1.80E+19
Exec. time (s)	0.5	0.5	0.4	0.6	1.80E+10
Area (%)	41.0	26.8	41.0	39.6	49.1
Power (%)	50.8	50.6	58.2	48.3	24.4
Delay (%)	7.6	11.6	8.6	5.4	1.2

However, the identity of camouflaged cells can be obscured so that they are indistinguishable from standard cells. Shakya et al. [11] proposed the novel technique termed as "covert gates." Under SEM imaging, an attacker would not be able to pinpoint which exact cells are camouflaged and would therefore need to assume all gates are camouflaged. This would require testing every gate in the netlist, an infeasible mode of attack, as the time required would be substantial. Figures 10.4 and 10.5 show the difference between regular camouflaging and "covert gates." Figure 10.4 shows how a NOR gate can be camouflaged as three potential functions in a black-box manner. This black box would be highly distinguishable from a regular standard cell. Figure 10.5 shows how standard logic gates can be reassembled as other standard logic gates. An inverter can be shown as a 2-input NOR gate if one of the inputs is equivalent to logic value zero. A 2-input NOR gate can be resembled by a 3-input NOR gate if the additional input is logic

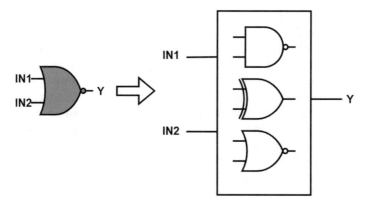

Fig. 10.4 Illustration of how a regular camouflaged gate is represented as a "black-box"

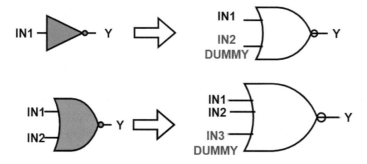

Fig. 10.5 Illustration of how "Covert gates" are represented as other standard gates

value zero. The dummy inputs in these reassembled gates can be used to obfuscate the output.

Many defenses against invasive and non-invasive attacks can be created by representing camouflaged cells as other standard logic cells. Because the adversary could not distinguish between true and covert cells, an attacker would have to use invasive techniques on every gate. This significantly increases the attack's complexity. Because of their low scalability, SAT-based and test-based non-invasive attacks would also suffer, greatly increasing the complexity and time of the attacks.

There are cases where covert gates could be more easily distinguishable to the attacker. For example, designers could choose many well-known procedures when choosing which gates to camouflage. If the attacker knew which procedure was chosen, they could easily know which gates are covert. Also, if the attacker had insider information from the design house, they might know that only certain types of gates are camouflaged. Table 10.2 shows the performance of covert gates against regular, dummy-contact camouflaging on SAT-based attacks [11]. It can be seen that the proposed methodology times out for each benchmark, except the first. Regular

178 10 Hardware Camouflaging in Integrated Circuits

camouflaging only times out for the last benchmark. Their initial claims that the complexity would drastically increase for SAT-based attacks are justified.

The area, delay, and power overheads of the "covert gates" methodology are shown in Table 10.3 [11]. It shows that covert gates have a minimal impact on the area and power overheads, with less than a 1% difference for almost all benchmarks. However, the delay overhead is fairly substantial for half of the benchmarks since they are greater than 10% different. A potential solution would be to top-lace the covert gates on non-essential paths. However, if the attacker knew that these cells were placed on non-essential paths only, it would shrink the search space for which gates to test. Therefore, it would be important to sparsely place covert cells along essential paths in a statistical manner so that the attacker could not predict the camouflaged cells.

10.5 Logic Locking and IC Camouflaging Schemes

Even after the implementation of gate camouflaging and design obfuscation, various methods of IC reversing exist. These attacks are intended to "deobfuscate" and resolve camouflaged logic. For example, the SAT-based attack, which employs Boolean Satisfiability solvers, is the type of attack investigated by Shamsi et al. [12]. The attack entails taking a known set of inputs and outputs from an unlocked IC and using a Boolean Satisfiability solver to solve the derived system of equations. This allows key bits to be resolved for logic locking or unknown gates to be resolved for camouflaging. Because an SAT-based attack requires input–output or Oracle access, it is referred to as an oracle-guided attack [5]. Large and complex circuits, such as multipliers, or circuits with high obfuscation/protection overhead, significantly slow down SAT attacks.

Regarding the implementation of an SAT-based attack, [12] proposed the use of AppSAT for maximum efficiency in circuit deobfuscation. The traditional SAT attack was considered precise deobfuscation, as once the algorithm found no more disagreeing inputs, the process was terminated. It was suggested that instead of a precise attack, an "approximate deobfuscation" would be a stronger attack. SAT-resilient circuits were good at protecting from exact deobfuscation since these kinds of SAT attacks were "agnostic to the corruptibility of the protection scheme."

Random input patterns were given to determine the most optimal time to finish querying in an approximate deobfuscation SAT-based attack. After that, for every d queries, a function was chosen from version space, and it was compared to r (patterns that were randomly selected) coming from the oracle to determine e_t. For a fixed number of iterations, the process should be terminated when e_t goes below a predetermined threshold. This is called the settlement threshold for the error estimation. In order to resolve the key bits, exact and approximate SAT attacks with anti-SAT and random logic locking were performed practically with different adjustments/methods on benchmark circuits [12]. Figure 10.6 shows various benchmark circuits used in the experiment to show the efficacy of AppSAT.

Table 10.2 Comparison of the performance for regular camouflaged gates and covert gates against SAT attacks

Benchmark	Gate/node count	Regular camouflaging 5% of NAND/NOR/XOR				Proposed camouflaging NAND+NOR+AND+OR			
		$|K|$	Attack time (s)	# Attack iterations	AppSAT time (h)	$|K|$	Attack time (h)	# Attack iterations	AppSAT time (h)
C1908	880	34	0.55	7	N/A	811	3.52 |5.91|	235 |191|	N/A
C2670	1193	26	0.65	11	N/A	1514	Timeout |Timeout|	2127 |4891|	Timeout
C3540	1669	28	0.68	11	N/A	2088	Timeout |Timeout|	28 |34|	Timeout
C5315	2307	46	3.58	25	N/A	3379	Timeout |4.27|	240 |459|	Timeout
C7552	3512	106	4.07	27	N/A	4454	Timeout |Timeout|	52 |91|	Timeout
Arbiter	11,839	1182	3815.00	855	N/A	23,678	Timeout |Timeout|	82 |141|	Timeout
Voter	13,758	1078	Timeout	33	Timeout	21,560	Timeout |Timeout|	51 |28|	Timeout

Table 10.3 Area, delay, and power overheads of covert gates on multiple different benchmarks

Benchmark	Area (μm^2)			Delay (ns)			Power (μW)		
	Covert	Original	%	Covert	Original	%	Covert	Original	%
AES	114098.90	113384.22	0.63	18.19	15.99	13.76	2689.2	2678.9	0.38
b12	9725.38	9646.59	0.81	2.98	2.88	3.47	154.9783	154.4319	0.35
b15	53432.06	53134.15	0.56	26.32	26.32	0.00	654.9308	657.4276	−0.38
b17	171193.62	170264.84	0.54	32.47	31.14	4.27	2015.7	2011.3	0.22
s35932	111402.38	111088.12	0.28	14.13	10.84	30.35	2290.2	2328.4	−1.67
s38417	107803.98	107349.70	0.42	20.48	16.69	24.87	1949.0	1949.6	−0.03
s38584	87647.35	87229.18	0.48	15.38	13.11	17.32	1572.1	1570.9	0.08

ISCAS sequential			
circuit	#inputs	#out	#gates
s382	24	27	392
s400	24	27	414
s641	54	42	459
s526n	24	27	494
s1488	14	25	843
s953	45	29	950
s3384	226	201	1966
s5378	214	228	5183
s13207	700	790	11248
s15850	611	684	13192
s35932	1763	1728	31833

ISCAS & MCNC combinational			
circuit	#inputs	#out	#gates
c432	36	7	160
c499	41	32	202
i4	192	6	338
c880	60	26	383
c1355	41	32	546
c1908	33	25	880
c2670	157	64	1193
i9	88	63	1315
i7	199	67	1581
c3540	50	22	1669
dalu	75	16	2298
c5315	178	123	2307
i8	133	81	2464
c7552	207	108	3512
dcs	256	245	6437

Fig. 10.6 The various benchmark circuits used in the experiment

Some protection methods against such approximate deobfuscation attacks have also been proposed in [12]. The diversified tree logic (DTL) technique is one promising scheme. One of the proposed advantages of such a technique is that the designer can adjust the protection to ensure a certain error per key value in SAT attacks. They can also be expanded to the point where the protection provides termination states, rendering bypass attacks useless.

The issue lies in the fact that there exists a trade-off between output corruptibility and the minimal number of queries needed to determine a locked circuit. Therefore, to increase the output corruptibility of the circuits, it is possible to combine DTL with the existing SAT-flexible techniques using these protection methods. By comparing the results and proposals with other existing schemes, it can be concluded that the proposed DTL method is not only more effective at improving corruptibility but also more lightweight than other methods such as those proposed in [18] and [19].

The use of cryptographic functions is another effective method of preventing SAT attacks, as they are difficult to reverse. They are resistant to SAT solvers because they introduce large integer multiplication or confusion layer repetition. The issue with this is that it adds a significant area overhead to the circuit, which may not be practical in some cases. As a result, iterative block ciphers can be used instead of these cryptographic methods because they are much more optimized and are a key-only transformation (less overhead). Although cyclic logic locking had previously been proposed, [12] advocated for its implementation alongside DTL techniques for additional obfuscation. Shamsi et al. [12] proposed an approximate SAT-based attack method and then proposed the novel technique of DTL in addition to existing SAT-resilient protection (logic locking methods) to further improve corruptibility and resilience to such attacks.

10.6 Security Analysis of IC Camouflaging

It is necessary to consider interference graphs when selecting gates for camouflaging. First, interference occurs when the output of one camouflaged gate crosses paths with the output of another. Each node in an interference graph represents a camouflaged gate, each edge represents interference, and the weight of each edge depends on the type of camouflage. Weights for isolated gates and fully resolvable gates are zero, two for partially resolvable gates, and three for non-resolvable gates. In other words, a heavier weight is preferable because it requires more work from an attacker. Figure 10.7 shows the interference graph of a dummy circuit [10].

From a defensive standpoint, activation and propagation must also be considered. Activation is the process of introducing ambiguity via input lines. A defender can ensure that the gate's identity is not revealed through the output by appropriately setting the input lines. Propagation is the process of taking an ambiguity from one gate and sending it to other outputs, forcing an attacker to determine the gate's identity. These two characteristics contribute to the corruptibility of the metric output. High output corruptibility is desired as a defender.

In [10], four different methods of camouflaging gates are considered to measure the effectiveness of camouflaged gates. These methods include random selection, output corruptibility (high output corruptibility gates), non-resolvable (gates whose functionality or identity cannot be resolved), and output corruptibility with non-resolvable (smart algorithm that considers both output corruptibility and gate non-resolvability) [10].

It was observed that random and output corruptibility selections resulted in significantly fewer required input patterns, requiring significantly less time to resolve the circuit. Non-resolvable and "output corruptibility + non-resolvable" selections, on the other hand, required significantly more input patterns and time to

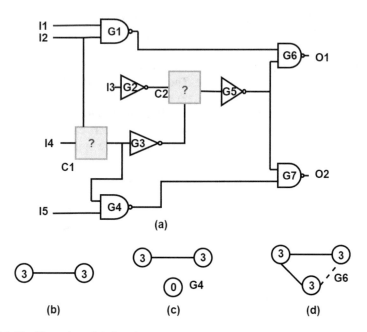

Fig. 10.7 The illustration of the interference graph

Table 10.4 The table shows the amount of power, delay, and area overheads associated with each camouflaged gate

| Function | Camouflaged gate | | | | | |
| | XOR+NAND+NOR | | | XNOR+NAND+NOR | | |
	Power	Delay	Area	Power	Delay	Area
NAND	5.5X	1.6X	4X	5.1X	1.8X	4X
NOR	5.1X	1.1X	4X	4.8X	1.4X	4X
XOR	0.8X	0	1.2X	N/A		
XNOR	N/A			0.7X	0	1.2X

resolve the functionality. Because the majority of the selected gates were resolvable or isolated, random camouflaging did not work well. Meanwhile, because most gates were isolated again, output corruptibility on its own was weaker than the other two selection methods [10].

One limitation of the camouflage strategy is overhead. Although camouflaging gates make it more difficult for attackers to determine the functionality of the circuit, they come at a high cost because they require more power, area, and delay. A camouflaged standard cell consumes approximately five times the power of a regular standard cell. As a result, the camouflaged gate selection process described in [10] is critical because unnecessary camouflaged gates cost power. Table 10.4 highlights the found overhead for different logical functions.

The assumption that an attacker must use a brute force method to resolve gate functionalities is another limitation of [10]. The number of inputs required to resolve gate functionalities can be significantly reduced using techniques demonstrated by

El Massad et al. [2] by carefully selecting a set of discriminating inputs. This set, which is much smaller than all possible inputs, allows a circuit to be completed in seconds rather than years.

10.7 Conclusions

While camouflaging standard cells provides some IP protection, it is far from a perfect solution. While [10]'s work shows that brute force attacks take years to resolve the circuit's functionality, other attacks, such as [2], show that there are faster ways to determine the identity of the camouflaged gates. According to the authors of [11], the true definition of camouflaging gates is that they are indistinguishable from standard cells. They demonstrate that using dummy contacts in certain gates will conceal the functionality of the gates, requiring the attacker to assume that all gates are camouflaged, increasing the time and complexity of the attack. Furthermore, as more camouflaged gates are added to the integrated circuit, the power, delay, and area overheads increase. This means spending more money and time trying to solve these problems. Time-to-market is critical in the fast-paced semiconductor industry because it can make or break a company. Thus, when considering cost, it is critical to choose which gates to camouflage.

References

1. Chen S, Chen J, Forte D, Di J, Tehranipoor M, Wang L (2015) Chip-level anti-reverse engineering using transformable interconnects. In: 2015 IEEE international symposium on defect and fault tolerance in VLSI and nanotechnology systems (DFTS). IEEE, Piscataway, pp 109–114
2. El Massad M, Garg S, Tripunitara MV (2015) Integrated circuit (IC) decamouflaging: reverse engineering camouflaged ICs within minutes. In: NDSS, pp 1–14
3. Guin U, Tehranipoor M (2013) Counterfeit detection technology assessment
4. Guin U, Huang K, DiMase D, Carulli JM, Tehranipoor M, Makris Y (2014) Counterfeit integrated circuits: a rising threat in the global semiconductor supply chain. Proc IEEE 102(8):1207–1228
5. Kamali HM, Azar KZ, Farahmandi F, Tehranipoor M (2022) Advances in logic locking: past, present, and prospects. Cryptology ePrint Archive
6. Karam R, Hoque T, Ray S, Tehranipoor M, Bhunia S (2016) Robust bitstream protection in FPGA-based systems through low-overhead obfuscation. In: 2016 international conference on ReConFigurable computing and FPGAs (ReConFig). IEEE, Piscataway, pp 1–8
7. Lu H, Wilson R, Vashistha N, Asadizanjani N, Tehranipoor M, Woodard DL (2020) Knowledge-based object localization in scanning electron microscopy images for hardware assurance. In: ISTFA 2020, ASM International, pp 20–28
8. Quadir SE, Chen J, Forte D, Asadizanjani N, Shahbazmohamadi S, Wang L, Chandy J, Tehranipoor M (2016) A survey on chip to system reverse engineering. ACM J Emerg Technol Comput Syst 13(1):1–34
9. Rahman MT, Dipu NF, Mehta D, Tajik S, Tehranipoor M, Asadizanjani N (2021) Concealing-gate: optical contactless probing resilient design. ACM J Emerg Technol Comput Syst 17(3): 1–25

10. Rajendran J, Sam M, Sinanoglu O, Karri R (2013) Security analysis of integrated circuit camouflaging. In: Proceedings of the 2013 ACM SIGSAC conference on computer & communications security, pp 709–720
11. Shakya B, Shen H, Tehranipoor M, Forte D (2019) Covert gates: protecting integrated circuits with undetectable camouflaging. In: IACR transactions on cryptographic hardware and embedded systems, pp 86–118
12. Shamsi K, Meade T, Li M, Pan DZ, Jin Y (2018) On the approximation resiliency of logic locking and IC camouflaging schemes. IEEE Trans Inform Forensics Secur 14(2):347–359
13. Shi Q, Vashistha N, Lu H, Shen H, Tehranipoor B, Woodard DL, Asadizanjani N (2019) Golden gates: a new hybrid approach for rapid hardware trojan detection using testing and imaging. In: 2019 IEEE international symposium on hardware oriented security and trust (HOST). IEEE, Piscataway, pp 61–71
14. Vashistha N, Lu H, Shi Q, Rahman MT, Shen H, Woodard DL, Asadizanjani N, Tehranipoor M (2018) Trojan scanner: Detecting hardware trojans with rapid SEM imaging combined with image processing and machine learning. In: ISTFA 2018: proceedings from the 44th international symposium for testing and failure analysis. ASM International, p 256
15. Vashistha N, Rahman MT, Shen H, Woodard DL, Asadizanjani N, Tehranipoor M (2018) Detecting hardware trojans inserted by untrusted foundry using physical inspection and advanced image processing. J Hardw Syst Secur 2(4):333–344
16. Woodard D, Tehranipoor MM, Asadi-Zanjani N, Wilson R, Lu H, Vashistha N (2022) Knowledge-based object localization in images for hardware assurance. US Patent App. 17/491,150
17. Yasin M, Mazumdar B, Sinanoglu O, Rajendran J (2016) CamoPerturb: secure IC camouflaging for minterm protection. In: 2016 IEEE/ACM international conference on computer-aided design (ICCAD). IEEE, Piscataway, pp 1–8
18. Yasin M, Sengupta A, Nabeel MT, Ashraf M, Rajendran J, Sinanoglu O (2017) Provably-secure logic locking: from theory to practice. In: Proceedings of the 2017 ACM SIGSAC conference on computer and communications security, pp 1601–1618
19. Zhou H (2017) A humble theory and application for logic encryption. IACR Cryptol ePrint Arch 2017:696

Chapter 11
Embedded Watermarks

11.1 Introduction

Aggressive time-to-market constraints have forced the global semiconductor industry to shift from a vertical to a horizontal model [50]. In a vertical model, a single entity is responsible for the entire process of designing, fabricating, testing, and assembling integrated circuits (ICs). In contrast, processes are distributed globally in a horizontal model, allowing companies to specialize in a specific process. Furthermore, system integrators found that obtaining licenses from different vendors for functional blocks and reusing them in their system-on-chip (SoC) were more cost-effective than designing every block from scratch [51]. These individual functional blocks are known as intellectual property (IP) blocks or IP cores. This shift in the semiconductor industry made IC manufacturing cheaper and faster, but it also made security and trust issues more pressing [22]. One such issue is intellectual property piracy and theft. The threat to IP owners is that an adversarial system integrator might use their IP in more ICs than they had licenses for. Similarly, an attacker could reverse engineer the IP core, extract design details, and resell unauthorized copies to system integrators as his own.

To address these concerns, IP owners can embed their own "signature" into the design, which can then be extracted and used to unequivocally prove their authorship in a legal context [7]. Watermarking is a technique inspired by similar practices in digital media. However, hardware IP watermarking presented its own set of challenges and requirements and was first proposed by the virtual socket interface alliance (VSIA) [2]. For example, the watermark must not interfere with IC or IP functionality, must be extremely difficult for an attacker to remove or copy, and provide irrefutable proof of authorship while incurring minimal area and performance overheads. IP watermarking techniques, according to [51], can be classified as additive, constraint-based, module-based, or side-channel-based. FPGA IPs have additive watermarks embedded in their functional core. Constraint-based watermarks embed the watermark by leveraging various design constraints. Soft IPs can be watermarked with module-based watermarks, which can duplicate or divide a hardware description language (HDL) module. Finally, watermarks based on side channels are extracted from or embedded in side-channel parameters such as power or electromagnetic (EM).

This chapter discusses some of the most recent and cutting-edge watermarking and metering approaches proposed in the literature, emphasizing their novelty and application and highlighting their underlying assumptions and limitations. They are also compared and contrasted with previous works in this domain throughout the chapter. These methods are classified as watermarking methods and can be extracted in various ways. Furthermore, each has its distinct method of incorporating the owner's signature into the design. An introductory summary of the items being focused on is provided for the reader's convenience as follows:

- FORTIS [23] is a novel approach employing forward trust among participating entities to prevent IP overuse, IP piracy, and IC overproduction.
- Hardware IP and Watermarking [7] is a digest of the recent trend in the research field of hardware watermarking and fingerprinting. It discusses the advantages and issues regarding different watermarking approaches and points out future research directions.
- Filling Logic and Testing Spatially (FLATS) [15] is an additive watermarking method that leverages unused LUTs in the post-synthesis phase in FPGAs to provide proof of authorship and tamper detection simultaneously.
- Side-channel-based watermarking [4] is a viable solution for protection against IP theft and cloning, where an artificial power side channel is used to embed a watermark.
- Through backside imaging, the embedded optical watermarking [59] technique helps to detect and locate hardware Trojans inserted at the fabrication stage.

11.2   Background

The current demand for integrated circuits is rapidly increasing. ICs are used in computers, mobile phones, automobiles, and medical equipment, among other things. In addition, new chip fabrication technologies are being developed for low-power consumption, better area constraint, and faster IC. However, setting up a fabrication facility that uses modern technology is extremely expensive, sometimes costing billions of dollars. Many IC design firms cannot afford the cost of establishing a fabrication facility. As a result, they contract out the physical fabrication of their chips to a foundry. Design firms are now purchasing third-party IPs (3PIPs) from various vendors worldwide to meet the aggressive time-to-market demand. In addition, to meet the market's ever-increasing IC demand, design facilities are shifting from a vertical to a horizontal business model. Many entities have access to IP or IC designs in a horizontal model. As a result, various attack scenarios on the IC supply chain have emerged. An untrustworthy design house may overuse third-party vendor IPs, insert Trojans into the design, or pirate the IPs [18, 47, 54, 55, 57]. A foundry can overproduce an IC and sell it for a profit, insert Trojans into the IC, or sell out-of-spec or defective ICs. Even in the field, ICs can be reverse-engineered, cloned, remarked, recycled, etc. [26, 56]. Research is being conducted to address these issues and protect the supply chain from untrustworthy entities. Various approaches have been proposed, focusing on protecting the IC from various attacks, such as hardware security primitives, hardware metering, watermarking, logic obfuscation, etc. Watermarking is one of these methods for identifying the designer. Watermarking research has gotten much attention because it can address IC cloning, unauthorized design modification, and IP piracy.

Watermarking is a procedure for embedding owner information in a design with a small overhead so that it has a minimal or no impact on the design's functionality [7]. The primary goal of watermarking was to use as proof of authorship in a design that identifies the owner to a third-party arbiter in a dispute over ownership. Watermarking can also help to ensure the design's integrity. Fingerprinting is another area of study related to watermarking [33]. Different IP buyers are provided with slightly different versions of the same design in the fingerprinting approach. If a violation of owner rights occurs, fingerprinting can be used to identify the dishonest buyer with high confidence.

Most tampering protection schemes proposed in the literature focus on pre-synthesis attacks and make designs resistant to such attacks. However, most watermarking methods proposed in the literature for FPGAs do not provide inherent proof of trust. As a result, there is a pressing need for a security solution that can simultaneously protect against all of these security threats.

IP theft is one of the most crucial hardware threats in the present multi-billion dollar market. In order to solve the problem, several watermarking approaches were proposed in the literature. However, since the existing methods had drawbacks, inspired by Lin et al. [37], Becker et. al. [4] proposed a technique for integrated circuits that used power side-channel-based embedded watermarking.

Although layout modification is one of the most common ways of hardware Trojan insertion, detection of Trojan at this stage is extremely difficult. The destructive method requires reverse engineering, which is expensive and very complicated. Other non-destructive methods proposed in the literature such as power analysis [41], timing-based analysis [36], thermal analysis [40] are time-consuming. In this regard, embedded optical watermarking fulfilled the need for a robust, rapid, and easy solution for hardware Trojan detection at the fabrication stage.

11.3   Hardware IP Security

Due to the current trend of aggressive time-to-market demand and cost cutting, design houses purchase various IPs from 3PIP vendors rather than designing them in-house. IPs are classified into three types: soft IP, firm IP, and hard IP. The

Table 11.1 Summary of characteristics differences between different IP cores

	Soft IP	Firm IP	Hard IP
Abstraction level	Behavioral	Structural	Physical
Optimization	Low	Medium	High
Technology dependency	Independent	Generic	Dependent
Flexibility	High	Medium	Low
Distribution risk	High	High (with RTL)	Low
		Medium (no RTL)	

characteristics of various IP forms are shown in Table 11.1. Soft IP is the most easily modified and thus the most vulnerable to attacks.

A third-party vendor can sell his design to many SoC integrators. However, if any integrator overuses or pirates the IP, watermarking is not capable of tracing it back to a specific integrator. Therefore, hardware IP fingerprinting is used to solve this problem. The specific SoC integrator who has misused the IP can be identified using fingerprinting. To ensure this, some buyer information, as well as owner information, is included in the watermark. As a result, each buyer receives the same functionality IP with a different fingerprint embedded in it. This makes it simple to distinguish between a guilty and an innocent buyer. The specifications for a good fingerprint are identical to those for a watermark. The following methods can be used to incorporate fingerprinting into a design:

1. Tile division [34, 35]: In this technique, the entire design can be partitioned into small tiles, which are then arranged differently for different buyers. However, it is possible to remove it through reverse engineering.
2. Iterative fingerprinting [5]: In this approach, specific nodes of the main design are assigned to a buyer. The nodes are assigned different weights than the other nodes, and optimization is carried out. As a result, different buyers receive different designs for the same functional IP. The drawback of this method is that the required optimization effort is not trivial.

Generally, three main types of attacks are performed on watermarks [7]. They are removal, masking, and forging attacks. In a removal attack, the attacker tries to remove the embedded watermarks. Masking of the watermark involves tampering with the design so that the watermark cannot be identified. Finally, in forging attacks, the attacker tries to embed his watermark into the IP and claim the design as his own.

11.3.1 Watermark Insertion Process and Requirements

Figure 11.1 shows a generic model of hardware IP watermarking [12]. The watermarking process is divided into two stages, as depicted in the figure. First, the

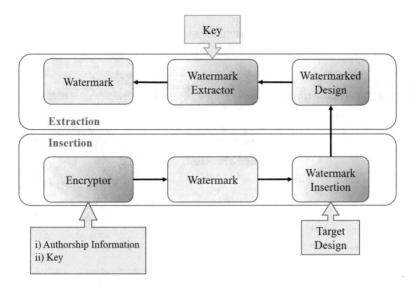

Fig. 11.1 Illustration of generic IP watermarking model

authorship information is embedded into the design during the watermark insertion phase. Then the embedded owner information is extracted during the watermark extraction phase for proof of authorship.

Figure 11.2 illustrates a generic pre-processing-based watermarking method. The watermark is converted into stego constraints and added to the optimizer along with the original design constraints. The pre-processing method produces a very strong watermark. However, this approach may result in unanticipated design overhead. Therefore, [58] proposed a three-phase watermarking approach to keep overhead to a reasonable level. First, the design is optimized using the original constraints. Then, the stego constraints are added in the second phase, and the design is re-synthesized in the third phase.

Watermark can be inserted into a design by two methods [7]:

1. **Additional functionality**: Adding additional logic function into IP to exhibit ownership information when queried
2. **Additional Constraint**: Adding additional constraints into the design that provides ownership information

11.3.1.1 Additional Functionality

The IP owner incorporates some ownership-related circuitry into the design in this watermarking approach. In most cases, this circuitry does not affect the original functionality of the design. A method for embedding a watermark generating circuit (WGC) is described in [16, 17], where the WGC is used with parallel-input serial-

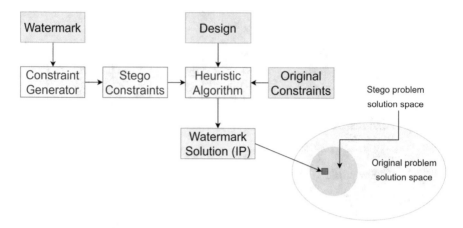

Fig. 11.2 Illustration of pre-processing constraint-based watermark for IPs

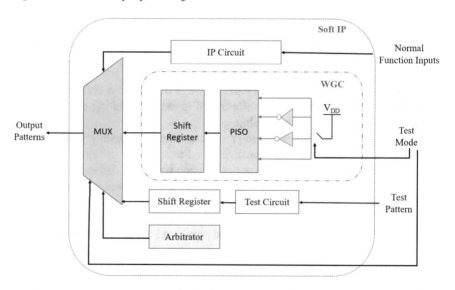

Fig. 11.3 Illustration of watermark generating circuit and test circuit

output (PISO) registers. Figure 11.3 depicts the proposed watermark generating circuit. The watermark bits are generated using the test mode signal and inverters. They are serially shifted out in the output to reveal ownership information.

The authors [29] describe another approach to additional functionality-based watermark insertion. This method employs side-channel information as a watermark information carrier. In this case, a small tag circuit is embedded that generates watermark bits based on the output of a timing module. This tag circuit powers a heat source. The embedded watermark is the heat source's power output. A similar approach is described in [4], where the authors extracted the watermark by correlation rather than directly detecting the power signal. Additional functionality-based watermarking is typically implemented at the behavior level, ensuring good ownership information protection in subsequent steps. However, the design is vulnerable during the behavioral stage because the watermark circuitry can be identified and removed.

11.3.1.2 Additional Constraint

Additional constraint-based watermarking entails incorporating some additional design constraints. During extraction, the design is checked to see if those constraints are present in the circuit to prove ownership. The additional constraints are used to reduce the design's solution space so that an owner can be identified. There are two ways to add constraints [7]:

1. Pre-processing: adding additional constraints as inputs into the optimizer
2. Post-processing: adding additional constraints to the outputs of the optimizer

Constraint-based watermarks can be embedded in a design at different levels [7]. For example, the insertion can be done at:

1. System Synthesis Level: At this method, the watermark insertion task is formulated as a graph partitioning problem [25].
2. Behavioral Synthesis Level: Watermark design is transformed into an RTL description to include in the design.
3. Logic Synthesis Level: Watermark is inserted by modifying the wire connection of the netlist.
4. Physical Synthesis Level: Watermark can be inserted by placing it on path-timing constraints, selecting placement with the specified row parity, or imposing cost in the underlying routing resource according to watermark bits [27].

The robustness of a constraint-based watermark is calculated by a metric called "probability of coincidence," P_c, which represents the probability of a design in which the watermarks are not added, satisfying the additional constraints. Therefore, the lower the P_c value, the more secure and robust the watermark. However, one issue with constraint-based watermarks is that the extraction of the watermark usually requires reverse engineering, which takes a long time.

The robustness of the watermarking schemes is prioritized in additional constraint-based watermarking techniques, but watermark verification is given less attention. Furthermore, because the watermark is embedded globally, it must be extracted from the entire IP indivisibly, making it more vulnerable to attackers, as an attacker's minor change in the design can destroy the watermark. To address this issue, [31] proposed inserting multiple local watermarks. Because these watermarks are embedded in different places on the IP and are smaller than a global one, they can be extracted individually. Another method for embedding watermarks is to insert multiple watermarks at different levels [8, 9] in a hierarchical manner. This method makes it more difficult for an attacker to remove the watermarking. The following conditions must be met for successful watermarking implementation:

- Maintenance of functional correctness
- Independence of the secrecy of the algorithm
- Strong authorship proof
- High reliability of extraction
- Low implementation overhead
- Ease of detection

11.3.2 Watermark Extraction

Watermarking schemes are ineffective if the watermark is difficult to detect and track from the owner's perspective. The watermark extraction method is divided into two parts: (i) physical processing and (ii) side-channel extraction [7]. Reverse engineering can be used to extract the watermark during physical processing. This is known as static detection. This approach detects all constraint-based watermark schemes. Another method, called dynamic detection, extracts a watermark as a challenge–response pair, with the watermark being found in the responses when specific challenges are given to the chip [27]. Typical dynamic watermarking techniques use the finite state machine's (FSM) state transition graph (STG) at the behavioral level or test structures such as scan chains at the design-for-testability (DFT) level as vehicles to carry the watermarks [14, 16, 30, 53]. However, FSM watermarks are challenging to extract, and test structure-based watermarks can be modified to destroy watermarks. To address these limitations, the authors [6] propose a scheme called synthesis-for-testability (SfT). Some approaches that use additional functionality mechanisms to insert the watermark use side-channel extraction.

11.4 FLATS: FPGA Authentication and Tamper Detection

FPGAs are the platform of choice for designers for many security applications since they possess unique reconfigurability capabilities. Modern FPGAs have the additional capacity of ensuring bitstream integrity through hashing and authentication through encryption [42]. However, recent post-synthesis attacks have been proposed in [10], which introduces two new attacks that can bypass traditional FPGA authentication and security protocols. The authors [15] introduce a novel methodology named Filling Logic and Testing Spatially (FLATS), a methodology for protection against post-synthesis tampering attacks, as well as provide proof of authorship.

11.4.1 Novelty of FLATS

- To detect runtime tamper events caused by hardware Trojans, pre-synthesis countermeasures such as formal verification, structural analysis, information flow tracking, etc., have been suggested [13, 18, 21, 38]. Similarly, proof of authorship is provided by embedding watermarks and extracting them in the field. Watermarking techniques for FPGA designs include using bitstream HMAC [1], using filler cells [24], or rely on FSM, power [61], EM [3] signatures, etc. However, tamper detection methods do not provide proof of authorship, while watermarking methods do not provide proof of trust. Therefore, FLATS methodology was proposed to provide both simultaneously.
- Watermarking methods proposed previously have limitations in terms of their ability to be unique to an FPGA serial number while also providing runtime protection and being an IP-level technique. FLATS methodology is free of these constraints.
- FLATS proposes a unique and novel watermarking embedding technique based on infrared imaging and unused FPGA LUTs that can be dynamically extracted during runtime through lock-in thermography.
- Two novel attacks are also demonstrated in [15] that are effective against ICAP- and JTAG-based bitstream authentication schemes used in modern FPGAs. In the first attack, a 3PIP with ICAP access is used to perform malicious FPGA configuration modification at runtime. Before the JTAG could verify the bitstream, the modified design was running in the FPGA for a while, and just before JTAG verification, the original configuration was restored. Through the loading of a malicious bitstream, the second attack forced the result of ICAP runtime verification to always appear as a pass to the user.

11.4.2 Methodology

From a design flow perspective, FLATS is represented as in Fig. 11.4. From a high-level viewpoint, the FLATS method consists of two phases: insertion and verification. First, the synthesized gate-level netlist is fed into the FLATS, which inserts the required modifications in circuits and initializations in the design and then performs the place and route stage. The following stages are similar to the current post-synthesis stages. When FLATS is embedded in the design phase of FPGA, it can be used to verify the design at every phase, including post-silicon stages such as runtime. The two major parts of FLATS are insertion and verification:

- Insertion phase: The first phase of FLATS is the insertion phase. It begins with the reservation of unused LUT inputs and outputs, which will be connected in a feedback configuration to create ring oscillators essentially.

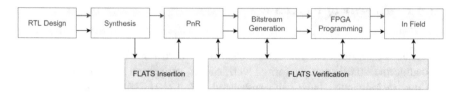

Fig. 11.4 Illustration of how FLATS methodology fits in the FPGA design flow

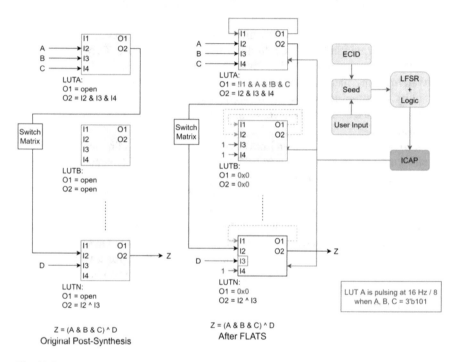

Fig. 11.5 FLATS implementation to convert LUTs into oscillators

The ICAP of the FPGA applies sequences to LUTs, which essentially determine three things:

− Which LUTs are to be used for FLATS verification?
− At which frequency to enable them?
− What outputs/inputs are to be selected to form the feedback?

Figure 11.5 shows the basic implementation method of FLATS. It highlights that when the sequence A = 1, B = 0, and C = 1 is applied, the LUTA oscillator should be pulsating at 16 Hz/8. Here, an LFSR leverages both ECID and User input to generate this sequence and feeds this to the ICAP. FLATs classified the oscillators implemented in FPGA LUTs into beacons, authenticators, and detectors, which can be defined as follows:

- Beacons are oscillators usually placed at corners or other spatially significant points.
- Authenticators are non-beacon LUTs placed randomly or according to a pattern.
- Detectors are LUTs that are integral to the functionality of the circuit and are placed at particular nodes of interest.

• Verification phase: The spatial distance between these three types of oscillators (beacons, authenticators, and detectors) provides evidence of authorship and trust. Authenticators are used to validating the authorship, and detectors are used to validate the trust. These distances are measured using infrared imaging.

During verification, the beacons would be activated first according to the sequences registered during the insertion phase, followed by the authenticator and detectors. The absence of beacon pixels or a mismatch in the distance between beacon pixels and authenticator pixels would indicate an inauthentic design for authorship.

The procedure for detecting tampering is similar. Rather than comparing the distances and positions of authenticator pixels, we would need to activate detectors and match the distances between beacons and detectors with distances registered during the insertion phase.

11.4.3 Experimentation

The authors [23] also demonstrate the implementation of the proposed FLATS method on a Xilinx Kintex-7 FPGA. First, the infrared imaging is done through a long-wave infrared camera with a resolution of 320 by 240. And after the image capture, lock-in analysis is done in the post-processing step to determine the pixel locations correlating to the LUT oscillators. Finally, from these images, combined amplitude and phase information is extracted. In experiments, a 1Hz square wave was used as a reference. Here, two observations have been made. A significant correlation between the reference and infrared magnitude is observed in the first case, which is for watermarked pixels. In the second case, for non-watermarked pixels, there is less correlation between the reference and the infrared magnitude. However, the bright area near the pixel coordinates correlates strongly to the reference waveform. The Laplacian of Gaussian (LoG) algorithm was used to detect the blob or bright area. An initial verification run with the golden design was done to provide a reference. Every fourth measurement in one of the plotted series demonstrates a considerable variation in distance, which is subsequently used to identify the design as unauthentic. The experiment that follows focused on four LUTs in a single slice. Each detector LUT was subjected to a total of 25 measurements. In all 25 cases, the LUT position was recorded within 1 pixel of the standard deviation.

11.4.4 Limitations

The principal limitation of the FLATS methodology is that since it uses unused LUTs to embed watermarks that essentially act as ring oscillators, the ROs dissipate power as they oscillate, which incurs power overhead. Also, using unused LUTs incurs an area overhead as well. The underlying assumption of the FLATS methodology is that the oscillations of ring oscillators will produce enough power or heat to be detectable by infrared cameras. The environmental setup required to extract this heat-induced IR signature has to be robust and minimize the impact of environmental heat, which may not always be possible. Furthermore, the watermark that FLATS implements is inserted at the post-synthesis level. As a general rule of thumb, the earlier a watermark can be inserted into the design flow, the more difficult it is to remove or tamper with.

11.5 Side-Channel-Based Watermarks

With the rapid growth of usage of integrated circuits in different applications, the protection of IP has become an important issue. Adopting watermarking of IP cores has been considered a viable solution to IP theft threat [39, 53]. With similar motivation, the authors [4] propose a technique for integrated circuits where a watermark is embedded using a power side channel.

The idea of implementing a side-channel-based watermark is not novel in the literature. The proposed method is highly inspired by Lin et al. [37], Ziener and Teich [62]. Lin et al. [37] proposed side-channel-based hardware Trojan to leak out secret information. On the contrary, the authors [4] use the power side channel to contain a watermark. The authors [62] also utilize side channel to embed watermark into IP core. But the method proposed by Becker et al. [4] is superior in terms of performance and can also be used for protection against cloning. Moreover, a thermal side-channel watermark is also proposed in [29], but such a method requires more energy and is slower in comparison to a power side-channel-based watermark. Besides these approaches, the constraint-based watermark is also used for IP protection [39].

11.5.1 Methodology

In [4], the following two different approaches of watermarking are proposed:

- Spread spectrum-based watermark
- Input-modulated watermark

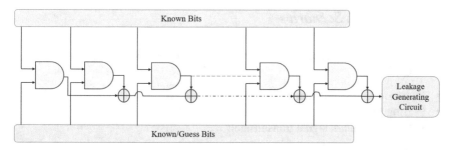

Fig. 11.6 Combinational and a leakage circuit illustrating the input-modulated watermark

Long linear shift registers (LFSR) are used as PRNG in the first approach to embedding the watermark. A secure stream cipher can be used instead of LFSR to make the produced bits unrecoverable. A leakage circuit then converts the PRNG output into physical power consumption. Large capacitances, toggling logic, pseudo-NMOS gates in ASIC design, and circular shift registers in FPGA implementation can all be used to implement leakage circuits. Differential power analysis is used to detect such embedded watermarking. The verifier begins by compressing the power trace by averaging measurement points over a clock cycle. The bit sequence is then simulated using PRNG, and the simulated sequence is finally correlated to the compressed power vector. Any significant correlation coefficient peak indicates the presence of a watermark. This type of detection mechanism makes the watermark resistant to environmental noise.

In the second approach, additional logic is added to the IC using a combinational function. Such a function uses known bits to compute the output bit. Next, a leakage circuit transmits the output bit mapping to power consumption. Figure 11.6 shows the diagram of such watermarking approach where the XOR gate is used as a combinational function. Such a combinational function introduces artificial data-dependent power consumption, which acts as a watermark. Differential power analysis is also used here as a detection mechanism. The main difference between these two approaches is that at the detection step, the first method uses a single measurement of multiple points, whereas the second method uses multiple measurements at a specific time instance with different input values.

It is critical that the watermarking confirms proof of ownership while also being easily detectable. The authors [4] describe how proposed methods can achieve proof of ownership. The company can generate a hash value of some design ID using a private key, which can then be transmitted using the proposed watermarking approaches. Because a signature can only be generated by the owner's private key, this method can prevent attackers from making false ownership claims.

11.5.2   Attack Scenarios

Three possible attack scenarios can be considered to remove the side-channel-based watermark:

- Reverse engineering attack
- Raising the noise
- Inverse watermark signal transmission

Since the proposed watermarks, chiefly the input-modulated watermarks, are very small, performing a reverse engineering attack is very difficult and costly. Moreover, such an attack becomes even more complex since the attack needs to perform such attack at the post-manufacturing level. The second attack deals with the signal-to-noise (SNR) ratio. Detectability of the watermark is related to SNR. The lower the SNR, the more challenging it is to detect watermark since more measurements are required. Therefore, additional noise sources can be added to remove the watermark in the second attack. Since measurements are averaged in the case of a spread spectrum watermark and additional measurements can be observed in the case of an input-modulated watermark, watermarks are resistant to this second type of attack.

In the third attack, an inverse watermark signal can be used to make the power consumption constant and make the original watermark undetectable. Such an inverse signal can be generated using an additional leakage source. In this regard, an experiment is also carried out in which an exact copy of the watermarking circuit, consisting of the same leakage circuit and PRNG, is added to the design, except that the PRNG output is inverted. The experiment revealed that the watermark can still be detected after measuring nearly 10,000 clock cycles. However, because of the differences in routing, the power consumption of the two circuits cannot be precisely inverse.

11.5.3   Experimentation

In [37], a first-order DPA-resistant AES implementation is used to evaluate the performance of a spread spectrum-based watermark. A 32-bit LFSR and a 16-bit circular shift register filled with alternative ones and zeros are used as PRNG and leakage circuits. Two separate experiments are carried out in this case. The AES core is idle in the first. The power trace is measured over 1000 clock cycles, compressed, and correlated to a simulated PRNG sequence. According to the findings, approximately 100 clock cycles are required to detect the watermark's presence. In the second case, AES is constantly encrypting, a longer power trace covering 250,000 clock cycles is used, and the same procedure described previously is followed. The watermark is easily detectable in this scenario as well.

11.6 Embedded Watermarking for Hardware Trojan Detection

Hardware Trojan is one of the most dangerous and significant hardware threats to electronic devices, especially in military and space applications. Detection of hardware Trojan is challenging due to its stealthiness in nature. Trojan can be inserted at any step of the IC chip development cycle. Among these phases, insertion of Trojan at layout is one of the most prevalent ways [48, 49]. Reverse engineering is a possible but difficult and costly approach to detect Trojan inserted at the fabrication stage. Thus, a robust, easy, rapid, and accurate methodology is needed to detect such Trojan. In order to fulfill this gap, the authors [59] provide an embedded optical watermarking-based solution, where backside optical imaging helps to detect and locate hardware Trojan inserted at the fabrication stage.

11.6.1 Methodology

Hardware Trojan detection technologies can be classified into two categories: destructive and non-destructive. Destructive approaches such as reverse engineering are costly [52]. On the other hand, non-destructive approaches such as power analysis [41], timing-based analysis [36], and thermal analysis [40] are time-consuming as these methods need to check a lot of challenge–response pairs. The method proposed in [59] is unique since to the best of our knowledge no other technique proposed optical watermark for the detection of hardware Trojan before.

At the design stage, the maximum amount of the M1 layer is embedded in the fill cells, resulting in highly reflective cells at the IC layout. When the backside of the IC is brought under the imaging system at a near-IR wavelength, these fill cells appear as a bright spot in the picture. The patterns created by the placement of these fill cells serve as a watermark. An optical watermark of this type is used to determine whether or not a Trojan has been inserted. Because inserting any hardware Trojan by modifying, replacing, or rearranging these cells will result in a watermark change, the location of the Trojan can be easily determined by comparing the tampered and original untampered layouts. The Euclidean distance or correlation coefficient is computed to compare and decide on the presence of a hardware Trojan.

11.6.2 Experimentation

For experimentation in [59], at first, hardware Trojans are inserted in the AES-T100 hardware block. After that, the following three changes are made to the layout (Table 11.2).

Table 11.2 Different watermarking approaches proposed in the literature

Paper	Type of water-marking	Novelty	Application	Strength	Limitation
Guin et al. [23]	Active metering	Novel	Establishes forward trust between entities	Comprehensive solution to prevent IP piracy, IP overuse, IC overproduction	Requires trusted EDA tools and may incur delay in test process
Chang et al. [7]	–	Review	–	Discussion of recent approaches	–
Duncan et al. [15]	Additive	Novel	Proof of trust and authorship	Can simultaneously provide runtime tamper protection and authentication	Area and power overheads incurred, time-consuming extraction method.
Becker et al. [4]	Power watermarking	Inspired by Lin et al. [37]	IP protection	Easily detectable even if SNR is low	Removable at RTL level
Zhou et al. [59]	Place and route	Novel	Hardware Trojan detection	Simple and quick method with negligible overhead	Low image resolution

- Fill cells are changed to functional gates.
- Bottom rows of gates are shifted to make room for Trojan.
- Functional gates are replaced by other types of functional gates.

Then, in response to the changes, reflectance maps are generated. The 2D correlation method is used to compute the difference between these two reflectance maps. The correlation coefficient found between two image matrices was 0.11. When this number is compared to the selected threshold value (0.65), it is obvious that the layout has been altered. Because process variation impacts the reflectance signal of fill cells, a variation in reflectance value is detected in the presence of 10% process variation to illustrate the method's efficacy over a wide range of wavelengths in the instance of two fill cells. Because of such variation, less than 5% change is observed on average, and such small change is easily tolerated. Multiple hardware blocks are evaluated in the proposed method: AES100, AES200, AES1000, PIC100, PIC200, and PIC300. The proposed method successfully detects hardware Trojans with negligible leakage power and area overhead for each test bench. Experiments with Monte Carlo simulations show that the Trojan detection error rate is zero if the SNR is greater than 7 and 4 in the case of AES and PIC circuits, respectively.

11.6.3   Strength and Weakness

The paper [59] has several benefits when it comes to detecting hardware Trojans. Compared to the state-of-the-art, the proposed method is simpler and faster. It necessitates very little overhead. The method has a wide field of view because it uses low-resolution images. As a result, this method can inspect billions of transistors in a matter of hours. However, there are some limitations to work as well. Because the imaging resolution used in this method is low, this method may fail to detect very small Trojans at the nanometer scale.

11.7   Enabling IP and IC Forward Trust

Although watermarking can provide proof of authorship, it is a passive technique because it does not actively prevent IP piracy and IC overproduction. For the sake of completeness of the discussion on hardware IP security, we discuss FORTIS, which is a proposed method to enable forward trust between various supply chain entities [23]. Several existing research provided a partial solution to IP piracy and IC overproduction. FORTIS provides a comprehensive solution to assure forward trust by utilizing existing SoC obfuscation techniques and a novel design flow to address IP piracy, overuse, and IC overproduction.

11.7.1 Related Works and Novelty

FORTIS relies on the existing techniques such as logic obfuscation, hardware watermarking, and IC metering that failed to address the problem entirely and brought novelty to provide a holistic solution. Logic obfuscation (or logic locking) encrypts a design of XOR/XNOR gates (typically) to hide the inner details, and a unique chip unlock key (CUK) is required to unlock the obfuscated circuit [28, 46, 60]. The early variants of logic obfuscation techniques were not resistant to reverse engineering and key sensitization attacks, which were addressed with improved logic encryption techniques [45]. In recent encryption approaches, a random symmetric session key was used to encrypt the IP, and then the session key was encrypted by EDA vendors with a public key and integrated with the IP to prevent unwanted modification to the IP. However, these approaches failed to provide integrity verification as additional features can be placed to an existing IP. The proposed solution added an IP digest header that contained the hashed output of the IP to prevent such unauthorized modification.

Hardware watermarking techniques use an IP's unique fingerprint as proof of authorship but fail to prevent IP overuse, piracy, and overproduction. IC metering techniques establish proof of trust between IC design house and foundry by providing control of the number of ICs fabricated. The passive metering techniques identify each IC using PUF-based unique chip IDs but cannot actively prevent overproduction as an untrusted foundry can falsify yield data. Active metering techniques solve this by locking every chip until the key is provided by the IC design house or OEM [32]. However, such active metering approaches have limitations as widely used test compression architectures cannot provide test responses without significant overhead. For example, the on-chip TRNG-based RSA encryption in EPIC metering [46] suffers from a large area overhead due to the complex algorithm of prime number checker, the vulnerability of man-in-the-middle attack, and key sensitization attacks. Secure split test (SST) [11] provides protection against overproduction, cloning, and defective/out-of-spec chips, but the back-and-forth communication between the IC design house and offshore foundry acts as a backlash and increases the delay in the test process.

FORTIS is a trusted authentication platform (TAP) that provides a communication protocol that activates chips at end-user location to prevent IC overproduction practice by an untrusted foundry. This approach improves the obfuscation techniques [45, 46] to make it attack resistant while removing on-chip key generation. It also enables the manufacturing tests without unlocking the chips. As the netlist is locked with a secret CUK, an IP encrypted header is added that is retrieved by the trusted EDA tool and enables simulation of the IP by the IC designer without accessing the CUK. IP piracy is prevented by IP integrity verification, which makes it resistant to malicious modification.

11.7.2 Methodology

The architecture of FORTIS and communication flow between the entities of FORTIS to provide forward trust between the IC design house and untrusted foundry are depicted in Fig. 11.7. FORTIS extends the existing IC design flows with lock insertion and functional activation steps. XOR/XNOR gates are used for lock insertion, using secure logic encryption that the chip can only unlock the key (CUK) to produce functionally correct output [45]. IC design house integrates the locked 3P-IPs and collects all the test vectors from the locked IPs using the trusted EDA tool that retrieves the CUK from the IP header. Each chip is unlocked after fabrication, packaging, and testing in the existing design flow.

In order to prevent untrusted foundry/assembly from fabricating defect-free ICs and hide the yield, the chips need to be tested before activation. In the proposed netlist shown in Fig. 11.8c, test patterns are generated by the ATPG tool without knowing the key. The locking key bit CUK[i] is connected to a scan flip-flop (FF_i) that is driven by the MUX output. When the scan enable (SE) signal is asserted, FF_i becomes a part of the scan chain and test pattern is generated for this modified netlist with $(n + 1)$ inputs rather than the original netlist with n inputs (Fig. 11.8a) or the locked netlist (Fig. 11.8b) with n inputs and CUK[i] $= 0/1$ [43, 44]. This modified netlist also prevents key sensitization attacks where key bits are treated as X_s and propagated to the output. Moreover, compressor logic structures suppress the effect of the Xs at the output if at least two key bits are input to the XOR gates in the compressor simultaneously, which will hide the individual effects [23].

The FORTIS's proposed communication flow shown in Fig. 11.9 prevents IC overproduction by securely transferring the CUKs to the chips in the foundry

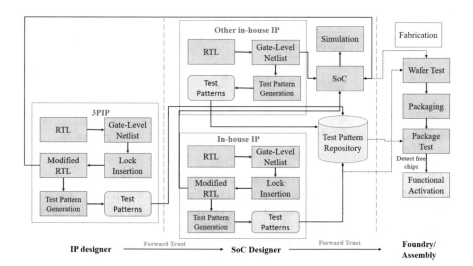

Fig. 11.7 FORTIS design flow

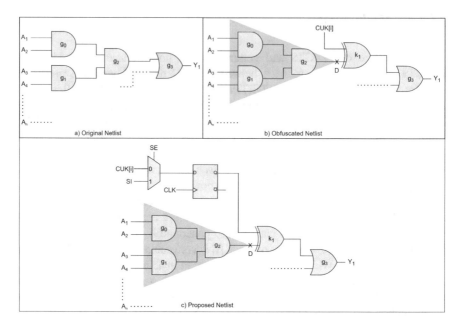

Fig. 11.8 Illustration of netlist obfuscation to enable manufacturing test without chip activation

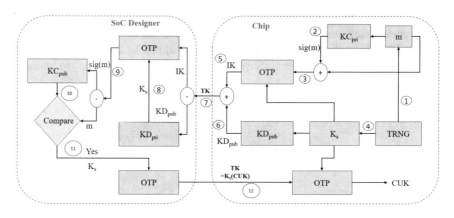

Fig. 11.9 Illustration of FORTIS communication flow for IC overproduction prevention

without any interception from an untrusted party. Then the protocol is extended to prevent 3PIP overuse by the IC design house in Fig. 11.10. In order to transfer the CUKs securely, the IC designer needs to ensure message integrity, end-point authentication, and confidentiality. RSA asymmetric encryption ensures integrity and end-point authentication, and a one-time pad (OTP) is used to ensure confidentiality. In Fig. 11.9, the public key of the IC designer (KD_{pub}) and the private key of the chip (KC_{pri}) are embedded in the design, and the SoC has the corresponding KD_{pri} and KC_{pub}. The communication protocol follows the steps below:

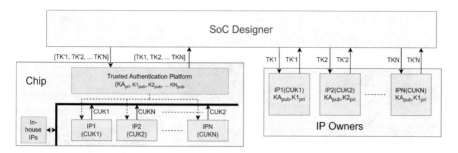

Fig. 11.10 Illustration of FORTIS flow for IP overuse prevention

- A message (m) is generated on-chip by TRNG, which is further encrypted with (KC_{pri}) to get sig(m). Then the sig(m) and message m are concatenated for padding.
- TRNG generates (K_S) that is a random session key, which is added for a one-time pad (OTP) to generate IK.
- (K_S) that denotes the session key is encrypted with KD_{pri} and concatenated with IK to get the transmission key (TK) that is sent to the IC designer.
- After receiving TK, separation of encrypted K_S and IK is done by the IC designer, by decrypting with KD_{pri}, and the session key K_S is retrieved. Then m and sig(m) are retrieved from IK using an OTP.
- Then IC designer decrypts the m from sig(m) using KC_{pub} and compares it with the original m to verify the integrity and end-point authenticity.
- Upon verifying that the TK is coming from the chip, not from an attacker, the IC designer encrypts the chip unlocking key (CUK) by using an OTP with K_S and sends it to the chip.
- The CUK is reconstructed inside the chip using its stored session key. The IP is unlocked without exposing the key to the foundry and preventing any opportunity for overproduction.

An IC design house's unauthorized overuse of 3PIP is prevented in FORTIS by a trusted authentication platform (TAP) introduced in the design flow. Figure 11.10 shows the architecture where each of the 3PIPs is obfuscated by the IP owners with individual CUK_i in a way that the IC designer cannot know the inner details and modify the design. The TAP holds all the public keys (Ki_{pub}) and its own private key (KA_{pri}) and generates the transmission keys (TK_i), which the IC designer forwards to each IP owner. Similarly, the IC designer receives the encrypted transmission keys (TK_i') from the IP owners and sends them to the foundry to unlock the IPs.

FORTIS prevents the issues of IP piracy, such as cloning and unauthorized modification of the IPs. Since the design netlist is locked, it is inherently protected against cloning because it cannot be unlocked without the key. However, the IPs need to be simulated without revealing CUKs to the IC designer. This is done by a modified design flow and trusted EDA tools, as shown in Fig. 11.11. Initially, the IP

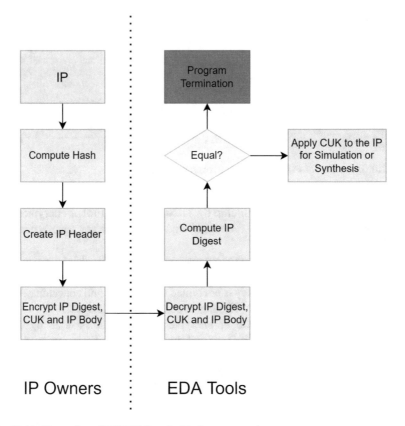

Fig. 11.11   Illustration of FORTIS flow for IP piracy prevention

owner computes the locked netlist's hash and creates an IP header that includes the CUK. Then, the header and IP body are encrypted using the symmetric encryption (AES-CBC) key of the EDA vendor. The EDA tool first decrypts the IP digest, CUK, and IP body and computes the hashed IP digest from the IP body. Finally, this newly calculated IP digest is compared with the retrieved IP digest to ensure that no modification to the IP has been done.

11.7.3   Experimentation

The FORTIS architecture was evaluated using large benchmark circuits from ITC99, opencores.org, and the OpenSPARC T1 processor core to demonstrate the impact of the obfuscation modification. A 128-bit CUK is used to lock each benchmark, which inserts 128 XOR/XNOR gates and 128 D-flip-flops into the synthesized netlist. Other than minor differences, there is no significant change in test pattern count or coverage when applied to secured and unsecured versions of benchmark circuits [19, 20].

The RSA module, OTP module, key gates, RSA keys, TRNG, and non-volatile memory all add to the design's area overhead. When an area-efficient RSA module is used, this overhead adds approximately 10K gates to the design and remains the same in all designs. As a result, while the overhead appears significant in small benchmark circuits ranging from 24.52 to 1.19%, it drastically decreases in industrial designs ranging from 0.77% in Xilinx Artix-7 to 0.15% in Kintex-7, which is negligible. The area overhead in designs that already include an RSA and TRNG module will be reduced even further.

Security is the most critical aspect of the proposed methodology to prevent IP piracy, overuse, and overproduction. The CUK length should be long enough to prevent brute force attacks, which is 80 bits of security and can be achieved with 1024-bit-long RSA key. The RSA key pairs reside in the chip, which prevents man-in-the-middle attacks. A new session key in each encryption prevents replay attacks. The secure logic obfuscation and unique session key for each chip make it almost impossible and economically infeasible to find the CUK by reverse engineering. Tampering of RSA keys, TRNG, or IP digest is also detected in the proposed methodology [23].

11.7.4 Strength and Limitations

The strength of the FORTIS methodology is that it provides a comprehensive solution that can be integrated into the existing design flow to establish forward trust between IP owners, IC designers, and foundry/assembly. However, the trusted authentication platform (TAP) and EDA tools must be trusted by all entities. Although the proposed protocol is excellent in terms of ensuring security and low overhead, the back-and-forth communication between the entities may incur some delays in the test process and time-to-market.

11.8 Conclusions

This chapter provided an overview of the fundamentals of hardware IP security through the use of embedded watermarks. We also discuss cutting-edge IP protection and watermarking techniques recently proposed in the literature, as well as their respective novelties, applications, strengths, and limitations. Hardware watermarking is a challenging security measure because it must be resilient enough that attackers cannot easily remove it, and owners can easily prove their authorship. It should also be subtle enough that it does not interfere with IP functionality, and is not easily detectable by malicious attackers. Unfortunately, these fundamental requirements frequently burden either the circuit's performance overheads or the degree of ease of extraction. The principal challenge and future work in this domain remain to find an approach that can balance all these trade-offs satisfactorily and provide strong proof of authorship. Currently, the research regarding hardware IP watermarking and fingerprinting is going in numerous directions. An interesting shift in research focus is from watermark embedding to IP active protection. This direction makes an active watermarking approach possible. With the advancement in this approach, a complete watermarking solution that integrates watermarking embedding, extraction, and enforcement of digital rights can be ensured.

References

1. (2018) Xilinx 7-series configuration user guide, ug470. https://docs.xilinx.com/v/u/en-US/ug470_7Series_Config
2. Alliance V (2000) Intellectual property protection white paper: schemes, alternatives and discussion version 1.1. Issued by Intellectual Property Protection Development Working Group, Ver 1
3. Balasch J, Gierlichs B, Verbauwhede I (2015) Electromagnetic circuit fingerprints for hardware trojan detection. In: 2015 IEEE international symposium on electromagnetic compatibility (EMC). IEEE, Piscataway, pp 246–251
4. Becker GT, Kasper M, Moradi A, Paar C (2010) Side-channel based watermarks for integrated circuits. In: 2010 IEEE international symposium on hardware-oriented security and trust (HOST). IEEE, Piscataway, pp 30–35
5. Caldwell A, Choi HJ, Kahng A, Mantik S, Potkonjak M, Qu G, Wong J (2004) Effective iterative techniques for fingerprinting design IP. IEEE Trans Comput-Aided Design Integr Circuits Syst 23(2):208–215. https://doi.org/10.1109/TCAD.2003.822126
6. Chang CH, Cui A (2010) Synthesis-for-testability watermarking for field authentication of VLSI intellectual property. IEEE Trans Circuits Syst I: Regul Papers 57(7):1618–1630
7. Chang CH, Potkonjak M, Zhang L (2016) Hardware IP watermarking and fingerprinting. In: secure system design and trustable computing. Springer, Berlin, pp 329–368
8. Charbon E (1998) Hierarchical watermarking in IC design. In: Proceedings of the IEEE 1998 custom integrated circuits conference (Cat. No. 98CH36143). IEEE, Piscataway, pp 295–298
9. Charbon E, Torunoglu I (1998) Intellectual property protection via hierarchical watermarking. In: Proc. international workshop on ip based synthesis and system design. Citeseer, pp 776–781
10. Chhotaray A, Nahiyan A, Shrimpton T, Forte D, Tehranipoor M (2017) Standardizing bad cryptographic practice: a teardown of the IEEE standard for protecting electronic-design intellectual property. In: Proceedings of the 2017 ACM SIGSAC conference on computer and communications security, pp 1533–1546
11. Contreras GK, Rahman MT, Tehranipoor M (2013) Secure split-test for preventing IC piracy by untrusted foundry and assembly. In: 2013 IEEE international symposium on defect and fault tolerance in vlsi and nanotechnology systems (DFTS). IEEE, Piscataway, pp 196–203
12. Cox IJ, Miller ML, Bloom JA, Honsinger C (2002) Digital watermarking, vol 53. Springer, Berlin
13. Cruz J, Farahmandi F, Ahmed A, Mishra P (2018) Hardware trojan detection using ATPG and model checking. In: 2018 31st international conference on VLSI design and 2018 17th international conference on embedded systems (VLSID). IEEE, Piscataway, pp 91–96
14. Cui A, Chang CH (2008) Intellectual property authentication by watermarking scan chain in design-for-testability flow. In: 2008 IEEE international symposium on circuits and systems. IEEE, Piscataway, pp 2645–2648

15. Duncan A, Skipper G, Stern A, Nahiyan A, Rahman F, Lukefahr A, Tehranipoor M, Swany M (2019) FLATS: filling logic and testing spatially for FPGA authentication and tamper detection. In: 2019 IEEE international symposium on hardware oriented security and trust (HOST). IEEE, Piscataway, pp 81–90

16. Fan YC (2008) Testing-based watermarking techniques for intellectual-property identification in SoC design. IEEE Trans Instrum Meas 57(3):467–479

17. Fan YC, Tsao HW (2005) Boundary scan test scheme for IP core identification via watermarking. IEICE Trans Inform Syst 88(7):1397–1400

18. Farahmandi F, Huang Y, Mishra P (2017) Trojan localization using symbolic algebra. In: 2017 22nd Asia and South Pacific design automation conference (ASP-DAC). IEEE, Piscataway, pp 591–597

19. Farahmandi F, Huang Y, Mishra P (2020) Automated test generation for detection of malicious functionality. In: System-on-chip security. Springer, Cham, pp 153–171

20. Farahmandi F, Huang Y, Mishra P (2020) SoC trust metrics and benchmarks. In: System-on-chip security. Springer, Cham, pp 37–57

21. Farahmandi F, Huang Y, Mishra P (2020) Trojan detection using machine learning. In: System-on-chip security. Springer, Cham, pp 173–188

22. Guin U, DiMase D, Tehranipoor M (2014) Counterfeit integrated circuits: detection, avoidance, and the challenges ahead. J Electron Testing 30(1):9–23

23. Guin U, Shi Q, Forte D, Tehranipoor MM (2016) FORTIS: a comprehensive solution for establishing forward trust for protecting IPs and ICs. ACM Trans Design Autom Electron Syst 21(4):1–20

24. Hazari NA, Niamat M (2017) Enhancing FPGA security through trojan resilient IP creation. In: 2017 IEEE national aerospace and electronics conference (NAECON). IEEE, Piscataway, pp 362–365

25. Hong I, Potkonjak M (1998) Techniques for intellectual property protection of DSP designs. In: Proceedings of the 1998 IEEE international conference on acoustics, speech and signal processing, ICASSP'98 (Cat. No. 98CH36181), vol 5. IEEE, Piscataway, pp 3133–3136

26. Hossain MM, Vashistha N, Allen J, Allen M, Farahmandi F, Rahman F, Tehranipoor M (2022) Thwarting counterfeit electronics by blockchain

27. Kahng AB, Lach J, Mangione-Smith WH, Mantik S, Markov IL, Potkonjak M, Tucker P, Wang H, Wolfe G (2001) Constraint-based watermarking techniques for design IP protection. IEEE Trans Comput-Aided Design Integr Circuits Syst 20(10):1236–1252

28. Kamali HM, Azar KZ, Farahmandi F, Tehranipoor M (2022) Advances in logic locking: past, present, and prospects. Cryptology ePrint Archive

29. Kean T, McLaren D, Marsh C (2008) Verifying the authenticity of chip designs with the DesignTag system. In: 2008 IEEE international workshop on hardware-oriented security and trust. IEEE, Piscataway, pp 59–64

30. Kirovski D, Potkonjak M (1998) Intellectual property protection using watermarking partial scan chains for sequential logic test generation. In: Proc. IEEE high level design, verification, and test conf. Citeseer

31. Kirovski D, Potkonjak M (2003) Local watermarks: Methodology and application to behavioral synthesis. IEEE Trans Comput-Aided Design Integr Circuits Syst 22(9):1277–1283

32. Koushanfar F (2011) Provably secure active IC metering techniques for piracy avoidance and digital rights management. IEEE Trans Inform Forensics Secur 7(1):51–63

33. Lach J, Mangione-Smith WH, Potkonjak M (1998) Fingerprinting digital circuits on programmable hardware. In: International workshop on information hiding. Springer, Berlin, pp 16–31

34. Lach J, Mangione-Smith WH, Potkonjak M (1998) FPGA fingerprinting techniques for protecting intellectual property. In: Proceedings of the IEEE 1998 custom integrated circuits conference (Cat. No. 98CH36143). IEEE, Piscataway, pp 299–302

35. Lach J, Mangione-Smith WH, Potkonjak M (2001) Fingerprinting techniques for field-programmable gate array intellectual property protection. IEEE Trans Comput-Aided Design Integr Circuits Syst 20(10):1253–1261

36. Li J, Lach J (2008) At-speed delay characterization for IC authentication and trojan horse detection. In: 2008 IEEE international workshop on hardware-oriented security and trust. IEEE, Piscataway, pp 8–14

37. Lin L, Kasper M, Güneysu T, Paar C, Burleson W (2009) Trojan side-channels: lightweight hardware trojans through side-channel engineering. In: International workshop on cryptographic hardware and embedded systems. Springer, Berlin, pp 382–395

38. Nahiyan A, Sadi M, Vittal R, Contreras G, Forte D, Tehranipoor M (2017) Hardware trojan detection through information flow security verification. In: 2017 IEEE international test conference (ITC). IEEE, Piscataway, pp 1–10

39. Narayan N, Newbould RD, Carothers JD, Rodriguez JJ, Holman WT (2001) IP protection for VLSI designs via watermarking of routes. In: Proceedings 14th annual IEEE international ASIC/SOC conference (IEEE Cat. No. 01TH8558). IEEE, Piscataway, pp 406–410

40. Nowroz AN, Hu K, Koushanfar F, Reda S (2014) Novel techniques for high-sensitivity hardware trojan detection using thermal and power maps. IEEE Trans Comput-Aided Design Integr Circuits Syst 33(12):1792–1805

41. Rad RM, Wang X, Tehranipoor M, Plusquellic J (2008) Power supply signal calibration techniques for improving detection resolution to hardware trojans. In: 2008 IEEE/ACM international conference on computer-aided design. IEEE, Piscataway, pp 632–639

42. Rahman F, Farahmandi F, Tehranipoor M (2021) An end-to-end bitstream tamper attack against flip-chip FPGAs. Cryptology ePrint Archive

43. Rahman MS, Nahiyan A, Amir S, Rahman F, Farahmandi F, Forte D, Tehranipoor M (2019) Dynamically obfuscated scan chain to resist oracle-guided attacks on logic locked design. Cryptology ePrint Archive

44. Rahman MS, Nahiyan A, Rahman F, Fazzari S, Plaks K, Farahmandi F, Forte D, Tehranipoor M (2021) Security assessment of dynamically obfuscated scan chain against oracle-guided attacks. ACM Trans Design Autom Electron Syst 26(4):1–27

45. Rajendran J, Pino Y, Sinanoglu O, Karri R (2012) Security analysis of logic obfuscation. In: Proceedings of the 49th annual design automation conference, pp 83–89

46. Roy JA, Koushanfar F, Markov IL (2010) Ending piracy of integrated circuits. Computer 43(10):30–38

47. Shi Q, Vashistha N, Lu H, Shen H, Tehranipoor B, Woodard DL, Asadizanjani N (2019) Golden gates: a new hybrid approach for rapid hardware trojan detection using testing and imaging. In: 2019 IEEE international symposium on hardware oriented security and trust (HOST). IEEE, Piscataway, pp 61–71

48. Stern A, Mehta D, Tajik S, Farahmandi F, Tehranipoor M (2020) SPARTA: a laser probing approach for trojan detection. In: 2020 IEEE international test conference (ITC). IEEE, Piscataway, pp 1–10

49. Stern A, Mehta D, Tajik S, Guin U, Farahmandi F, Tehranipoor M (2020) SPARTA-COTS: a laser probing approach for sequential trojan detection in cots integrated circuits. In: 2020 IEEE physical assurance and inspection of electronics (PAINE). IEEE, Piscataway, pp 1–6

50. Tehranipoor M, Peng K, Chakrabarty K (2011) Introduction to VLSI testing. In: Test and Diagnosis for small-delay defects. Springer, New York, pp 1–19

51. Tehranipoor MM, Guin U, Forte D (2015) Hardware IP watermarking. In: Counterfeit integrated circuits. Springer, Berlin, pp 203–222

52. Tehranipoor MM, Shen H, Vashistha N, Asadizanjani N, Rahman MT, Woodard D (2021) Hardware trojan scanner. US Patent 11,030,737

53. Torunoglu I, Charbon E (2000) Watermarking-based copyright protection of sequential functions. IEEE J Solid-State Circuits 35(3):434–440

54. Vashistha N, Lu H, Shi Q, Rahman MT, Shen H, Woodard DL, Asadizanjani N, Tehranipoor M (2018) Trojan scanner: detecting hardware trojans with rapid SEM imaging combined with image processing and machine learning. In: ISTFA 2018: proceedings from the 44th international symposium for testing and failure analysis. ASM International, p 256

55. Vashistha N, Rahman MT, Shen H, Woodard DL, Asadizanjani N, Tehranipoor M (2018) Detecting hardware trojans inserted by untrusted foundry using physical inspection and advanced image processing. J Hardw Syst Secur 2(4):333–344

56. Vashistha N, Hossain MM, Shahriar MR, Farahmandi F, Rahman F, Tehranipoor M (2021) eChain: a blockchain-enabled ecosystem for electronic device authenticity verification. IEEE Trans Consum Electron 68(1):23–37
57. Vashistha N, Lu H, Shi Q, Woodard DL, Asadizanjani N, Tehranipoor M (2021) Detecting hardware trojans using combined self-testing and imaging. IEEE Trans Comput-Aided Design Integr Circuits Syst 41(6):1730–1743
58. Yuan L, Qu G, Ghouti L, Bouridane A (2006) VLSI design IP protection: solutions, new challenges, and opportunities. In: First NASA/ESA conference on adaptive hardware and systems (AHS'06). IEEE, Piscataway, pp 469–476
59. Zhou B, Adato R, Zangeneh M, Yang T, Uyar A, Goldberg B, Unlu S, Joshi A (2015) Detecting hardware trojans using backside optical imaging of embedded watermarks. In: 2015 52nd ACM/EDAC/IEEE design automation conference (DAC). IEEE, Piscataway, pp 1–6
60. Zhuang X, Zhang T, Lee HHS, Pande S (2004) Hardware assisted control flow obfuscation for embedded processors. In: Proceedings of the 2004 international conference on compilers, architecture, and synthesis for embedded systems, pp 292–302
61. Ziener D (2010) Techniques for increasing security and reliability of IP cores embedded in FPGA and ASIC designs. Friedrich-Alexander-Universität Erlangen-Nürnberg (FAU)
62. Ziener D, Teich J (2008) Power signature watermarking of IP cores for FGGAs. J Signal Process Syst 51(1):123–136

Chapter 12
Lightweight Cryptography

12.1 Introduction

Cryptography has become indispensable in protecting sensitive communication, data, and personal information [3, 28, 31]. However, most conventional encryption algorithms were designed for use in devices with sufficient resources, such as servers, personal computers, and smartphones. These algorithms primarily optimize the trade-off between security, performance, and resource requirements in the afore-mentioned platforms with high computational complexity and resources. However, it makes these algorithms challenging to run on resource-constrained devices [25].

Today, many smart devices are connected through the Internet, creating a network known as the Internet of Things (IoT) [22]. It is proliferating due to ubiquitous applications such as wearable technology, autonomous vehicles, smart grids, smart supply chain management, smart home appliances, and so on. These applications are quickly becoming part of our daily life. Resource-constrained embedded processors, sensors, and short-range communication functions also strengthen the base of communication between IoT devices. These devices usually have access to sensitive data from their sensors, and the data is processed after being transferred to a cloud-based remote server. Such data transactions and processing in a resource-constrained system can be vulnerable to new attacks and require encryption. Thus, with the progress of the Internet of Things (IoT), embedded systems, RFID tags, and sensors, lightweight cryptographic algorithms are needed to prevent these vulnerabilities [4].

Many attempts have been made to standardize lightweight cryptographic algorithms by calling for submissions of existing algorithms to consider for standardization [25]. Necessary feedback is requested from stakeholders and industry members for lightweight cryptographic algorithms. After that, NIST developed a method of evaluating cryptographic algorithms based on the profile so that they can recommend specific algorithms for different hardware and different use cases [25]. This chapter explores various lightweight cryptographic algorithms suitable for resource-constrained devices and highlights the characteristics and design considerations for such cryptographic implementations.

12.2   Background

This section discusses the characteristics of lightweight cryptography and various design considerations.

12.2.1   Characteristics of Lightweight Cryptography

A cryptographic algorithm requires an optimum balance between performance, resources, and the desired security level. Power consumption, latency, and throughput are significant attributes that represent a system's performance. In contrast, gate area, gate equivalents (GEs), or slices represent the resource requirements for a hardware implementation [25]. Slices are the reconfigurable units of an FPGA consisting of LUTs, flops, and multiplexers. For current CMOS technology, gate equivalents represent the area required by a two-input NAND standard cell for a specific technology library. The number of slices for FPGA implementation, or the number of GEs for ASIC for the given technology library, specifies the area of the algorithm in hardware.

Additionally, the number of registers, size of RAM, and ROM signify the resource requirements for a software implementation [25]. These hardware and software platform resource requirements can be termed hardware-specific and software-specific metrics, respectively. For a resource-constrained application, a cryptographic algorithm should require limited resources to illustrate the smaller value of the metrics. Thus, lightweight cryptography is characterized by performance and metrics related to resource constraints.

Microcontrollers (e.g., NXP RS08, TI COP912C, Microchip PIC, etc.) are restricted to use a minimal amount of RAM and ROM [1, 2, 34]. Furthermore, embedded systems with a wide range of performance attributes have a limited set of instructions. Therefore, running traditional cryptographic algorithms on such systems takes an extensive number of clock cycles. Since the throughput is low with increased power overhead, traditional algorithms are not feasible for such power and resource constraint real-time systems. Similarly, some RFID tags require an EM signal to run the internal circuitry where battery power is not allowed. Therefore, a lightweight cryptographic algorithm is required for these devices with a minimal number of logic gates, low power, and timing constraints [33]. For instance, using 200–2000 gates for security purposes out of 1000–10,000 gate counts can reduce both areas and power consumption along with production cost for an RFID tag [20].

In addition, specific real-time applications (e.g., automotive, aerospace, etc.) need low latency and demand swift response times. Lightweight cryptography seeks to achieve this quick response by lowering the latency of an encryption process. A cryptographic algorithm defines latency as the time between the initial request (plaintext sent for encryption) and the response (received ciphertext) [25]. On the other hand, throughput is the pace at which new replies (such as authentication tags or ciphertext) are generated. As a result, in order to achieve adequate performance, even the most resource-constrained machine requires a lightweight cipher.

12.2.2 Lightweight Cryptography's Design Considerations

Along with application-specific requirements, some general considerations are necessary for a cryptographic algorithm to be considered lightweight [25]. NIST uses these desired considerations for evaluating a design and is described as follows:

1. **Security strength**: Any lightweight cryptographic algorithm must have appropriate security strength, with more than 112 bits.
2. **Flexibility**: In order to enable a wide range of applications, an algorithm should be able to support many implementations on the same platform. Algorithms that allow to adjust design features (such as state and key sizes) are more versatile because they provide more options for conserving resources.
3. **Low overhead for multiple functions**: Compared to functions with fully independent implementations, multiple functions (e.g., encryption and decryption) sharing the same logic circuit incur less overhead. For example, a block cipher that employs the same logic for encryption/decryption rounds is more practical for a lightweight application than having separate logic circuits for encryption and decryption rounds. The requirements for resources are further lowered in the case of a resource-constrained platform by allowing numerous primitives (e.g., block cipher and hash function) to share the same fundamental logic [25].
4. **Ciphertext expansion**: Algorithms that manage ciphertext size to minimize data capacity incur lower storage and transmission costs. These algorithms are ideal for use in lightweight cryptography.
5. **Side channel and fault attacks**: Cryptographic algorithm implementations in hardware and software are prone to various side-channel and fault injection attacks. Due to implementation flaws, a design can leak sensitive information as various side-channel attributes, such as time, power, electromagnetic radiation, etc. An attacker takes advantage of these physical leakages to obtain the key or plaintext. An attacker can also deliberately introduce faults to a design to change the system's runtime behavior. The altered behavior may also leak sensitive information from the system. For example, faulty ciphertexts obtained through fault injection attacks are analyzed to determine a cipher's key, known as differential fault analysis. Since an attacker may have direct physical access to the pervasive devices used in IoT-based systems, algorithms with side-channel and fault injection attack mitigation techniques are more viable [25].

6. **Plaintext–Ciphertext pairs limit**: An upper margin of plaintext/ciphertext pairs can be considered during the design of an algorithm. This assumption is acceptable for a lightweight application where the same key processes a limited amount of data or the message formats are well defined [25].
7. **Related-key attacks**: Cryptographic protocols where the relation between random and independent keys poses a security threat called a "related-key attack." Based on a known relation among multiple keys, an adversary can guess the information about an unknown key through this attack. This attack still threatens the implementations that permanently burn keys into hardware with no way for a replacement. So, a good cryptographic algorithm should have resiliency against this threat for a lightweight application.

Not all the requirements mentioned above can be satisfied simultaneously in a lightweight cryptographic algorithm. Some requirements may not be necessary depending on the use case. However, all lightweight cryptographic implementations must ensure the desired security level. Therefore, lightweight cryptography does not mean weak cryptography but a minimal version that can be implemented on resource-constrained devices. On top of that, lightweight cryptography is not designed to prevent attacks from powerful adversaries and is not supposed to replace traditional cryptography [30].

12.3 Lightweight Cryptographic Primitives

Scientists and researchers have introduced many cryptographic primitives, such as block and stream ciphers, message authentication codes (MACs), hash functions, etc. Application-specific usage and restriction on an attacker's ability are the significant differences between these primitives with traditional algorithms. For instance, a lightweight cryptographic primitive can limit the data available to an attacker under a single key. However, these primitives are intended to provide a better balance between resource requirements, performance, and security, unlike conventional algorithms [25]. This section presents some of these primitives and their various aspects.

12.3.1 Lightweight Block Ciphers

To improve performance over NIST's AES-128, several lightweight block ciphers have been introduced. A conventional block cipher is simplified to design these more efficient ciphers. For instance, DESL [21], a variant of DES, provides minimal hardware size, removing initial and final permutations using a single Sbox [25]. Some lightweight block ciphers (e.g., PRESENT [8]) have been developed especially for platforms with limited resources. SIMON and SPECK ciphers target both hardware and software with simplicity, flexibility, and better performance [6]. For resource-constrained software platforms, some algorithms such as RC5 [32], TEA [36], and XTEA [26] can be used, consisting of simple round operations. Some characteristics of lightweight block ciphers provide advantages over standard block ciphers, as indicated below:

- **Smaller block sizes**: Instead of 128 bits of block size, a lightweight block cipher may use a reduced block size of 64 bits or 80 bits to reduce the memory overhead.
- **Smaller key sizes**: To increase efficiency, a lightweight block cipher can use smaller key sizes (less than 96 bits). For instance, a PRESENT block cipher can have a key size of 80 bits.
- **Simpler rounds**: A simplified version of the round operations and related hardware components are maintained during the design and implementation of a lightweight cipher. It helps to reduce the area overhead of the overall design. For instance, a 4-bit Sbox used in PRESENT instead of an 8-bit one saves the area of 367 gate equivalents (GEs) compared to an AES Sbox requiring 395 GEs [15]. In addition, instead of complex-linear layers, bit permutations or recursive MDS matrices can be used to save area in some hardware implementations [25].
- **Simpler key schedules**: A lightweight block cipher uses a simplified version of the key scheduling process to save power, timing, and resource overhead. Some implementations may prefer generating sub-keys during runtime. To ensure resiliency against attacks related to chosen keys, known keys, weak keys, or related keys, independent generation of keys using a secure key derivation function can be followed [25].

AES and Triple DES (TDES) are NIST-approved lightweight block ciphers for resource-constrained environments. For instance, AES-128 is well-suited for 8-bit AVR microcontrollers, and TDES is appropriate for Renesas RL78 16-bit microcontrollers (limited ROM and ROM).

12.3.2 Lightweight Stream Ciphers

Stream ciphers can also have a minimal design for specific environments. For example, some lightweight stream ciphers designed for resource-constrained hardware platforms are as follows:

- **Grain**: This stream cipher is flexible for implementation and supports authentication [18].
- **Trivium**: This widely explored stream cipher also provides flexibility in design and supports a key size of only 80 bits [9].
- **Mickey**: This stream cipher is less flexible compared to Grain and Trivium and vulnerable to some attacks (timing and power side channel) [5]. Exhaustive analysis is necessary to ensure its security.

12.3.3 Lightweight Hash Functions

Lightweight hash functions are suitable for resource-constrained platforms as they have smaller internal state sizes and low power consumption requirements [25]. Modern lightweight hash functions (e.g., PHOTON [17], QUART SPONGENT [7], and Lesamnta-LW [19]) have several features, unlike conventional hash functions. These features are outlined as follows:

- **Smaller internal state and output size**. A lightweight hash function demands a smaller output size along with a smaller internal state where collision resistance is not a requirement [25].
- **Smaller message size**. A lightweight hash function supports much smaller input sizes (e.g., maximum 256 bits) compared to a conventional hash function. So, it is appropriate for a lightweight application where short messages as inputs are one of the requirements.

12.3.4 Lightweight Message Authentication Codes

To verify whether a message is authentic or not, a tag from a message and a secret key is generated by a message authentication code (MAC) [25]. 64 bits of tag size is typically used for conventional MACs. However, some applications (voice over IP) can occasionally accept inauthentic messages, which can have a limited impact on security. These applications demand lightweight MACs (e.g., Chaskey [27], TuLP [16], LightMAC [23], etc.) where shorter tags are carefully chosen so that the system's security is not degraded. For constrained environments, NIST approved some block ciphers (e.g., CCM and GCM) that provide encryption as well as authentication at the same time [25]. Some other NIST-approved MACs include CMAC, GMAC, and HMAC for message authentication in lightweight applications.

12.4 Standard for Lightweight Cryptography

This section describes the standardization of state-of-the-art lightweight cryptographic algorithms. NIST developed profiles corresponding to a specific use case to standardize cryptographic algorithms. The most recent lightweight cryptographic algorithms are recommended for different profiles or use cases to guide people in selecting a lightweight cryptographic algorithm for their platform.

A single algorithm cannot meet all the requirements of an application. For this reason, different algorithms will be used for different applications described by the profiles NIST developed. Of course, any lightweight design will have trade-offs. However, the profiles NIST developed help minimize impacts on cryptographic security by utilizing an appropriate algorithm for the use case.

Table 12.1 Characteristics of the profiles

Physical characteristics	Performance characteristics	Security characteristics
Area (in GEs, logic blocks or mm^2)		Minimum security strength (bits)
Memory (RAM/ROM)	Latency (in clock cycles or time period)	Attack models (e.g., related key, multi-key)
Implementation type (hardware, software, or both)	Throughput (cycles per byte)	Side-channel resistance requirements
Energy (J)	Power (W)	

Table 12.2 Profile template

Profile <profile name>	
Primitive	Type of primitive
Physical characteristics	Name physical characteristic(s), and provide acceptable range(s) (e.g., 64–128 bytes of RAM)
Performance characteristics	Name performance characteristic(s), and provide acceptable range(s) (e.g., latency of no more than 5 ns)
Security characteristics	Minimum security strength, relevant attack models, side channel resistance requirements, etc.
Design goals	List design goals

Table 12.1 describes some of the characteristics used to develop a profile. Physical characteristics include area and energy consumption. Performance includes latency and throughput, and security features minimum security strength and possible attack vectors. The developed profile is intended to capture the limitations of the cryptographic algorithms. It includes the purpose/functionality of the cryptographic algorithms, design goals, and the algorithms' physical, performance, and security characteristics. Table 12.2 shows the items included in the profile template.

12.5 Lightweight Ciphers for IoT Devices

As IoT technology continues to develop, so does the need for a variety of cryptographic primitives that require low production costs and little resources [29]. Therefore, researchers have set out to create a survey of already existing lightweight ciphers, choosing to highlight existing implementations made in lightweight cryptography to explore and compare their differences. This section explains two varieties of lightweight ciphers, block ciphers and stream ciphers, in detail. We also provide a side-by-side comparison for each one.

Block ciphers are incredibly versatile and open to lightweight modifications since designers can utilize and combine many elementary operations. These elementary operations can establish unique encryption based on their application's constraints. The operation of lightweight block ciphers depends on a fixed length of bits, includ-

Table 12.3 Comparison between different lightweight block ciphers

Block ciphers	Parameters					
	Block size	Key size	Area (GEs)	Throughput	No. of rounds	Made of
PRESENT	128	80/128	1000	High	32	SPN
KATAN	32/48/64	80	802	Low	254	Feistel
Humming bird	16	256	2159	High	4	SPN
SIMON	32 48 64 96 128	64 72/96 98/128 96/144 128/192 256	739 809 958 955 1234	High	32 36 42/44 52/54 68/69/72	Feistel
RECTANGLE	64	80	1600	High	25	SPN
DESXL	64	184	2168	Low	144	Feistel

ing encryption and decryption of each package as it is transmitted and received. The elementary operations of a lightweight cipher consist of a substitution–permutation network (SPN) or Feistel network [29]. Symmetric structure of the networks is responsible for identical encryption and decryption operations that help reduce the resources of hardware or software implementations. This section describes some block ciphers intended to be incorporated into IoT platforms. Table 12.3 highlights the comparisons between the lightweight block ciphers.

- **PRESENT**: It is credited as a milestone for research on lightweight ciphers and also stands as a model for comparison against new lightweight ciphers. It is well-suited for a resource-constrained application for its ability to be implemented within 1000 GEs. It uses a block size of 128 bits and a key size of 80/128 bits along with the combination of substitution and permutation network for 32 rounds of operation [29]. 4-bit Sbox and bit permutations form the substitution and permutation layers, respectively. To summarize, it combines the features of both AES and DES and provides optimized hardware with no compromise for security. These features improve the efficiency of PRESENT implementation on hardware and software platforms [8].
- **KATAN**: In comparison with PRESENT, the KATAN block cipher has less area overhead (minimal size), requiring only 802 GEs during implementation. The hardware design of KATAN is based on the Feistel network and is well-known for its simplicity in key scheduling processes. It can use a variety of block sizes (e.g., 32 bits, 48 bits, or 64 bits) along with a key size of 80 bits. In addition, 254 rounds of operations are based on two invertible nonlinear Boolean operations [29]. The significant drawbacks of this cipher for lightweight applications include low throughput, increased energy consumption, and inefficient software implementation [10].
- **Humming Bird**: Humming Bird, another promising lightweight cipher, provides better security than KATAN by increasing key size (256 bit). It also uses a

combination of block ciphers and stream ciphers along with 16 bits of block size. It is aimed to be used for control system applications since it has a smaller block size. Its substitution–permutation network (SPN) consists of Sboxes and linear transformations, which control four regular rounds and one final round of operations [29]. Compared with PRESENT, it provides higher throughput and less area overhead in both hardware and software platforms. However, since its encryption and decryption functions are not the same, it requires larger GEs for implementing encryption and decryption altogether in a typical hardware [24].

- **SIMON and SPECK**: SIMON and SPECK are new lightweight ciphers initiated by the National Security Agency (NSA) to improve existing lightweight algorithms compatible with different platforms [6]. The existing algorithms worked very well on specific platforms, but with the increase in IoT devices, there is a need for cryptographic algorithms that can work on a wide range of hardware and software. For instance, one of the most widely used lightweight ciphers, PRESENT, is very compact when running in ASIC implementations but not efficient if running on constrained software or an FPGA. In addition, the conventional lightweight block ciphers have a single block size and, at most, two key sizes, and these limitations need to be resolved. SIMON and SPECK can utilize five different block sizes, ranging from 32 to 128 bits, as well as multiple key sizes for providing flexibility in a wide range of applications [6]. Table 12.4 shows only the block and key sizes available in both SIMON and SPECK.

The types of operations used in SIMON and SPECK are limited since they are intended to run on various lightweight platforms. For this reason, they avoid using substitution–permutation networks (SPNs) and instead use Feistel permutations. Another design decision was to use rotations for the bit permutations to run efficiently on both software and hardware. In SIMON, the design decisions are generally optimized for hardware, while with SPECK, the design choices favor software implementations over hardware. For example, in SPECK, modular addition is used for nonlinearity and achieving cryptographic strength. Nevertheless, it is not efficient in hardware implementations. In contrast, SIMON has two rotation operations, along with a bitwise AND, which is very easy to implement on hardware with reduced security strength [6].

SIMON and SPECK use straightforward rounds for extremely compact implementation compared to AES but require more iteration to gain desired security. SPECK can also run without performing any move operations, allowing

Table 12.4 Available block and key sizes

Block size	Key size
32	64
48	72, 96
64	96, 128
96	96, 144
128	128, 192, 256

it to run very efficiently in software and using fewer rounds than SIMON. In addition to performing well in resource-constrained environments, the two perform exceptionally well on more typical computing devices and run quicker than any other algorithm. SIMON and SPECK are designed to be secure, however, a significant breakthrough in cryptanalysis is highly likely to render both less effective as cryptographic algorithms [6].

- **RECTANGLE**: In implementing a RECTANGLE cipher, bit-slice techniques help achieve less area overhead and higher throughput [37]. Substitution–permutation network (SPN) is the foundation of its 25 intermediate rounds of operation, where the key length of 80 bits and block size of 64 bits are used. In the hardware of this cipher, 16 4×4 Sboxes form the substitution layer, and XOR with rotation operations forms the permutation layer [29]. For both hardware and software platforms, it performs efficiently.

- **DESXL**: Discarding 7 Sboxes along with multiplexer and thus reducing the area overhead (e.g., reduced GEs), a lightweight version of DES can be implemented. This lightweight version, known as DESL, provides sufficient security by showing resiliency against linear and differential cryptanalysis and David-Murphy attack [13]. Furthermore, the security strength can be further improved by key whitening. This more secure version is DESXL, requiring 144 clock cycles for encryption. Although it provides better security, it suffers from low throughput and high area overhead compared to other aforementioned lightweight ciphers.

Lightweight stream ciphers differ from lightweight block ciphers due to their method of encrypting and decrypting plaintext on a bit-by-bit level, utilizing a key having an equal size to the plain text. Since only bit operations are required to implement a stream cipher, they are typically more compact and simpler to implement, with faster hardware performance than block ciphers. Due to these characteristics, bit operations can mostly be found in cell phone and wireless communication applications [29]. This section presents some lightweight stream ciphers and compares the elementary function elements that make up the encryption/decryption algorithm, GE area consumption, initialization vector size, key size, and hardware vs. software optimization. Table 12.5 highlights the comparisons between the lightweight stream ciphers.

Table 12.5 Table illustrating parameters of different lightweight stream ciphers

Stream ciphers	Parameters				
	Initialization vector	Key size	Area (GEs)	Throughput	Made of
TRIVIUM	80	80	2600	High	LFSR
GRAIN	96	128	1300	Low	Both
WG 8	80	80	1984	High	LFSR
ESPRESSO	96	128	1500	High	NLFSR
CHACHA	128	256	750	High	Both

- **TRIVIUM**: TRIVIUM is designed to provide optimum speed, security, and flexibility. It uses linear feedback shift registers (LSFRs) as part of its structure. It employs an initialization vector of 80 bits and a key length of 80 bits [29]. TRIVIUM is suited better for resource-constrained devices, although it requires a large GE area for implementation. One of its major drawbacks is that it uses too many sequential cells (flops) for hardware implementations. On top of that, it is prone to cube attacks [35].
- **GRAIN**: GRAIN is an appropriate lightweight cipher for resource-constrained devices. It uses a key of length 128 bits, an intermediate stage of 256 bits, and an initialization vector of 96 bits. It improves GE area consumption due to its combination of LFSR, NLFSR (non-linear feedback shift register), and Boolean function [29]. Nevertheless, it suffers from high propagation delay and is susceptible to cryptanalysis attacks [18].
- **WG 8**: The Welch Gong stream cipher family introduced the WG-8 lightweight stream cipher that consists of a series of LFSRs with an 80-bit key and an initialization vector-like TRIVIUM [29]. However, implementations of WG-8 provide more significant throughput than other designs and require less memory. In addition, it has provable randomness property and consumes less energy; however, it is susceptible to key recovery attacks [14].
- **ESPRESSO**: This cipher is credited for being the fastest implementation of lightweight stream ciphers below 1500 GE due to its unique combination of NLFSRs. It uses a key size of 128 bits, an initialization vector size of 96 bits, and a short propagation delay that allows it to minimize its hardware footprint while maximizing its throughput. Furthermore, since it provides low latency, it is well-suited for 5G applications [12].
- **CHACHA**: The round function of a variant of a stream cipher, Salsa20, is modified to form a new lightweight cipher, CHACHA. It supports a key length of 256 bits and an initialization vector of 128 bits and uses a combination of both LFSR and NLFSR for eight-round operations [29]. It is well-suited for software-constrained platforms since it uses the same hardware for encryption and decryption operations. A detailed description and implementation of this stream cipher in combination with a Poly1305 Authenticator are presented in the next section.

12.6   CHACHA20-Poly1305 Authenticated Encryption

When used together, the CHACHA20 stream cipher and the Poly1305 Authenticator are intended to provide increased security and performance on many systems. To build an Authenticated Encryption with Associated Data (AEAD) scheme, the adoption and standardization of the CHACHA20 stream cipher and the Poly1305 authenticator are encouraged. It also contributes to the increased secrecy and integrity of permitted data.

In the CHACHA stream cipher, a 256-bit key is expanded into 264 random streams of 264 randomly accessible 64-byte blocks [11]. They are encrypted by a round number suffix, recommended at 20 rounds, which controls the number of times the round module is applied to the cipher's internal state (it must be an even number). The round module applies a sequence of constant shifting operation, rotating operation, and XOR operation. These operations provide high randomness in the encryption process and rely on the mechanical sympathy of how CPUs physically implement XOR and shifting instructions [11].

Poly1305, a cryptographic message authentication code (MAC), is used to validate a communication's data integrity and authenticity. A 16-byte tag is created by combining a 32-byte key and a message. This tag is used to validate the message, and each key can only be used once. The function Pad1305 chops the input message into 17-byte chunks where each block is padded [11]. Then, the function Clamp clears some bits so that some costly carry propagation can be saved with a minimal loss of a few security bits. Finally, the evaluation of the polynomial defined by the coefficients is exploited to generate the authentication tag.

The Poly1305 authenticator and CHACHA20 stream cipher must be combined to implement the CHACHA20-Poly1305 authenticated encryption with the AEAD scheme specified with RFC7539. The inputs of CHACHA20-Poly1305 AEAD are a 12-byte nonce, 32-byte secret key, and a variable-length plaintext. The return value of it is a ciphertext and a 16-byte authentication tag.

Implementing Poly1305 and CHACHA20 on an ARM Cortex-M4 processor aims at both lightweight and high-speed applications. Thus, an assembly level of optimizations is required for CHACHA20 and Poly1305. CHACHA20's 20 rounds are optimized in two ways. To begin, the double rounds are optimized by altering the quarter round order. Second, by utilizing the "flexible second operand" feature, the quarter rounds are optimized. This optimization technique reduces the encryption round from 20 to 4 rounds, making a comparable large encryption algorithm into a light encryption algorithm [11].

The Poly1305 authentication algorithm also necessitates Horner's method optimization. The authentication algorithm can be implemented using addition, multiplication, and reduction instructions. In ARM Artex-7, five add instructions are used to perform addition instructions. Next, the product scanning method is used to implement multiplication instructions. Finally, several shifts are used to apply reduction instructions, and the final reduction is computed to reduce the result.

The advantages of combining the CHACHA20 and Poly1305 authentication algorithms are overwhelming other algorithms with their speed and code size [11]. Therefore, by optimizing the two encryption algorithms, a lighter encryption algorithm with a lighter code size and better speed can be implemented. However, the code size for encryption algorithms is not controllable because it aims to make the encryption process complex enough to be unbreakable. Furthermore, after optimizing the algorithm, there must be a cost for its security. So, the limitation of this implementation is that it has security issues compared to other encryption methods [11]. To overcome these limitations, a method evaluating the security and complexity of the encryption algorithm needs to be implemented. Therefore, along with evaluating the speed and code size, security is also a parameter in choosing an encryption algorithm.

12.7 Conclusions

This chapter discussed lightweight cryptography and presented several lightweight cryptographic algorithms. The demand for resource and performance versatility in IoT applications increases, making lightweight cryptography more advantageous than existing cryptographic solutions. Since lightweight cryptography is an emerging field, researchers seek to establish lightweight cryptographic standards with profiles that tailor specific device and application resource requirements. To create these profiles, research in existing lightweight ciphers is illustrated, focusing on block and stream ciphers such as SIMON, SPECK, PRESENT, TRIVIUM, ESPRESSO, etc. This chapter defines and highlights characteristics used to compare lightweight ciphers. Notably, there is a trade-off when using sophisticated algorithms to increase the lightweight cipher application's security, like an increase in GE area consumption. Because of this, SIMON and SPECK ciphers are introduced to create a single algorithm that works well on all resource-constrained platforms. However, the developers created two different algorithms optimized separately for hardware and software, which illustrates the difficulty of designing an efficient algorithm for all resource-constrained environments. Finally, we present the high-speed lightweight encryption implementation of the CHACHA20 stream cipher, Poly1305 authenticator, and CHACHA20-Poly1305 authenticated encryption scheme on an ARM Cortex-M4 processor. The advantages CHACHA20 has over other encryption algorithms made it a promising candidate as a light encryption algorithm for IoT devices that supports TLS secured communication on ARM Cortex-M4 devices.

References

1. (2020) 8-bit legacy MCUs—NXP semiconductors. https://www.nxp.com/products/processors-and-microcontrollers/legacy-mpu-mcus/8-bit-legacy-mcus:8-BIT-LEGACY-MCUS
2. (2021) Parametric-search—microchip technology. https://www.microchip.com/en-us/parametric-search.html/627. Accessed 26 Dec 2021
3. Ahmed B, Bepary MK, Pundir N, Borza M, Raikhman O, Garg A, Donchin D, Cron A, Abdel-moneum MA, Farahmandi F et al. (2022) Quantifiable assurance: from IPS to platforms. arXiv preprint arXiv:220407909
4. Alaba FA, Othman M, Hashem IAT, Alotaibi F (2017) Internet of things security: a survey. J Netw Comput Appl 88:10–28
5. Babbage S, Dodd M (2008) The mickey stream ciphers. In: New stream cipher designs. Springer, Berlin, pp 191–209
6. Beaulieu R, Shors D, Smith J, Treatman-Clark S, Weeks B, Wingers L (2015) The SIMON and SPECK lightweight block ciphers. In: Proceedings of the 52nd annual design automation conference, pp 1–6

7. Beyne T, Chen YL, Dobraunig C, Mennink B (2020) Dumbo, jumbo, and delirium: parallel authenticated encryption for the lightweight circus. IACR Trans Symmetric Cryptol 2020:5–30

8. Bogdanov A, Knudsen LR, Leander G, Paar C, Poschmann A, Robshaw MJ, Seurin Y, Vikkelsoe C (2007) Present: an ultra-lightweight block cipher. In: International workshop on cryptographic hardware and embedded systems. Springer, Berlin, pp 450–466

9. De Canniere C, Preneel B (2008) Trivium. In: New stream cipher designs. Springer, Berlin, pp 244–266

10. De Canniere C, Dunkelman O, Knežević M (2009) KATAN and KTANTAN—a family of small and efficient hardware-oriented block ciphers. In: International workshop on cryptographic hardware and embedded systems. Springer, Berlin, pp 272–288

11. De Santis F, Schauer A, Sigl G (2017) Chacha20-poly1305 authenticated encryption for high-speed embedded IoT applications. In: Design, automation & test in europe conference & exhibition (DATE), 2017. IEEE, Piscataway, pp 692–697

12. Dubrova E, Hell M (2017) Espresso: a stream cipher for 5g wireless communication systems. Cryptogr Commun 9(2):273–289

13. Eisenbarth T, Kumar S, Paar C, Poschmann A, Uhsadel L (2007) A survey of lightweight-cryptography implementations. IEEE Design Test Comput 24(6):522–533

14. Fan X, Mandal K, Gong G (2013) Wg-8: A lightweight stream cipher for resource-constrained smart devices. In: International conference on heterogeneous networking for quality, reliability, security and robustness. Springer, Berlin, pp 617–632

15. Feldhofer M, Dominikus S, Wolkerstorfer J (2004) Strong authentication for RFID systems using the AES algorithm. In: International workshop on cryptographic hardware and embedded systems. Springer, Berlin, pp 357–370

16. Gong Z, Hartel P, Nikova S, Tang SH, Zhu B (2014) TuLP: a family of lightweight message authentication codes for body sensor networks. J Comput Sci Technol 29(1):53–68

17. Guo J, Peyrin T, Poschmann A (2011) The photon family of lightweight hash functions. In: Annual cryptology conference. Springer, Berlin, pp 222–239

18. Hell M, Johansson T, Meier W (2007) Grain: a stream cipher for constrained environments. Int J Wirel Mobile Comput 2(1):86–93

19. Hirose S, Ideguchi K, Kuwakado H, Owada T, Preneel B, Yoshida H (2010) A lightweight 256-bit hash function for hardware and low-end devices: Lesamnta-lW. In: International conference on information security and cryptology. Springer, Berlin, pp 151–168

20. Juels A, Weis SA (2005) Authenticating pervasive devices with human protocols. In: Annual international cryptology conference. Springer, Berlin, pp 293–308

21. Leander G, Paar C, Poschmann A, Schramm K (2007) New lightweight des variants. In: International workshop on fast software encryption. Springer, Berlin, pp 196–210

22. Li S, Xu LD, Zhao S (2015) The internet of things: a survey. Inform Syst Front 17(2):243–259

23. Luykx A, Preneel B, Tischhauser E, Yasuda K (2016) A mac mode for lightweight block ciphers. In: International conference on fast software encryption. Springer, Berlin, pp 43–59

24. Manifavas C, Hatzivasilis G, Fysarakis K, Rantos K (2013) Lightweight cryptography for embedded systems–a comparative analysis. In: Data privacy management and autonomous spontaneous security. Springer, Berlin, pp 333–349

25. McKay K, Bassham L, Sönmez Turan M, Mouha N (2016) Report on lightweight cryptography. Tech. rep., National Institute of Standards and Technology

26. Moon D, Hwang K, Lee W, Lee S, Lim J (2002) Impossible differential cryptanalysis of reduced round XTEA and TEA. In: International workshop on fast software encryption. Springer, Berlin, pp 49–60

27. Mouha N, Mennink B, Herrewege AV, Watanabe D, Preneel B, Verbauwhede I (2014) Chaskey: an efficient mac algorithm for 32-bit microcontrollers. In: International conference on selected areas in cryptography. Springer, Berlin, pp 306–323

28. Park J, Anandakumar NN, Saha D, Mehta D, Pundir N, Rahman F, Farahmandi F, Tehranipoor MM (2022) PQC-SEP: power side-channel evaluation platform for post-quantum cryptography algorithms. IACR Cryptol ePrint Arch 2022:527

29. Philip MA, et al (2017) A survey on lightweight ciphers for IoT devices. In: 2017 International conference on technological advancements in power and energy (TAP Energy). IEEE, Piscataway, pp 1–4
30. Poschmann A (2007) Lightweight cryptography from an engineers perspective. In: Workshop on elliptic curve cryptography (ECC 2007)
31. Pundir N, Aftabjahani S, Cammarota R, Tehranipoor M, Farahmandi F (2022) Analyzing security vulnerabilities induced by high-level synthesis. ACM J Emerg Technol Comput Syst 18(3):1–22
32. Rivest RL (1994) The rc5 encryption algorithm. In: International workshop on fast software encryption. Springer, Berlin, pp 86–96
33. Saarinen MJO, Engels D (2012) A do-it-all-cipher for RFID: design requirements (extended abstract). Cryptology ePrint Archive, Report 2012/317. https://ia.cr/2012/317
34. Semiconductor N (2021) Cop912c datasheet pdf—8-bit microcontrollers. http://www. datasheetcatalog.com/datasheets_pdf/C/O/P/9/COP912C.shtml. Accessed 26 Dec 2021
35. Srinivasan C, Pillai UU, Lakshmy K, Sethumadhavan M (2015) Cube attack on stream ciphers using a modified linearity test. J Discrete Math Sci Cryptogr 18(3):301–311
36. Wheeler DJ, Needham RM (1994) Tea, a tiny encryption algorithm. In: International workshop on fast software encryption. Springer, Berlin, pp 363–366
37. Zhang W, Bao Z, Lin D, Rijmen V, Yang B, Verbauwhede I (2015) Rectangle: a bit-slice lightweight block cipher suitable for multiple platforms. Sci China Inform Sci 58(12):1–15

Chapter 13
Virtual Proof of Reality

13.1 Introduction

Virtual Proof of Reality, more commonly known as VP, pushes the boundaries further in the field of cryptography by providing a safer, more secure way of communication between parties. Virtual security has become increasingly important. It is the driving force for technological advancements in this digital age. In this chapter, the concept of VP is reviewed to provide an informative insight into this novel approach for ensuring security. The VP process involves sharing physical statements of hardware attributes through digital communication lines and implementing well-developed key-less cryptographic methods. This new security framework gives way to improved hardware security defenses by using physical attributes to confirm system verification and accuracy.

In order to present the concept of VP protocols and the challenges it faces in modern technology, the following topics have been discussed in this chapter:

- Physical hardware statements shared between a "prover" and a "verifier"
- A key-less cryptographic security primitive known as Simulation Possible but Laborious (SIMPL) system
- Artificial Intelligence (AI) integration in cyber-physical systems for the protection of privacy
- Symmetric cryptography protocols influenced by virtual proof
- Secure wireless sensing through applications of physical unclonable function (PUF) [15]

These initiatives come as impressive steps toward hardware security, such as avoiding vulnerable secret key sharing protocols through security sensors and converting physical system attributes into digital data for verification. However, these improvements have a few application challenges, such as precise sensor accuracy, system stability, and extreme complexity.

© The Author(s), under exclusive license to Springer Nature Switzerland AG 2023

M. Tehranipoor et al., *Hardware Security Primitives*,

https://doi.org/10.1007/978-3-031-19185-5_13

13.2   Background

Virtual proofs of reality (VPs) are entirely new forms of security solutions. VPs go above and beyond traditional security sensors in a multitude of ways. For example, VPs do not require secret keys stored in trustworthy and tamper-resistant sensors. As a consequence, VPs differ significantly from conventional sensors. They also increase the range of physical claims that may be verified compared to conventional sensors, such as an item's irreversible alteration or annihilation. The underlying security architecture makes no assumptions about secret keys in the prover's system or conventional tamper-resistant sensor hardware that the verifier may trust. Instead, virtual proofs are performed between the "prover" and the "verifier."

VPs may have a private or a public setup phase. VPs with a private setup phase allow the verifier to prepare several physical objects before initiating the actual proof. Consequently, it allows the verifier to measure and store some characteristics of those objects without the prover knowing what was stored. The objects are then transferred to the prover's system for later use. However, in virtual public proofs, the prover may still use several objects in the proof, but there is no secure setup phase or object transfer before the proof begins.

Blockchain   In a blockchain, multiple transactions/entries are combined together into data blocks, which, when chained together, build the entire blockchain [18]. The new blocks are added to the blockchain after being validated by the network using some cryptographic proof. Each block carries its timestamp, the previous block's hash, and a random nonce to verify the hash. Because of this architecture, the integrity of the entire blockchain is ensured [16]. Even a one-bit manipulation in the data can lead to a completely different hash. And since blockchain follows a consensus mechanism, meaning the majority of nodes in the blockchain should agree to add the block to the blockchain; therefore, it is almost impossible for a single adversarial node to cause harm to the blockchain.

Smart Contract   A contract is a binding agreement between two parties to execute/fulfill certain things when certain conditions are met. However, if two parties are human, then one may try to back out of the contract due to malicious intent. In contrast, a smart contract is a programming piece of code on the blockchain, which is bound to execute when the contract conditions are fulfilled [19]. And since smart contract lies on the blockchain, no parties can back out once the contract is signed.

13.3   Illustrative Overview of Virtual Proof of Reality

13.3.1   Physical Implementation of Virtual Proof of Reality

The methodology used in the research on virtual proofs of reality focuses on proving a physical statement over a digital communication channel between two parties [14].

The physical statements could be any physical characteristics of the system, such as temperature, sensor data, etc. The primary goal was to address how one party (the "prover") could prove a physical statement to another party (the "verifier") over a digital channel. The prover should convince the verifier with a high conviction that the claim is valid, while the verifier should notice if the claims are false with a high conviction. It was ultimately determined that there needed to be a way to bypass the need for using secret keys and hardware security primitives. It led to an idea to use the physical domains to process variations and extract electrical/optical structures (similar to physical unclonable functions (PUFs)) and generate challenge-and-response pairs (CRPs) [1, 8, 9, 12, 15, 17]. The primary application for this research would be for virtual proofs of physical statements for closed systems. This includes bank vaults, ATMs, and other forms of mutual authentication via security tokens.

An outline of one of the implementations is as follows. There is an initial setup phase where the verifier creates a strong PUF. Then, the verifier generates and stores a private CRP list, and a PUF is transferred to the prover. During the virtual proof stage, the prover claims that a PUF is a witness object (WO) at a particular value. The verifier will select a random challenge, apply it, and measure the response. From there, the verifier will receive the values and either accept or reject the statement. Figure 13.1 shows an architectural diagram of the process.

For the procedure to be valid, the system must use a strong PUF. Ideally, the PUF should have complex input–output behavior, many possible challenges, a publicly accessible interface, and appropriate intra-Hamming or inter-hamming distances. In addition, to measure physical quantities and verify them, it is necessary to make the PUF response a function of the challenge and the witness object, such as the temperature. At the same time, it should be resistant to other variations, such as the voltage and ambient noise variation [9–11].

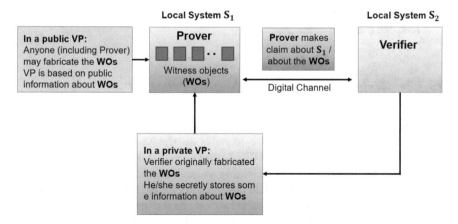

Fig. 13.1 Block diagram illustrating the public and private virtual proof of reality architecture

There are several underlying assumptions in the VP architecture proposed in [14]. It is assumed that the system's keys or classical sensor hardware are tamper-resistant for the underlying model. Essentially, the system itself should not be trusted by the verifier since secret keys can be attacked. The prover and the verifier are anticipated to be in separate physical locations. The only communication between the two will be through one channel, known as the digital channel. Aside from that channel, the two systems are closed off to any type of interaction.

The advantages of virtual proofs are that they can be applied in a variety of ways. Virtual proofs of location or distance can be effectively used for location-dependent encryption or decryption. Virtual proofs of destruction prove that a particular object in the system of the prover was modified irreversibly or, more precisely, "destroyed." An advantage of virtual proofs of destruction is that they can be utilized to guarantee that any digital object (i.e., a non-fungible token for a piece of crypto-art) was provably destroyed, or a media owner can guarantee that a licensee is no longer capable of utilizing a copyrighted piece of digital media once their usage rights have ended. While this technology offers plenty of advantages, there are a few disadvantages. One disadvantage is that significant optimizations would need to be made for these specific virtual proof styles (i.e., virtual proofs of location or distance and virtual proofs of destruction) to become feasible in the mainstream.

The results of this work displayed the usefulness and feasibility of three different types of virtual proofs. Virtual proofs of temperature can prove a temperature of a witness object in a system to a verifier. Moreover, virtual proofs of location can prove the position of multiple physical objects. Lastly, virtual proofs of destruction can prove that a witness object in a system was irreversibly changed or destroyed. Shortcomings associated with virtual proofs of reality that still need to be addressed are as follows:

- The design is still in its infancy and needs many optimizations for mainstream use.
- The design does not necessarily have a secure setup phase before a proof begins, which can cause potential problems.

13.3.2 Keyless Cryptographic Security Primitive

A novel cryptographic primitive called Simulation Possible, or SIMPL, has been introduced in [13]. SIMPL acts as a lucrative alternative to the commonly used Physical Unclonable Function (PUF). Like PUFs, SIMPL systems are highly disordered and unclonable with a highly complicated input–output behavioral relationship. Having so many possible inputs makes the SIMPL system very hard to model or predict the relation between the inputs and the outputs. A comprehensive analysis of the SIMPL system has shown its advantages compared to PUFs and potential as a cryptographic and security primitive.

The central focus of current research is the feasibility of SIMPL systems in practice. These systems may outperform PUFs in certain applications and provide further options in particular scenarios where PUFs were insufficient previously. Since this is a relatively new topic, the analysis provided in this chapter can help advance and encourage further research in the area by increasing awareness of it.

SIMPL systems work by producing publicly known and computable functions faster than other systems, which is how SIMPL systems remain secure without keeping as many secrets as PUFs. To compute functions and address challenges faster than other systems (and therefore for the systems to be considered SIMPL), the following circumstances must be true. First, the system must be a partly disordered physical system, meaning that a particular challenge should be applied as an input to the system, producing a stable response. The system also needs to derive an individual numeric description and an algorithm for the system, with which anyone would be able to produce the correct responses to any given challenge. Additionally, any numeric or physical attempt to predict the system's response to a challenge must be slower than the system's response time. Finally, it must not be easy to physically copy or clone the system, including for someone with knowledge of the system's numeric description, algorithm, and internal characteristics.

The architecture of an SRAM-based SIMPL system is presented in Fig. 13.2. The system has four main components: a skew SRAM (SS) block, a voltage control (VC) block, a feedback and output control (FOC) block, and a challenge control (CC) block. SIMPL systems generate responses R_i when they have challenges C_i as their inputs. The skew SRAM is "scrambled" by the challenge control block CC, generating voltage select (SEL), write data (D_{IN}), read or write address (ADR), and other control (CTRL) signals, such as the write-enable signal. The challenge control block can be implemented using a hash function. The skew SRAM block SS is designed similarly to normal SRAM blocks. The main distinction is that different types of cells (normal cells and skew cells) are spread throughout the array, and depending on the current supply voltage, write operations conducted in those specially engineered skew cells may fail to alter their previously stored contents.

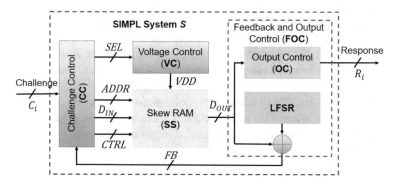

Fig. 13.2 Block diagram illustrating the SRAM-based SIMPL system's architecture

Skew cells are distinguished from normal cells (i.e., typical SRAM cells) by the size of their transistors and, as a result, their electrical properties, particularly their write behavior [2].

Since emulators or simulators cannot beat skew SRAM behavior, they can tell the difference between authentic and counterfeit SIMPL systems. The performance advantage of the SIMPL system over emulators or simulators can be enhanced if the data read from the skew SRAM in a cycle is directly fed into the challenge control. This can, of course, be done several times. The feedback loop can be implemented using a linear feedback shift register (LFSR) that XORs the skew SRAM blocks' read-out data D_{OUT} [2].

The public key (i.e., the simulator component) implements a function that calculates a response R_i when supplied the description, D(S), of a certain SIMPL system, S, and the challenge C_i in public-key applications (the verifier should give the challenge C_i). The simulator should also indicate SIMPL system S's reaction time limit t_{max} so that a verifier may validate both the speed and accuracy standards. To construct a public simulation program for a particular SIMPL system, one must first understand the specific logic features of each block of the system, including the configurations of the skew SRAM block, which include normal, skew, and fixed cell allocations. The simulation program has to implement the SIMPL system's logic in software and depending on the requirements, estimate the legal SIMPL system's response time limit t_{max} [2].

Many features of the SIMPL system are unique compared to a PUF for specific applications. A SIMPL system comes with a publicly known numeric description that permits its simulation and prediction of outputs. However, for a PUF, such a description is not publicly available. The output of a SIMPL system can be found slowly. However, only the actual holder of the SIMPL can determine the output quickly via physical measurement. Secondly, secret information should not be stored in a SIMPL system to ensure cryptographic security. This feature of the SIMPL system makes it unique compared to a PUF and makes it much more attractive compared to PUFs. In the case of PUFs, secret information is hidden inside random and analog features of the device. As a result, security provided by the SIMPL system depends on two essential things: physical unclonability and the computational complexity of simulating the output of a SIMPL system.

Conversely, a drawback of using a SIMPL system is the lack of secret information and that one can slowly generate the correct output for a prompt. Another disadvantage of a SIMPL system is its implementation. PUFs have already been the standard in many situations, and in many cases, Public PUFs (PPUFs) are functionally similar to SIMPL systems. Relatively few known applications for the novel SIMPL system may not be enough to convince people to adopt them in practical applications as an alternative to PUFs. In summary, research challenges that still need to be addressed are as follows:

- SIMPL systems are still relatively new, designs may not be as fleshed out, and they could have unforeseen flaws that have not yet been realized due to lack of testing.

- Implementing SIMPL systems may have a higher cost than what they are worth, considering they overlap with PPUFs in terms of functionality. Though some applications are unique to SIMPL systems, they are still being explored and discovered.

13.3.3 Smart Contract Privacy Protection: AI in Cyber Systems

Blockchain (BC) and Cyber-Physical Systems (CPSs) applications are proliferating. However, due to their complexities, framing robust and appropriate smart contracts (SCs) for these intelligent systems is a difficult task. SCs are modernizing traditional industrial, technological, and business processes. Moreover, they are self-executing, self-verifiable, and integrated within the BC, removing the need for trustworthy third-party systems and, as a result, saving both administrative and service expenditures. SCs also increase a system's efficiency, lower security threats, and encourage new technological reforms in contemporary industry. However, SCs still have several security and privacy issues to be resolved.

Cyber-Physical Systems (CPSs) incorporate networking, computing, and physical processes with advanced sensors that efficiently handle essential components (e.g., cybersystems and physical processes). However, as the connection between these two aspects grows, physical systems become increasingly vulnerable to security breaches. To address these issues, blockchain (BC) is a game-changing technology that secures and executes transactions in an open network environment without relying on a centralized third-party system (TTS). TTS is a distributed ledger in which all transactions are recorded in a chain of blocks. Before being stored in the block in the chain, a transaction is authenticated by all of the BC's participants.

A solidity, Go, Kotlin, or Python-based SC is a software code. The software is both enforceable and immutable and is in charge of developing, compiling, and deploying digital asset logic for automated execution. Once written, SCs are difficult to alter, and through their involvement, the trust, credibility, and complexities of transactions in a decentralized system may be appropriately handled and maintained. SCs also facilitate real-world concerns of low-cost and high-accuracy design (such as banking systems). However, one challenge of an SC is that it cannot fix any flaws detected after it has been deployed.

Smart contracts are still in their infancy regarding advancement and development. Most organizations will soon adopt blockchain technology and be regulated by smart contracts. Because of the rising use of SCs, more efficient and comprehensive security processes are needed. Developing a secure and bug-free SC is a challenging task for developers. Re-entrancy, transaction-ordering reliance, forcefully delivering ether to a contract, DoS using revert, and integer overflow and underflow are all possible security issues in SCs. When constructing SCs, SC developers must be conscious of such weaknesses. These vulnerabilities can result in million-dollar losses for companies.

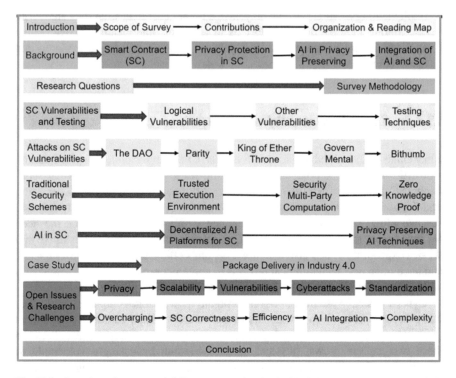

Fig. 13.3 Overview of survey and different approaches in the blockchain and smart contract field as presented in [7]

In such cases, AI algorithms may be used to evaluate the SC as well as the behavior of the participants who have consented to the transaction's execution. It would help in monitoring and identifying dangerous SC transaction execution behavior. In addition, AI coupled with natural language processing (NLP) might be used to analyze SC patterns in greater depth. The power of NLP may be put to use in semantic parsing and entity recognition. Both AI and NLP technologies help analyze and develop a sophisticated and compelling SC. As more data becomes available, the more precise the analysis can be. Consequently, merging AI with the BC and SC can be an excellent way to ensure privacy and security. Figure 13.3 provides the overview of security research in the domain of BC and SC as presented in [7].

The results of the state of the research are the findings of different vulnerabilities, testing options, and AI techniques for mitigation or optimization [3, 4]. A comprehensive view of the vulnerabilities and testing options found by current research efforts [7] can be seen in Fig. 13.4. A host of AI techniques and platforms for smart contracts have been found outside of this. Eight decentralized AI platforms for intelligent contracts have been explicitly discussed, broken down, and analyzed in how they can be used to revamp smart contracts' efficiency, flexibility, and intelligence in particular applications.

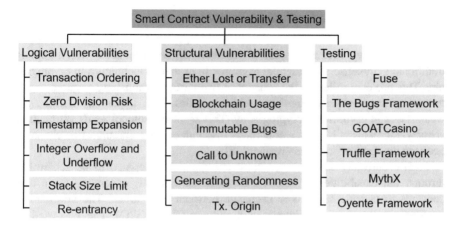

Fig. 13.4 Overview of different smart contract vulnerabilities and testing approaches

There are a couple of benefits that intelligent contracts bring to the table. One significant benefit is that in conventional blockchain systems, data privacy can be affected based on the nature of its decentralized distributed design; however, involving intelligent contracts can resolve this issue as they auto-execute themselves across a decentralized blockchain network. Smart contracts also improve existing blockchain technology, reducing the transaction cost, execution time, and increasing security strength in blockchain systems. Finally, there are many applicable fields for this technique to grow into, such as smart homes, asset management, and e-commerce.

One major issue with smart contracts is that fixing bugs discovered after deployment is entirely impossible. Like any software system, it is exceptionally challenging to create a secure and bug-free smart contract. Smart contracts have a few possible security vulnerabilities. The most notable vulnerabilities are re-entrancy, transaction-ordering dependence, and integer overflow and underflow. Another significant disadvantage to smart contracts is that the diversity in clever contract designs is relatively low, making it more difficult to catch a malicious user.

The main takeaway is that a handful of existing techniques using AI in smart contracts can solve existing issues and overhead costs incurred in intelligent contract code by traditional security schemes. It is claimed that AI techniques and integrating them into smart contracts will result in a much more widespread implementation of smart contracts across a large variety of intelligent applications in the future. Finally, there are some open challenges in this domain to be overcome by future research efforts. Some of them are listed as follows:

- Smart contracts are entirely unable to be adjusted to fix any bugs discovered after deployment.
- May contain other possible software security vulnerabilities like integer overflow, integer underflow, and re-entrancy.

13.3.4   Secure Key Exchange Protocol

Now, a novel key exchange security protocol will be introduced, which applies traditional symmetric and essential asymmetric exchange techniques in cryptography. This protocol relies on the concept of two parties establishing a shared secret developed from physical hardware characteristics, thus bypassing well-known weaknesses in asymmetric cryptography. This security protocol sends verification acknowledgments of hardware conditions over vulnerable communication channels between the "prover" and the "verifier" [5].

Such secure key sharing over untrusted channels is argued to be achievable through the capabilities of sensor PUF methods and VP, expanding these two concepts to provide secure key exchange between two parties. The information between the two parties is known as a Witness Object (WO). The WO used in this protocol is a strong PUF since it carries the necessary features for secure communication. This experimentation process begins with the verifier preparing the WO, sending it through the vulnerable communication channel, and confirming whether the response from the prover is accurate. See Fig. 13.5 for more detail on the communication process.

Key characteristics of the WO, or Strong PUF in this communication link, are dependent upon two hardware features: temperature and position. With these dependencies, this protocol is built upon two phases known as the "enrollment phase" and the "key exchange phase." The enrollment phase is responsible for setting up the WO for delivery to the prover and creating a set of challenge–response pairs based on the challenges created by the temperature and position features.

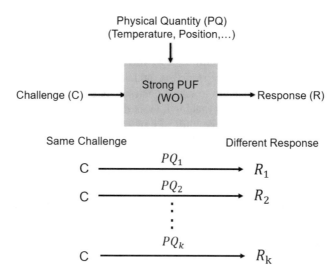

Fig. 13.5 Block diagram illustrating the verification process using CRPs from a strong PUF

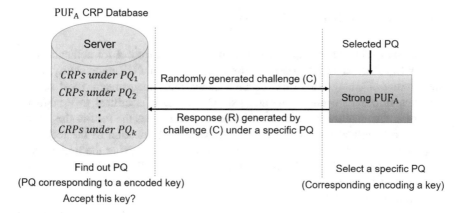

Fig. 13.6 Block diagram illustrating the key exchange protocol

	PQ	BER	SC-DPQ-FHD	WO
Table 13.1 Average BER and SC-DPQ-FHD performance under different PQs [5]	Temperature	1.0%	6.2%	4 XOR BR-PUF
	Position	8.7%	36.3%	Optical PUF

The key exchange process allows the verifier to select a challenge to deliver to the prover and the prover then responds by applying the challenge to the PUF. This response carries the necessary information on the temperature and position hardware features and is the key between the prover and the verifier. Then, the verifier accepts the key once the response is compared and matched with the verifier's stored responses (Fig. 13.6).

This protocol depends on two pre-conditions to prove that this key exchange method is both reliable and secure. This first condition is the security of the strong PUF being sent over the communication line. From this process, the challenge–response repairs generated by this PUF are too abundant to be taken advantage of by a perpetrator and have the response and challenge exploited. Second, the hamming distance of each response pair (termed "SC-DPQ-FHD") is utilized to supersede the Bit Error Rate (BER). The BER ensures that the verifier can view the temperature and position features based on the key sent from the prover. Table 13.1 shows the general results from the experimentation presented in [5].

This specific experiment supports the security of sending hardware physical statements over an untrusted communication line. The verifier can verify secret hardware characteristics via the PUF-based WO. This model can be further improved by enhancing the performance of the XOR BR-PUFs concerning temperature confirmation. This objective can be accomplished by increasing the WO's receptibility to temperature. Few limitations of the technique proposed in [5] that still need to be addressed are as follows:

- The performance of XOR BR PUFs needs to be enhanced with the temperature confirmation.
- Sensitivity to WO needs to be increased.

13.3.5   Secure Wireless Sensing

The primary focus of this approach was to develop a protocol for authentication, similar to a Virtual Proof, based on the exploitation of PUF unreliability. Since the Internet of Things (IoT) has become increasingly mainstream, most sensors and data information are interconnected. Therefore, it is critical to establish a form of authentication that eliminates or at least minimizes the man-in-the-middle attack across insecure channels. This effort was a continuation of the technique proposed in [14], which expanded on these results by removing the limitation of only using strong PUFs. This endeavor aimed to accomplish cryptographic tasks, even in resource-constrained environments, while still maintaining a proper level of protection against threats.

Sensitivity to parameter variation is undesirable for most PUF applications. In contrast, process variation is a desirable trait that can be harnessed to measure specific operating conditions. It was assumed that secret keys stored in non-volatile memory (NVM) could be attacked through various invasive and non-invasive attacks. Therefore, the devices were expected not to use such keys since digital keys may not be safe. Any signal converted into an electrical bitstream, such as a transducer, would be an acceptable technique to generate a PUF challenge and response pair (CRP) [6]. This implies that any physical quantity (PQ) or environmental parameter can be generalized. There were in total four primary requirements for the PUF analyzed in [6]:

- The PUF should be a function of the challenge and current sensor value.
- Identical PUF signatures of two different sensors cannot be made.
- The response to a challenge should be stable for a specific PQ value.
- Given a response to a PQ value, it should not be possible to predict the challenge or any other outcome.

Another assumption was that the inter-PQ distance and intra-PQ distance follow a binomial distribution in the measurement process. The primary approach used experimental data from a ring oscillator PUF (RO-PUF). An example setup of an RO-PUF can be seen in Fig. 13.7.

The central arguments surrounding the findings established a protocol that would detect man-in-the-middle tampering. By converting the parameters into electrical signals, it would be more scalable to other features. The main reasons for using PUFs are that PUFs would eliminate the need for NVM storage for keys, the PUF will obfuscate the voltage value read in the PQ, easy detection of tampered responses, and the prevention of replaying attacks (challenges only used once or through a reverse fuzzy extractor). Pre-selecting a more significant number of unreliable response bits may ultimately be sensitive to environmental parameter changes.

An authenticated sensing protocol has been proposed in [6]. First, a trusted entity (the user) measures and securely stores various PQ responses under given challenges in the enrollment phase. When data collection is requested in the authentication

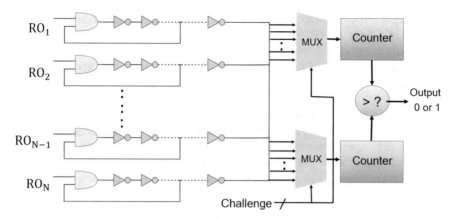

Fig. 13.7 Circuit illustrating the RO-PUF configuration

phase, the user will select a random challenge and send it to the PUF, where a response will be generated. The final step is to compare the response with the stored responses and ultimately reject or accept the CRP. In selecting the RO-PUF pairs, it is critical to have frequencies that differ by less than a certain threshold to guarantee responsive bits.

The main benefit of using this technique is that it can be applied to resource-constrained devices. Instead of being limited to using strong PUFs, weak PUFs and techniques other than the RO-PUFs can be used in the case of high resource constraints. Integrating PUF sensors with the PQ mitigates attacks on digital keys. Since electrical signals are used, it implies that almost any measurable response can be authenticated. If an adversary were to delayer the device layer of the chip or attempt an invasive attack, the PUF behavior would be destroyed. In addition, PUF sensors do not require any calibration since the enrollment phase measurements take this into account based on environmental changes.

However, there are several limitations in this approach. The use of ring oscillators and measurement of output frequency in the experiment consumes a great amount of power and area. However, this can be countered with the use of reverse fuzzy extractors. Reverse fuzzy extractors use a secure sketch to eliminate noise from the collected data and a randomness extractor to guarantee uniform distributions (random distributions) of keys, meaning that other applications (e.g., SRAM PUFs and Arbiter PUFs) can be used. Another drawback of this approach is that feeding the sensed signal to the power of the RO-PUF can limit the dynamic range of the measurement. Localized electromagnetic attacks are also possible to implement against RO-PUFs.

Experimental results have suggested that intra-PQ and inter-PQ distance vary from less than 10% to more than 40%. The best responses were generated when the frequency variation was less than 0.3 MHz between ring oscillators. This consequently had the lowest false acceptance rate (FAR), defined as mistakenly accepting a response under a false PQ, and the lowest false rejection rate (FRR), defined as falsely rejecting an accurate response. The FAR indicates the risk of incorrectly accepting false values, which could raise security concerns. FRR values indicate the robustness of the authentication. The total number of bits required for practicality was reduced by increasing responsive bits, making it highly practical for lightweight applications.

The experimental results presented in [6] show how increasing the number of unreliable response bits or sensitivity of a PUF increases the effectiveness of proving physical statements. Increasing sensitivity reduces the number of bits required for a PUF, making it useful for resource-constrained applications. However, there are some shortcomings of the approach proposed in [6] presented below which need further improvement:

- Limited CRP space in specific applications requires the use of weak PUFs.
- Some PUF Virtual Proof implementations, like RO-PUFs, require more power and area, which is resource-consuming.
- Feeding sensed signals directly to the power rail can result in dynamic range measurement limitations.
- The methodology employed is still prone to localized electromagnetic attacks.

13.4   Conclusions

Overall, Virtual Proof of reality is a viable strategy to remove the necessary keys in cryptographic exchanges. Exploiting PUF properties and making the CRP a function of the physical quantity in question make it possible to avoid such threats. Although the system is still being perfect, it is evident that more work needs to be done to increase the sensitivity to parameter changes. It is also imperative that this protocol is extended to lightweight applications.

References

1. Amsaad F, Pundir N, Niamat M (2018) A dynamic area-efficient technique to enhance ROPUFs security against modeling attacks. In: Computer and network security essentials. Springer, Berlin, pp 407–425
2. Chen Q, Rührmair U, Narayana S, Sharif U, Schlichtmann U (2015) MWA skew SRAM based SIMPL systems for public-key physical cryptography. In: Conti M, Schunter M, Askoxylakis I (eds) Trust and trustworthy computing. Springer International Publishing, Cham, pp 268–282
3. Farahmandi F, Huang Y, Mishra P (2020) Automated test generation for detection of malicious functionality. In: System-on-chip security. Springer, Cham, pp 153–171
4. Farahmandi F, Huang Y, Mishra P (2020) Trojan detection using machine learning. In: System-on-chip security. Springer, Cham, pp 173–188
5. Gao Y (2015) Secure key exchange protocol based on virtual proof of reality. Cryptology ePrint Archive

6. Gao Y, Ma H, Abbott D, Al-Sarawi SF (2017) PUF sensor: exploiting PUF unreliability for secure wireless sensing. IEEE Trans Circuits Syst Regul Pap 64(9):2532–2543

7. Gupta R, Tanwar S, Al-Turjman F, Italiya P, Nauman A, Kim SW (2020) Smart contract privacy protection using AI in cyber-physical systems: tools, techniques and challenges. IEEE Access 8:24,746–24,772. https://doi.org/10.1109/ACCESS.2020.2970576

8. Pundir N, Amsaad F, Choudhury M, Niamat M (2017) Novel technique to improve strength of weak arbiter PUF. In: 2017 IEEE 60th international midwest symposium on circuits and systems (MWSCAS), IEEE, Piscataway, pp 1532–1535

9. Rahman MT, Forte D, Fahrny J, Tehranipoor M (2014) ARO-PUF: an aging-resistant ring oscillator PUF design. In: 2014 design, automation & test in Europe conference & exhibition (DATE), IEEE, Piscataway, pp 1–6

10. Rahman MT, Xiao K, Forte D, Zhang X, Shi J, Tehranipoor M (2014) TI-TRNG: technology independent true random number generator. In: 2014 51st ACM/EDAC/IEEE design automation conference (DAC). IEEE, Piscataway, pp 1–6

11. Rahman MT, Forte D, Rahman F, Tehranipoor M (2015) A pair selection algorithm for robust RO-PUF against environmental variations and aging. In: 2015 33rd IEEE international conference on computer design (ICCD). IEEE, Piscataway, pp 415–418

12. Rahman MT, Rahman F, Forte D, Tehranipoor M (2015) An aging-resistant RO-PUF for reliable key generation. IEEE Trans Emerg Topics Comput 4(3):335–348

13. Rührmair U (2012) SIMPL systems as a keyless cryptographic and security primitive. Springer, Berlin, pp 329–354. https://doi.org/10.1007/978-3-642-28368-0_22

14. Rührmair U, Martinez-Hurtado J, Xu X, Kraeh C, Hilgers C, Kononchuk D, Finley JJ, Burleson WP (2015) Virtual proofs of reality and their physical implementation. In: 2015 IEEE symposium on security and privacy, pp 70–85. https://doi.org/10.1109/SP.2015.12

15. Suh GE, Devadas S (2007) Physical unclonable functions for device authentication and secret key generation. In: 2007 44th ACM/IEEE design automation conference. IEEE, Piscataway, pp 9–14

16. Vashistha N, Hossain MM, Shahriar MR, Farahmandi F, Rahman F, Tehranipoor M (2021) eChain: a blockchain-enabled ecosystem for electronic device authenticity verification. IEEE Trans Consum Electron 68(1):23–37

17. Xiao K, Rahman MT, Forte D, Huang Y, Su M, Tehranipoor M (2014) Bit selection algorithm suitable for high-volume production of SRAM-PUF. In: 2014 IEEE international symposium on hardware-oriented security and trust (HOST). IEEE, Piscataway, pp 101–106

18. Zheng Z, Xie S, Dai HN, Chen X, Wang H (2018) Blockchain challenges and opportunities: a survey. Int J Web Grid Serv 14(4):352–375

19. Zou W, Lo D, Kochhar PS, Le XBD, Xia X, Feng Y, Chen Z, Xu B (2019) Smart contract development: challenges and opportunities. IEEE Trans Softw Eng 47(10):2084–2106

Chapter 14
Analog Security

14.1 Introduction

Hardware Trojan insertion, cloning, forgery, overproduction, recycling, and reverse engineering (RE) problems are increasing yearly in the semiconductor supply chain [12, 15, 23, 39, 40]. According to statistics, the annual loss caused by the leakage of intellectual property (IP) in the semiconductor domain is about \$4 billion [35]. Security researchers and engineers are trying to solve these security vulnerabilities through different ways, such as hardware metering [20, 21], remote IC activation [3], physically unclonable function(PUF) [33], reconfigurable logic barrier [6], security split test [9], etc. Currently, most of the research in the field of hardware security is aimed at digital circuits. However, the analog circuit is an unavoidable part of the modern computing system. Amplifiers, filters, analog-to-digital converters (ADCs), voltage regulators, digital-to-analog converters (DACs), and RF integrated circuits (ICs) are nearly universally utilized in integrated circuits used in various sectors such as medical care, energy, space, defense, and Internet of Things.

14.2 Background

When it comes to hardware security, analog and mixed-signal circuits are typically neglected by researchers and designers. There can be various reasons; first, AMS is a custom transistor-level design that involves multiple design iterations and reviews to meet particular specifications. Second, analog circuits, compared to digital ICs, cover a small fraction of the semiconductor industry and IC supply chain. However, these downsides of analog IC do not stop an adversary from exploiting vulnerabilities to inject any malicious circuit or perform counterfeiting IC. For an adversary, the widespread use of analog ICs (e.g., physical interfaces,

© The Author(s), under exclusive license to Springer Nature Switzerland AG 2023
M. Tehranipoor et al., *Hardware Security Primitives*,
https://doi.org/10.1007/978-3-031-19185-5_14

sensors, actuators, wireless communications) can be a lucrative opportunity to clone, recycle, or overproduce these ICs.

Although analog and digital designs face similar threats, they are attacked differently. The proposed logic encryption and camouflage for digital systems can protect them, but they cannot be directly extended to analog systems. This chapter will introduce some security methods to deal with the threats faced by analog design.

14.3   Stochastic All-Digital Weak PUF for AMS Circuits

Like digital, analog-mixed signal (AMS) ICs also face counterfeiting threats. It can harm the OEM's interests and raise security concerns for end users through industry and consumers by pushing counterfeit IC into the supply chain [15, 40].

Various security primitives such as physically unclonable function (PUF) circuits have been intensely investigated for counterfeiting detection and avoidance [13, 14, 30]. PUF is implemented at the silicon level and takes the specific challenge as the input to generate a stable specific response. Random process changes and equipment mismatches ensure the uniqueness of PUF response. Therefore, it is almost impossible to reproduce PUF entirely and accurately. In addition, PUF must be stable and reliable in various environments; its response should not be changed due to environmental variations. Among them, environmental factors include noise, voltage, and temperature. PUF itself needs to be measured from two aspects: uniqueness and reliability. Uniqueness is measured by inter-chip Hamming distance (HD), while reliability is measured by intra-chip HD.

Generally speaking, PUF has the difference between strong PUF and weak PUF, which is divided according to the number of challenge–response pairs (CRPs) contained in PUF. The number of CRPs in strong PUF grows exponentially as the PUF scale grows, but the number of CRPs in weak PUF grows linearly. Sometimes there is only one CRP in a PUF. It is called physical obfuscation key (POK) [22].

This section will introduce the stochastic all-digital weak Physically Unclonable Function [8] for AMS circuits. This PUF uses a dynamic latch comparator to generate stable and reliable chip-specific identifiers. The schematic representation of the weak PUF is shown in Fig. 14.1 [7]. We can see that it uses n comparator chains to create n-bit unique IDs. It can be implemented using two different approaches, such as double-tailed and three-latched extended range comparators [8]. The double-tail dynamic comparator can work stably under low power supply voltage [5]. At the same time, this design can make the PUF circuit to produce stable output when the power supply voltage changes.

The $\sum$ symbol in Fig. 14.1 represents a digital averager, which can reduce the influence of noise. Its working principle is to average the output of each comparator in a given cycle and then determine the corresponding unique ID bit according to most of the outputs. Due to the existence of a digital averager, the users can shield those unreliable bits to ensure the reliability and stability of PUF output.

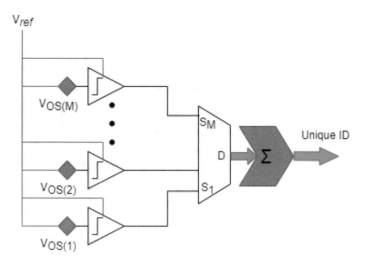

Fig. 14.1 Block diagram of analog PUF circuit

Experiments show that the two different PUFs have stable bit outputs of 81.9% and 69.9%, respectively, in a given voltage range (0.8 V–1.4 V). Furthermore, when only the temperature is concerned, the given temperature range is 0°C-80 °C, and the two PUFs have stable bit outputs of 91.2% and 83.0%, respectively.

As mentioned earlier, the reliability of each PUF is measured by intra-HD. The uniqueness of PUF is measured by inter-HD. The experimental results show that the normalized intra-HD of the two PUFs under the voltage of 0.8 v–1.4 v is about 0.9%, which is lower than 0.89% and 0.96%, respectively. When the temperature factor is considered, the intra-HD of both PUFs is less than 0.15% at temperatures up to 80 °C. The inter-HD achieved by both PUFs is close to 50%, and 50% is the ideal value of inter-HD.

14.4 Chaogate

Chaogate is a dynamic circuit that shows highly nonlinear behavior, called chaos [10]. The characteristic of chaos is that the nearby trajectories in the dynamic system diverge exponentially with time, but they are deterministic simultaneously. This property has been applied to cryptography and computing. At the same time, chaotic systems are also sensitive to initial conditions and similar random behaviors. These properties of the chaotic system are customized according to the requirements of each module in the cryptosystem, including compression, encryption, and modulation schemes [18, 24, 32]. Chaogates are also considered as a low-cost and potential technology.

Fig. 14.2 A three-transistor
(3T) chaogate

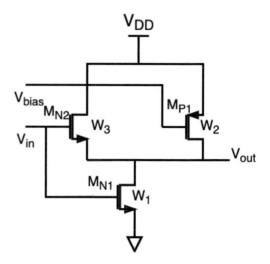

It is a promising method to design with the help of relevant artificial intelligence technologies. A chaogate made of three MOSFET transistors is shown in Fig. 14.2 [19]. The input voltage V_{in} is the same for both transistors M_{N1} and M_{N2}. These three transistors can form a huge search space, a huge task for designers. In this regard, several automatic circuit design methods were proposed in the 2000s. Artificial intelligence is used to optimize the parameters of analog IC [11, 34].

The dynamics of the chaogate can be described by this equation:

$$V_{n+1} = f(V_n; \alpha) \ n = 0, 1, 2, 3, \ldots, N, \qquad (14.1)$$

where V_n is the input voltage of the $n th$ iteration, and α is the parameter list of the specified chaogate; then the output voltage of this iteration is V_{n+1}. The following sequence is generated when α is iterated in a fixed manner.

$$\{V_n\}|_\alpha = \{V_0, f(V_0; \alpha) = V_1, f(V_1; \alpha) = V_2, \ldots\} \qquad (14.2)$$

Even if α changes slightly near a critical point may be enough to make the chaogate enter and exit chaos.

Python scripting can be used to build a netlist for the circuit as shown in Fig. 14.2 [1]. Furthermore, through transistor parameter optimization and simulation (using the new width value), the optimal W for the chaogate design can be determined using Bayesian Optimization (BO) and the genetic algorithm (GA). BO and GA use the same number of iterations to produce more convincing results. In addition, the parameter range of search optimization is also fixed. They use these two optimization algorithms to obtain width values, respectively. Although the results achieved by these optimization techniques appear comparable, GA can converge to one solution, where *lambda* is positive on a greater scale.

The results show that no matter what transistor technology is used, BO and GA can converge to a similar width. However, the time complexity could be considered to decide which optimization algorithm to use in the end. GA takes slightly less time than BO to complete the task through experiments. Similar trends were observed in other transistor models.

14.5 Key-Based Parameter Obfuscation

This section covers an obfuscation strategy for preventing the counterfeiting of analog ICs. Potential overproduction threats mainly come from an untrusted foundry [38]. In addition, an adversary can reverse engineer the analog IC design for cloning purposes [37].

The obfuscation of digital circuits needs to mask Boolean functions and logic values [16, 36]. In contract for analog circuits, depending on the continuous range of input/output values and the setting of various bias parameters increases the complexity of realizing the obfuscation. Furthermore, this analog obfuscation obscures crucial analog circuit characteristics such as the gain of the amplifier, filter cut-off frequency, and PLL operating frequency. Because analog circuits are usually designed and developed in stages, the impact on circuit parameters will be reduced if the key bits are distributed on the critical circuit parameters in each stage.

Analog ICs are more responsive to environmental factors like noise and temperature than digital circuits. Furthermore, the function and performance of the circuit are directly dependent on the set bias voltage and current. Therefore, setting the appropriate bias point in the analog circuit to produce optimal operating conditions is critical. Only when the correct key is used, it further activates the specific transistors inside the analog circuit to set the correct bias conditions at the target node for normal functioning of the analog integrated circuit.

Figure 14.3 [27] depicts a typical voltage bias circuit. When the correct key is applied, it can work normally and obtain the required V_{out}. A typical current bias circuit is shown in Fig. 14.4 [27]. The proper transistor width generates the desired current bias I_{ds} after applying the right KEY1.

Taking the RF front-end circuit as an example, an amplifier, mixer, filter, PLL, and demodulator make up a standard RF superheterodyne receiver. The obfuscation technology is applied to each stage of the superheterodyne receiver, as shown in Fig. 14.5. The receiver's settings are protected by a 512-bit key, with 40 bits utilized to obfuscate the PLL frequency.

A phase frequency detector (PFD) is the first essential element of PLL (see Fig. 14.7), which is made up of D flip-flop architecture based on an edge trigger. This block is followed by a low-pass filter (LPF) to clean noisy reference signals and a crystal-based voltage-controlled oscillator (VCO). Compared with other VCO topologies, LC-based VCO can provide excellent phase noise performance [29] (Fig. 14.6).

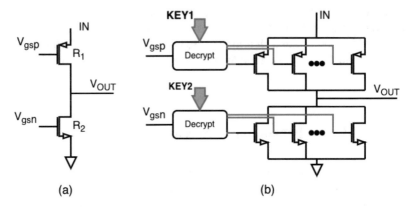

Fig. 14.3 (**a**) Unencrypted and (**b**) obfuscated voltage bias circuit

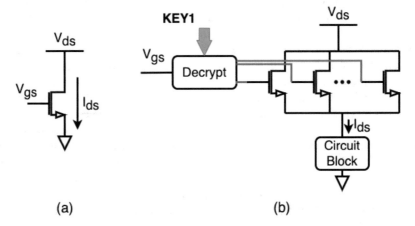

Fig. 14.4 (**a**) Unencrypted and (**b**) obfuscated current bias circuit

It is worth noting that the bias parameters are (1) the control voltage range, (2) the size of the varactor, and (3) the size of the negative resistance circuit. The output frequency of the voltage-controlled oscillator (VCO) is obfuscated by 40 transistors. The VCO will deliver the desired output frequency only if the correct key is utilized. Figure 14.7 [27] shows the connection of the key.

The experimental results show that the key-based parameter obfuscation can effectively improve the security of analog IC, with an increase of the area of 6.3%, a power consumption of 0.89%, and a phase noise of 5 dBc/Hz. By brute force attack, the likelihood of identifying the correct key sequence is only 9.095×10^{-13}. This technology is an effective countermeasure to prevent IP piracy, forgery, and overproduction of analog ICs.

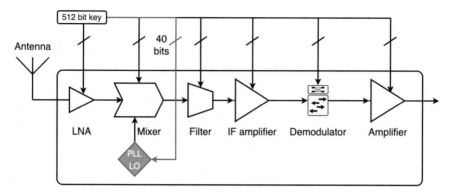

Fig. 14.5 Block diagram of a superheterodyne receiver with an integrated PLL local oscillator (l_o)

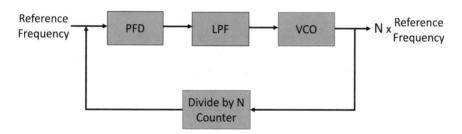

Fig. 14.6 Detailed block diagram of PLL-LO [28]

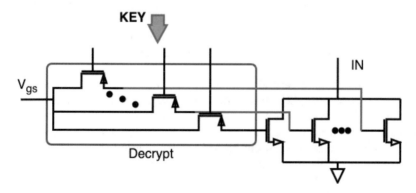

Fig. 14.7 Schematic of the decryption block

14.6 Combinational Locking

Combinational locking is a type of logic locking, originally used to secure digital integrated circuits [2, 17, 25, 26, 31]. However, the locking of an analog IC design is substantially different from that of a digital chip, as analog IC design

Fig. 14.8 Overview of
locking system

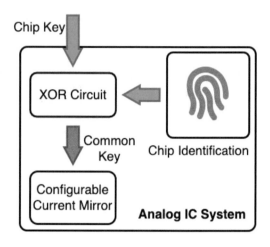

is far more complicated. A single bit error in the key in a digital circuit might easily cause the entire system to fail. In contrast, a minor fault in analog circuit configuration can result in significant performance variations. Therefore, wrong key inputs can lead to major performance degradation of circuits. This section will introduce a combinational locking technology, which ensures that most incorrect analog inputs result in considerable output deviations or poor performance. Without the correct key, reverse engineering or recycled chips are difficult to work normally. Furthermore, if the manufacturing test is carried out in a trusted facility, this technology will make piracy and overproduction in the foundry more difficult.

Figure 14.8 depicts an overview of the analog IC locking system. Locking with ordinary keys is usually not enough. The attacker can unlock all chips in the design if they have the public key. Chip identification technology may be used to provide a unique key for each chip.

Figure 14.9 shows a general locking architecture with a large design space to generate security locks [42]. The architecture is designed by $R \times N$ transistor array and the key lines. It contains a control transistor array of $3 \times N$ transistors. Some transistors in the array can be removed to increase flexibility. For example, there is no transistor in the second row and second column. Furthermore, a single key line may connect multiple control transistors; q_3 connects two transistors as shown in Fig. 14.9. The goal of locking is to find a configurable current mirror (CCM) design so that only one key can make the whole analog IC system work normally, while all other keys will lead to a large output deviation or failure of system performance. The control matrix of Fig. 14.9 is shown below. This matrix shows which key is linked to which transistor in row i and column j. There is no transistor at the corresponding place if the element is zero (see orange squared part of Fig. 14.9).

$$X^{3 \times N} = \begin{bmatrix} x_{11} & x_{12} & \dots & x_{1N} \\ x_{21} & x_{22} & \dots & x_{2N} \\ x_{31} & x_{32} & \dots & x_{3N} \end{bmatrix} = \begin{bmatrix} 1 & 3 & \dots & 2 \\ 3 & 0 & \dots & 4 \\ 3 & 4 & \dots & 3 \end{bmatrix}. \tag{14.3}$$

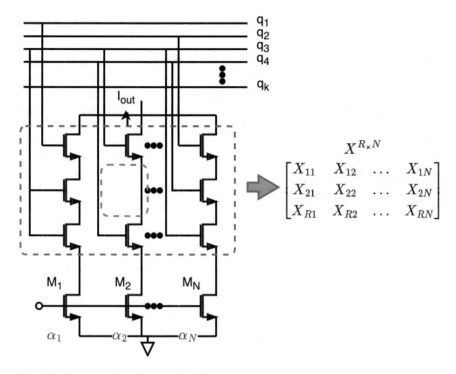

Fig. 14.9 Example of combinational locking

The following equation determines the bias current for the specific design using this architecture:

$$I_{out} = \sum_{j=1}^{N} \alpha_j \prod_{i=1}^{R} \Phi\left(x_{ij}\right) I_{REF}, \tag{14.4}$$

where α is the size of each column of transistors, and the transistor's control signal is $\phi(x_{ij})$ and expressed by:

$$\Phi\left(x_{ij}\right) = \begin{cases} q_k & \text{if } x_{ij} = k \neq 0, \\ 1 & \text{else } x_{ij} = 0. \end{cases} \tag{14.5}$$

This approach has been tested on the following analog IC designs: (1) bandpass filter, (2) quadrature oscillator, (3) LC oscillator, and (4) class-D amplifier. After simulation, only one key can meet the requirements, while other keys will significantly impact the performance [42].

14.7   Obfuscation with Analog Neural Network

IC recycling causes the semiconductor industry to lose billions of dollars annually. In addition, many discarded ICs after their usage may often contain sensitive information that could be extracted after recycling. This section describes how to utilize an analog neural network to safeguard analog integrated circuits from illegal usage and to lock their performance. This method does not directly obtain the input to analog modules (amplifier, ADC/DACs) from the bias circuitry, but it relies on an analog neural network to bias IC to its normal working point (transistor's saturation region). When the correct key is applied to its input, the IC will be supplied with the required bias voltage. The trained neural network serves as a lock during the procedure. In other words, only the neural network input matching to the right key will cause the IC to function properly. The analog neural network's programmability is achieved through the use of an analog floating gate transistor (FGT). It is used as permanent storage of synaptic weight and continuous input domain, generating a substantial analog key space so that this method can resist brute force attacks and model approximation attacks.

Once other parameters such as transistor size are determined in the design process, the bias value can also be determined for the required performance. One strategy for locking analog IC performance is to utilize the bias as the key because the IC can only function well if the bias value is accurate. Let us look at how to lock analog IC performance by preventing applying bias directly. This technique depends on an on-chip analog neural network (ANN) to generate the bias necessary for IC functioning following the specification, as shown in Fig. 14.10. The analog input serves as the key, while the trained ANN serves as the lock. A pair of correct locks and keys can make the IC reach the bias value required by its operating point. At the same time, any incorrect key will significantly deviate from the performance of the IC. In addition, to prevent finding the key through the model approximation

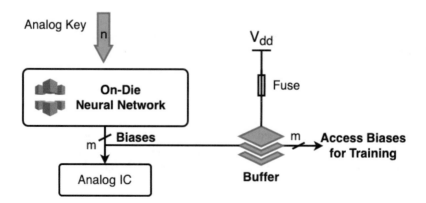

Fig. 14.10 General architecture of bias locking method based on neural network

algorithm, the wrong key should also reduce the leakage of the information of the working point of the analog IC as much as possible.

This design approach must program the ANN to accept a certain key before selling the IC to the end user. The end user is subsequently given the key together with the IC. Finally, this key must be applied to the IC by the end user to supply the correct bias value so that the entire IC can function normally.

The following four facts support this technology's capacity to withstand attacks:

1. The substantial labor/workload and time cost of applying and verifying analog keys can prevent brute force assaults due to the big enough space of analog input.
2. Because the neural network's learnt function has high entropy, when the incorrect analog key is applied, the corresponding performance of the analog IC cannot disclose the information of the correct key.
3. As described earlier, each integrated circuit can have its unique key to resist the key sharing attack.
4. The inclusion of a fuse basically eliminates the observability of the neural network's output (i.e., bias voltage) once it has been trained.

This analog performance locking approach is evaluated on a custom-designed and manufactured experimental platform that includes two custom-designed and manufactured ICs, an LNA chip with three bias voltages, and an analog neural network prototype chip. The experiments show that this method can effectively protect the analog IC [41].

14.8 Multi-threshold Design

Compared with digital circuits, analog design has fewer transistors, which makes it more vulnerable to attacks such as reverse engineering (RE). In reverse engineering, malicious attackers unpack the IC and take each layer's images [37]. The metal layer image can offer connection information, whereas the transistor configuration is identified using the base layer image. Finally, the malicious attacker splices the information obtained from the image together to unlock the IP netlist. This section will introduce a multi-threshold voltage(V_{TH}) countermeasure to prevent RE-based attacks. The idea is to use Low V_{TH} (LVT) and/or High V_{TH} (HVT) transistors to replace a few normal V_{TH}(NVT) transistors while keeping the performance of analog IC unchanged, such as gain, bandwidth. If the attacker cannot correctly obtain the V_{TH} of the transistor, it can result in the poor performance of the reverse-engineered circuit. For example, there will be a significant performance difference after the amplifier applies the correct V_{TH} and the incorrect V_{TH}. This example shows that if the attacker fails to recognize the correct V_{TH}, the analog IC cannot normally work, thus protecting the analog IC. However, this workload is huge if the attacker only relies on guess and verification to identify the transistor's V_{TH}.

This approach can also be used with the previously discussed key-based parameter obfuscation technique.

This method can protect against the counterfeiting of analog IC. Even when the attacker obtains the netlist through RE, if he/she does not know the correct threshold voltage, the analog IC cannot work normally. An adversary can often use brute force attacks to obtain the threshold voltage; that is, the attacker will attempt all combinations of V_{TH} values with the aim of evaluating the response of each transistor to estimate the circuit and repeat this process until the IC works with acceptable performance. However, each transistor in the design has three V_{TH} options. As a result, if the design contains n transistors, the adversary must evaluate 3^n combinations. For smaller analog designs, brute force attacks may find the correct V_{TH} combination in a limited time. However, as the transistors in the design increase, the time required for a brute force attack increases exponentially. Therefore, analog designs with higher transistors are clearly more appropriate for this method. For small-scale analog design, an effective way to increase the workload of attackers is to split into parallel transistors.

For example, Fig. 14.11a shows a self-biased reference current generator. The reference current is given by the following equation [4]:

$$I_{REF} = \frac{1}{8.R^2 \mu_n C_{OX} \left(\frac{W}{L}\right)_2} \tag{14.6}$$

In this equation, the size of MB2 is $(W/L)_2$. Four transistors $(MB1 - 4)$ and resistors (R) in the circuit primarily determine the reference current and bias voltage V_{biasp}. Three different threshold voltages are possible for each of the four transistors.

This scalable cascaded operational amplifier topology requires five separate voltages ($V_{bias1-4}$ and V_{MCM}) to bias different transistors. At the same time, more than two bias voltages (V_{pcas} and V_{ncas}) are necessary for these transistors. $V_{bias1-4}$, V_{pcas}, and V_{ncas} are generated by the bias network, as shown in Fig. 14.11b.

In this test case, there are 55 transistors. However, considering the existence of pairing, there will be 47 effective transistors, so the attacker needs to guess and verify that the number of combinations is less than the theoretical maximum. If the V_{TH} modification is not performed symmetrically to paired transistors, the attacker can find abnormal transistor size and destroy the design simultaneously. However, if V_{TH} changes are applied in pairs, the overall key space will be less than the theoretical maximum ($3^{47} < 3^{55}$). Assuming that the opponent only needs 0.1s to verify each key using automated scripts and simulators, the RE workload may be 10^{13} years, which is a long time.

For smaller analog IC designs, the workload of RE can be increased by other methods such as transistor splitting. This way, the effective number of transistors can be increased to 49. Therefore, RE workload increases to $0.1sec \times 3^{49} \sim 10^{14}$ years (10x higher). If all transistors in the design are split similarly, the workload of RE will increase by several orders of magnitude. Therefore, even for small-scale analog design, this method can effectively prevent reverse engineering.

Fig. 14.11 (**a**) Self-biased
reference current circuit. (**b**)
Bias voltages generator for
op-amp. (**c**) Wide-swing
telescopic cascode op-amp
with output stage [4]

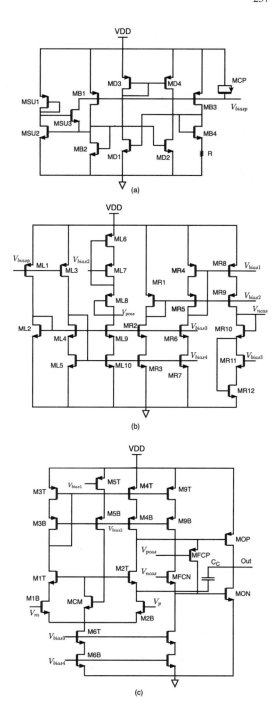

14.9   Conclusions

Analog integrated circuits (ICs) are widely employed in a wide range of industrial applications, including telecommunications, military systems, automotive, and space. However, the security of analog circuits security is somewhat "ignored" as compared to digital circuits for various reasons such as challenging design cycles and lower market share. At the same time, it is very lucrative for an adversary to inject malicious circuits to cause reliability issues and counterfeit analog ICs. Hence analog world needs special attention by researchers toward the authenticity and security of analog ICs. We discussed various obfuscation techniques for analog logic locking to prevent analog IC overproduction and cloning through reverse engineering.

References

1. Acharya RY, Charlot NF, Alam MM, Ganji F, Gauthier D, Forte D (2021) Chaogate parameter optimization using Bayesian optimization and genetic algorithm. In: 2021 22nd international symposium on quality electronic design (ISQED). IEEE, Piscataway, pp 426–431
2. Alkabani Y, Koushanfar F (2007) Active hardware metering for intellectual property protection and security. In: USENIX security symposium, vol 20, pp 1–20
3. Alkabani Y, Koushanfar F, Potkonjak M (2007) Remote activation of ICS for piracy prevention and digital right management. In: 2007 IEEE/ACM international conference on computer-aided design. IEEE, Piscataway, pp 674–677
4. Ash-Saki A, Ghosh S (2018) How multi-threshold designs can protect analog IPS. In: 2018 IEEE 36th international conference on computer design (ICCD). IEEE, Piscataway pp 464–471
5. Babayan-Mashhadi S, Lotfi R (2013) Analysis and design of a low-voltage low-power double-tail comparator. IEEE Trans Very Large Scale Integr VLSI Syst 22(2):343–352
6. Baumgarten A, Tyagi A, Zambreno J (2010) Preventing IC piracy using reconfigurable logic barriers. IEEE Des Test Comput 27(1):66–75
7. Bryant T, Chowdhury S, Forte D, Tehranipoor M, Maghari N (2016) A stochastic approach to analog physical unclonable function. In: 2016 IEEE 59th international midwest symposium on circuits and systems (MWSCAS). IEEE, Piscataway, pp 1–4
8. Bryant T, Chowdhury S, Forte D, Tehranipoor M, Maghari N (2017) A stochastic all-digital weak physically unclonable function for analog/mixed-signal applications. In: 2017 IEEE international symposium on hardware oriented security and trust (HOST). IEEE, Piscataway, pp 140–145
9. Contreras GK, Rahman MT, Tehranipoor M (2013) Secure split-test for preventing IC piracy by untrusted foundry and assembly. In: 2013 IEEE international symposium on defect and fault tolerance in VLSI and nanotechnology systems (DFTS). IEEE, Piscataway, pp 196–203
10. Ditto WL, Miliotis A, Murali K, Sinha S, Spano ML (2010) Chaogates: morphing logic gates that exploit dynamical patterns. Chaos An Interdiscip J Nonlin Sci 20(3):037,107
11. Fakhfakh M, Cooren Y, Sallem A, Loulou M, Siarry P (2010) Analog circuit design optimization through the particle swarm optimization technique. Analog Integr Circuits Signal Process 63(1):71–82
12. Farahmandi F, Huang Y, Mishra P (2017) Trojan localization using symbolic algebra. In: 2017 22nd Asia and South Pacific design automation conference (ASP-DAC). IEEE, Piscataway, pp 591–597

13. Gassend B, Clarke D, Van Dijk M, Devadas S (2002) Silicon physical random functions. In: Proceedings of the 9th ACM conference on computer and communications security, pp 148–160
14. Herder C, Yu MD, Koushanfar F, Devadas S (2014) Physical unclonable functions and applications: a tutorial. Proc IEEE 102(8):1126–1141
15. Hossain MM, Vashistha N, Allen J, Allen M, Farahmandi F, Rahman F, Tehranipoor M (2022) Thwarting counterfeit electronics by blockchain. https://scholar.google.com/citations?view_op=view_citation&hl=en&user=n-I3JdAAAAAJ&citation_for_view=n-I3JdAAAAAJ:9ZlFYXVOiuMC
16. Juretus K, Savidis I (2016) Reducing logic encryption overhead through gate level key insertion. In: 2016 IEEE international symposium on circuits and systems (ISCAS). IEEE, Piscataway, pp 1714–1717
17. Kamali HM, Azar KZ, Farahmandi F, Tehranipoor M (2022) Advances in logic locking: past, present, and prospects. Cryptology ePrint Archive
18. Keuninckx L, Soriano MC, Fischer I, Mirasso CR, Nguimdo RM, Van der Sande G (2017) Encryption key distribution via chaos synchronization. Sci Rep 7(1):1–14
19. Kia B, Lindner JF, Ditto WL (2016) A simple nonlinear circuit contains an infinite number of functions. IEEE Trans. Circuits Syst. Express Briefs 63(10):944–948
20. Koushanfar F, Qu G (2001) Hardware metering. In: Proceedings of the 38th annual design automation conference, pp 490–493
21. Koushanfar F, Qu G, Potkonjak M (2001) Intellectual property metering. In: International workshop on information hiding. Springer, Berlin, pp 81–95
22. Li J, Seok M (2016) Ultra-compact and robust physically unclonable function based on voltage-compensated proportional-to-absolute-temperature voltage generators. IEEE J Solid State Circuits 51(9):2192–2202
23. Lowry RK (2007) Counterfeit electronic components-an overview. In: Military, Aerospace, Spaceborne and Homeland Security Workshop (MASH)
24. Mishkovski I, Kocarev L (2011) Chaos-based public-key cryptography. In: Chaos-based cryptography. Springer, Berlin, pp 27–65
25. Rahman MS, Li H, Guo R, Rahman F, Farahmandi F, Tehranipoor M (2021) LL-ATPG: logic-locking aware test using valet keys in an untrusted environment. In: 2021 IEEE international test conference (ITC). IEEE, Piscataway, pp 180–189
26. Rajendran J, Pino Y, Sinanoglu O, Karri R (2012) Security analysis of logic obfuscation. In: Proceedings of the 49th annual design automation conference, pp 83–89
27. Rao VV, Savidis I (2017) Protecting analog circuits with parameter biasing obfuscation. In: 2017 18th IEEE Latin American test symposium (LATS). IEEE, Piscataway, pp 1–6
28. Razavi B (2001) Design of analog CMOS integrated circuits. McGraw-Hill, Spain
29. Razavi B (2012) Design of integrated circuits for optical communications. John Wiley & Sons, Hoboken
30. Rosenblatt S, Fainstein D, Cestero A, Safran J, Robson N, Kirihata T, Iyer SS (2013) Field tolerant dynamic intrinsic chip ID using 32 nm high-k/metal gate SOI embedded dram. IEEE J Solid-State Circuits 48(4):940–947
31. Roy JA, Koushanfar F, Markov IL (2010) Ending piracy of integrated circuits. Computer 43(10):30–38
32. Stojanovski T, Pihl J, Kocarev L (2001) Chaos-based random number generators. Part II: practical realization. IEEE Trans. Circuits Systems I Fund. Theory Appl 48(3):382–385
33. Suh GE, Devadas S (2007) Physical unclonable functions for device authentication and secret key generation. In: 2007 44th ACM/IEEE design automation conference. IEEE, Piscataway, pp 9–14
34. Taherzadeh-Sani M, Lotfi R, Zare-Hoseini H, Shoaei O (2003) Design optimization of analog integrated circuits using simulation-based genetic algorithm. In: International symposium on signals, circuits and systems, 2003. SCS 2003, vol 1. IEEE, Piscataway, pp 73–76
35. Tehranipoor M, Wang C (2011) Introduction to hardware security and trust. Springer Science & Business Media, Berlin

36. Tehranipoor MM, Forte DJ, Farahmandi F, Nahiyan A, Rahman F, Rahman MS (2022) Protecting obfuscated circuits against attacks that utilize test infrastructures. US Patent 11,222,098
37. Torrance R, James D (2009) The state-of-the-art in IC reverse engineering. In: International workshop on cryptographic hardware and embedded systems. Springer, Berlin, pp 363–381
38. Torrance R, James D (2011) The state-of-the-art in semiconductor reverse engineering. In: Proceedings of the 48th design automation conference, pp 333–338
39. Vashistha N, Lu H, Shi Q, Rahman MT, Shen H, Woodard DL, Asadizanjani N, Tehranipoor M (2018) Trojan scanner: detecting hardware trojans with rapid SEM imaging combined with image processing and machine learning. In: ISTFA 2018: proceedings from the 44th international symposium for testing and failure analysis. ASM International, Novelty, p 256
40. Vashistha N, Hossain MM, Shahriar MR, Farahmandi F, Rahman F, Tehranipoor M (2021) eChain: a blockchain-enabled ecosystem for electronic device authenticity verification. IEEE Trans Consum Electron 68(1):23–37
41. Volanis G, Lu Y, Nimmalapudi SGR, Antonopoulos A, Marshall A, Makris Y (2019) Analog performance locking through neural network-based biasing. In: 2019 IEEE 37th VLSI test symposium (VTS). IEEE, Piscataway, pp 1–6
42. Wang J, Shi C, Sanabria-Borbon A, Sánchez-Sinencio E, Hu J (2017) Thwarting analog IC piracy via combinational locking. In: 2017 IEEE international test conference (ITC). IEEE, Piscataway, pp 1–10

Chapter 15
Tamper Detection

15.1 Introduction

Hardware is the root of trust for electronics security [5]. Unfortunately, from intellectual property theft and exploitation of hardware vulnerabilities to steal sensitive data, problems in hardware security can cost billions of dollars annually [10, 11]. Tampering is an example of one of these attack vectors, where physical or environmental modifications are used to alter integrated circuits or printed circuit board (PCB) systems [1, 12]. The exploitation of these vulnerabilities may be used to modify the expected performance of systems or recycle to resale integrated circuits [23, 24]. For example, unauthorized modifications of video game consoles cost billions to video game developers in 2008 as console modifications enabled owners of the tempered hardware to play pirated games, which could not be done with an off-the-shelf console [25]. Since then, hardware tampering techniques have become more advanced, so tamper detection and prevention techniques need to be developed.

This chapter analyzes five novel methods to detect systems or ICs tempering, specific to five different attack vectors. Each method has its unique capabilities and, perhaps, addresses unique vulnerabilities. Starting with FLATS that stands for Filling Logically and Testing Spatially is a design architecture to provide for authentication and tamper detection against post-synthesis attacks in FPGA designs [8]. This architecture can be designed using programmable feedback oscillators, dynamically inserted into the design as beacons, authenticators, and detectors. Feedback oscillators will output more heat than the rest of the chip. This heat can be captured through an IR imaging device, and distances between points of interest can be measured and compared to reference values. Any unauthorized design deviations will change the target oscillators' locations or unique characteristics.

The second design uses a method to detect the use of a recycled IC using physical unclonable functions (PUFs) and encryption [2]. As IC ages, the transistors slow down, and thus the frequency of a ring-oscillator (RO) PUF decreases. Storing the

M. Tehranipoor et al., *Hardware Security Primitives*,
https://doi.org/10.1007/978-3-031-19185-5_15

initial frequency at the time of semiconductor fabrication in memory and comparing it to the RO frequency at usage can give an accurate age reading. This technique can be used to detect an aged IC. Furthermore, this technique can be combined with encryption of the initial frequency and various IC parameters to detect if the memory has been tampered with to forge the initial RO frequency to sell a recycled device as new.

In the third design, a method to prevent Local ElectroMagnetic Attacks (L-EMA) uses an LC-based oscillating sensor circuit that can detect an incoming microprobe [16]. The circuit is designed to protect a cryptographic engine, the component responsible for encrypting data in the device. Furthermore, the technique involves creating an auxiliary detector circuit that uses two oscillators to create a mutual inductance with the attacker's microprobe, resulting in the chip entering a locked state. Hence, the secret key that the engine encrypts is safe.

The fourth design prevents physical attacks such as nano-probing by surrounding the circuit with a three-layer conductive mesh (CM) [22]. This circuit monitors invasive probing by regularly sending a known series of pulses through the CM via an active tamper detection circuit and registering mismatches from the expected threshold. In case of a mismatch, data is removed from an internal memory device by removing power from memory.

The final design depicts a conductive grid-based circuit that works as a shield for sensitive data, and an active tamper detecting unit against physical intrusions [21]. Aside from physical attacks, side-channel attacks utilize faults created by logic circuits working under severe conditions [15, 17]. When logic circuits, for example, are exposed to temperatures that exceed their operational limitations, they may malfunction, creating excellent conditions for cryptographic assaults. This tamper detection approach uses a grid with unique resistance and capacitance values, often known as the "RC constant," and checks for variations most likely caused by severe temperatures.

Section 15.2 of this chapter covers the underlying threat model and challenges to be encountered while addressing the tamper detection problem. Section 15.3 of this chapter discusses various tamper prevention and detection methods in detail. Finally, Sect. 15.4 summarizes all methods.

15.2 Background

15.2.1 Threat Model

The threat of tampering with hardware has existed for as long as hardware has existed. The idea behind tampering with an integrated circuit is to gain unauthorized access to a device by circumventing the security features to leak information or sabotage functionality [4]. The tempering poses a massive risk to devices that can serve as system-critical components or devices that hold system-critical information. The attempt to prevent these attacks in the first place is one field of study in hardware security [18]. Still, this chapter focuses on tampering detection as a prerequisite to temper prevention. Most tamper detection techniques must be built into the IC as an effective countermeasure against malicious attacks on the circuitry. Furthermore, tampering can also break through the package to probe the silicon and attack the bitstream of an FPGA to change the functionality [7]. Once the detection mechanism has detected these attacks, it can alert the security circuitry to take preventive actions such as erasing critical information or shutting down the power supply on the chip.

15.2.2 Challenges in Tamper Detection

The primary challenge with creating a universal tamper detection method is that there is no single method to detect all forms of tampering. There are many attacks to develop countermeasures for, and VLSI technologies come with new vulnerabilities. Thus, previously unseen attacks make it nearly impossible to detect them until it happens in the field. Often tamper detection algorithms and added circuitry work toward identifying general tampering rather than focusing on a specific attack.

Other difficulties involve the physical implementation of the solution on the chip itself. If a designer attempts to create a low-overhead design, fitting an effective temper detection mechanism could be difficult due to silicon space or power constraints. Furthermore, additional time and on-chip resources are required to detect specific attacks effectively. If the IP designer cannot set aside those resources, they may have no choice but to leave their chip vulnerable in the field. Therefore, low-cost and less resource-consuming solutions are desirable for effective tamper detection.

Finally, it is necessary to keep those tamper detection methods as tamper-proof and robust as possible. In particular, an intelligent adversary can plan an attack on integrated circuits by disabling detection methods or circumventing on-chip security. Typically, if the detection method halts its function, that is a significant clue that something malicious may have been inserted into the circuit or some connection has been cut, and the circuitry can still sound an alarm. The challenges can come when an attack implants false "safe" values to the tamper detection mechanism (rather than just shutting it off), thus preventing tamper detection even though the system appears active. The ability to effectively detect attacks like these proves a challenge going forward.

15.3   Various Tamper Prevention and Detection Mechanisms

15.3.1   FLATS

FLATS, which stands for Filling Logic and Testing Spatially, is a tamper detection authentication method for FPGAs. It leverages the heat-generating property of feedback oscillators to identify the positions and relative distances between key look-up tables (LUTs) in the design. The process is divided into two phases—insertion and verification. This process assumes that no malicious IP is incorporated during the design phase. It also assumes that EDA tools are not compromised and do not generate or simulate the design with undesired malicious code [8].

The FLATS method draws inspiration from Built-in Self Authentication (BISA) [26] test methodology and Golden Gates Circuits (GGC) [19, 20] technique, which fills unused areas in an ASIC with logic cells to insert testable logic cells for hardware Trojan detection. These ASIC-based methods were further expanded to FPGAs that fill unused logic space and are used to detect Trojans in earlier stages, such as synthesis and RTL design [6, 13, 14]. However, FLATS is unique compared to the previously mentioned methods as it looks at filling the unused logic in later processes in the FPGA flow, such as place and route and programming. So while similar to a handful of methods in numerous ways for detecting Trojans, it is novel in its methods for analyzing circuits post-synthesis. This research could be continued to apply it to pre-synthesis methods or combine it with compatible pre-synthesis techniques for an FPGA end-to-end tamper detection method. Another way to continue would be to see if similar methods could be created for monitoring other types of devices, perhaps systems of ICs or FPGAs on a printed circuit board. The insertion phase begins with the addition of configurable oscillators to every LUT in the design. Each LUT must have one output directly connected to the input of the same LUT. Any floating inputs created as a consequence of this design consideration are set to "1" (high). Next, additional logic is added to enable or disable each particular oscillator. Additionally, a Linear Feedback Shift Register is integrated with the system to obfuscate the input and oscillator selection/configuration relationship. Seed input for this register is generated from a concatenation of device ECID and some portion of user input. Next, clock gating circuitry is added to the design to enable the design to be locked in a particular state for observation and clock dividers for LUT oscillator pulsing. The insertion phase also creates an Internal Configuration Access Port to generate frames for each LUT. Finally, the "place and route" step creates a boundary, places, routes the design, identifies and converts unused logic blocks into programmable feedback oscillators and completes routing for the modified LUTs.

At this stage, the design is considered mature and can be registered in preparation for subsequent verification passes. Some oscillators (typically found in the corners of the design) are assigned as reference beacons. Several other oscillators can be selected for a pattern or, at random, may be identified as authenticators. Oscillators at particular nodes of interest are referred to as detectors. The distances between

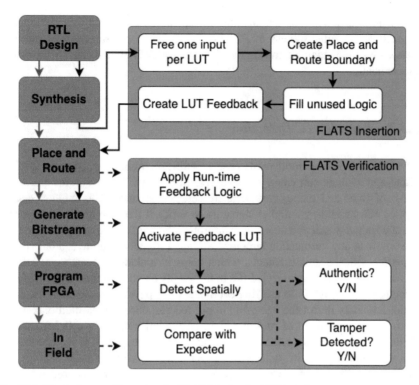

Fig. 15.1 Block diagram illustrating the FLATS during the FPGA design stages

beacons, authenticators, and detectors are generated from simulation and treated as the "golden model." The verification phase consists of capturing an infrared image of the chip to identify the locations of beacons, authenticators, and detectors. The distances are compared against the golden design, and any deviations indicate a tamper. Both phases are presented in Fig. 15.1 [8].

This implementation is relatively inexpensive, with a low area overhead and no discernible effect on power use or speed. It also does not require specialized manufacturing hardware as it is directly implemented on an FPGA. However, this method is only relevant to FPGAs as it leverages the unique architecture of these devices. FLATS also requires specialized testing equipment to verify the locations of beacons, authenticators, and detectors. This methodology may also require exposing the bare die of the chip for better-infrared imaging capabilities, depending on-chip packaging characteristics. FLATS is also limited to verifying Trojan insertion or tamper that occurs after the design has been synthesized. Using a long-wave infrared imaging device (LWIR), the functionality of FLATS architecture could be verified. Using the frequency of the RO and the principle of lock-in thermography, the beacons could be separated from the other logic of the circuit. The beacons and authenticators could be physically located on the device to verify the spacing. The authenticator to beacon distance versus authenticator to beacon number for an authentic device and an unauthentic device can be compared to a reference to demonstrate the effectiveness of FLATs. Every fourth measurement of one plot series shows a considerable shift in the distance, which is used to classify the design as unauthentic.

15.3.2   Recycled IC Detection

Recycled ICs from electronic waste pose a significant problem as using these ICs in critical applications can cause massive system failures. These failures can result in loss of lives and money due to decreased performance. This method detects recycled ICs by adding a digital signature to verify if the chip has been tampered with and forged to sell as a new chip. Unfortunately, this forging can make aged ICs appear new to any verification circuitry. Adding a digital signature can help detect tampering with the measurements, which helps in making systems more reliable and secure by identifying recycled ICs in the supply chain. The fundamental idea of detecting recycled ICs is to use ring-oscillator (RO) physical unclonable functions to approximately detect the age of a circuit. RO response is measured once when the circuit is manufactured and stored in non-volatile memory (NVM) such as EEPROM. As the chip ages, the transistors are generally expected to become slower. Thus, the frequency of a RO will decrease over time. With further usage of chips, RO frequency significantly decreases compared to its initial frequency (stored in NVM), and it can be used to detect the age of the chip. The IC is flagged as recycled if the difference is more significant than a predefined threshold value. The main concern with this approach is that the contents of the NVM can potentially be tampered with, thus enabling the initial RO frequency measurement to be overwritten with a value more closely matching the aged RO frequency. The final product of this tampering is the IC ultimately classifying itself as a new IC despite having encountered some significant aging. This method uses a digital signature to verify the NVM contents to detect tampering (Fig. 15.2) [3].

This recycled IC detection method assumes that the chip has an electronic chip id (ECID) assigned to it, which is a valid assumption as many chips available in the market have an ECID or a similar on-chip identifier such as chip DNA. Once ECID is programmed into the chip, it cannot be easily changed. An RO and an NVM are the two main components of the on-chip sensor structure (see Fig. 15.3).

Multiplexing an existing primary output can be used to make the RO output available (PO). To determine the frequency of the RO, a counter and a timer are necessary. The NVM is then programmed with the registration data and a data signature. A test access port and a boundary-scan architecture can be used to access the NVM content. This sensor necessitates the use of a single small non-volatile memory capable of storing data even when the power is turned off. The NVM will contain data and a signature of that data to prevent manipulation. The data is obtained by concatenating the electronic chip ID (ECID), measurement conditions (e.g., voltage supply, duration, and temperature), and the counter value. That data is

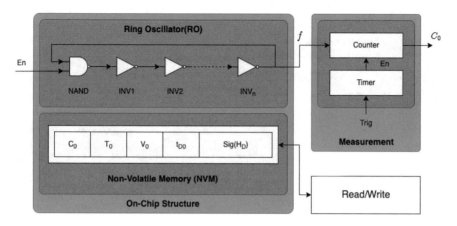

Fig. 15.2 Circuit diagram of an on-chip sensor for aging detection

used to create a fixed-length hash using a hashing algorithm, and finally, a digital signature is produced using a private key. The data and the digital signature are stored on the NVM on the chip. Whenever the chip age is evaluated, the data string and the ECID are used to create a new hash, and the signature is decrypted using a private key to retrieve the original hash. If the new hash and original hash are not equal to each other, then it is highly likely that NVM tampering has occurred, and the stored frequency data should not be trusted. ECID is easily accessible via the test access port (TAP).

After the digital signature generation, the chips go through the registration process. It is a five-step process (see Fig. 15.3) as follows:

1. A measurement unit is used to measure the ring-oscillator frequency. The counter value (C_0), supply voltage (V_0), and the temperature (T_0) are recorded for a fixed time interval (t_{D0}). These values are concatenated to form a data value called ring oscillator (RD). The test access port (TAP) can be used to read the chip's ECID values.
2. ECID value and RD are concatenated to generate data (D). Further additional information can be added in D, such as manufacturer and production site identification codes.
3. A fixed-length hash (H_D) is generated using a cryptographically secure hash algorithm such as SHA-2/3.
4. A digital signature ($Sig(H_D)$) is created by encrypting the hash output (H_D) with the private key of the original component manufacturer (OCM). Only the OCM has access to this protected private key.
5. The NVM on the chip is used to store the oscillator data, digital signature $Sig(H_D)$, and the RD.

ICs can be validated at any step in the supply chain using the RO-based, low-cost on-chip sensor. To measure the cycle count of the ring oscillator over a specified

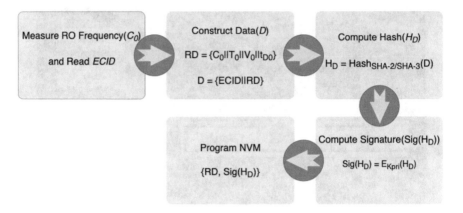

Fig. 15.3 Block diagram illustrating the steps for creating the digital signature of the IC for registration

time interval, a timer IC and a counter can be employed. A trigger input (T_rig) starts the measuring process. The authentication technique is straightforward and may be performed at any point along the supply chain with low-cost measuring equipment. A counter and a timer were formerly required in the setup. It must also be able to read the chip's ECID and NVM content. The suggested authentication procedure consists of six steps (see Fig. 15.4) as follows:

1. The NVM contents of the chip that need to be authenticated are read using the measurement setup described earlier. The data consists of ring-oscillator data (RD) and digital signature ($Sig(H_D)$). The ECID value must also be read. Concatenating ECID and RD results in the data (D).
2. On the data (D), a hash (H_D) is computed. And, signature ($Sig(H_D)$) is used to recover another hash (H_D^*).
3. Mismatches are checked between the computed hash (H_D) and hash (H_D^*). Mismatches between the two hashes indicate that the NVM contents of the chip have been tampered with by an adversary.
4. During registration, if both hashes match, the measurement parameters ((T_0), (V_0), and (t_{D0})) are extracted from the ring-oscillator data (RD).
5. During registration, the RO clock cycle count (C_0^*) for the fixed time interval (t_{D0}) is measured at the same condition (T_0, V_0).
6. The difference in measured and registration clock cycle counts is computed as (C_0). If the discrepancy exceeds the precision of the counter (measurement error), the chip is marked as recycled. Aside from that, the chip is marked as new.

There are a couple of limitations of this design. First, there is no protection against unauthorized modifications of the NVM if the private key is leaked. While leaking a private key from a trusted source is highly unlikely, the risk remains. The second limitation is that this design is vulnerable to improper registration by the

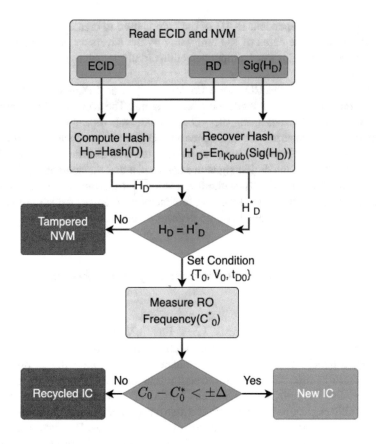

Fig. 15.4 Steps for NVM tamper detection followed by recycled IC detection

foundry; however, it is assumed that there is no motive to undertake such action, especially since it has no clear benefit for the foundry. Even if such an act would occur, it could be mitigated by an external active hardware metering. Finally, a demerit of this method is the requirement of additional circuitry and NVM. Along with the RO, the additional circuitry to perform hashing and encryption should be present (though it can be offloaded to a CPU unit present in the design), and the NVM is recommended to be at least 1kB in size. The additional circuitry increases the area overhead in the overall design. If the overall design is large enough to where this extra area is only a small percentage or can even fit between other modules on the design, this is not a considerable problem.

Besides limitations, some of the advantages of this design include elevating security to prevent recycled ICs. Since the data produced by the initial RO measurement is hashed and encrypted to create a signature, it is exceptionally resilient to brute force attacks and is at little to no risk of cracking. It is also resilient to replacing the initial RO measurement with a measurement from a brand-new chip, as the device-dependent ECID of the chip is used to create the unique digital signature. Even in the cases of RO tampering by the adversary, the circuitry itself can be protected by additional tamper detection circuitry, such as an active shield on a higher metal layer.

To test the design, the ROs were implemented in an FPGA and subjected to various temperatures to accelerate the effect of aging. The accuracy of the approach was demonstrated by measuring the RO at set intervals during a simulated aging experiment and measuring the difference in RO frequency at various times. The results are significant because aging can be detected with as little as one day of use, and the generation of the digital signature has shown that the contents of the NVM cannot be modified externally without detection. Furthermore, the experimental results justify the designer's claims that this method of temper detection improves the security and reliability of electronic systems by detecting recycled ICs.

15.3.3   Digital-Oscillator-Based Sensor for EM Probing Detection

This technique can detect and resist Local ElectroMagnetic Attacks (L-EMA), a type of side-channel attack, on a cryptographic engine (CE) [16]. Many chips are equipped to handle other forms of side-channel attacks externally to prevent an attacker from gaining necessary information that can lead to the extraction of a secret key. For example, an attacker can perform power analysis on the device's cryptographic engine, but with a current equalizer, [16], the power consumed by the engine can be monitored and masked to prevent the analysis. However, these same chips are not necessarily equipped to handle an L-EMA attack. Instead, an attacker uses a microprobe to analyze the radiation induced by the cryptographic engine and gain confidential key information. Therefore, the probe used by an attacker for L-EMA would need to be very close to the CE.

The tamper detection solution creates a circuit that can detect a Local-Electromagnetic Attack and prevent it by counteracting the attacker's probe to secure the secret key information further. The outcome of a prevented attack would be to output a dummy key or lock the crypto-engine entirely.

The cryptographic engine used in the method is outfitted with an LC-oscillator-based dual coil-sensor system. The idea behind the detector is to use two distinct circuits in tandem as a countermeasure to incoming EM probes that approach the coils routed around the cryptographic engine: the sensor core and the oscillator core (see Fig. 15.5 [16]). The coils around the CE form two LC oscillators that would use the mutual inductance to generate two different frequencies when approached by the EM microprobe. The difference between both frequencies, f_{LC}, is constantly monitored by a digital counter in the sensor core. To detect the attack, the LC-oscillator core, made up of a series of power-gated CMOS inverters, is fed a signal

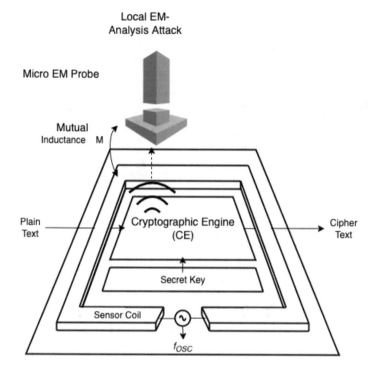

Fig. 15.5 An oscillator-based tamper-access sensor as a countermeasure to local EM analysis attack

LC_{en} to check if a probe has approached the device. A trigger pulse enables an oscillation if the oscillator detects a probing attack.

The proposed sensor circuit is depicted in Fig. 15.6. Dual-sensor coils L_1 and L_2 are pulled over CE to generate two LC oscillators. Key-related circuit blocks (such as S-BOX in AES) are positioned beneath the dual coils, whereas key-unrelated blocks (such as 110 interface) are positioned between the coils. To capture local EM radiation, the tiny EM probe must approach one of the dual coils, which causes an LC oscillation frequency f_{LC} difference between two LC oscillators. Control logic's digital counters can further detect this frequency difference (f_{LC}).

In order to fine-tune the detection method, a one-step digital calibration sensor was implemented within the device to parse out PVT variation and ensure the difference in frequency between the two oscillators, f_{LC}, is as accurate as possible. It is important to note that the accuracy of the f_{LC} did not seem to be the intended goal of this method. Instead, it sought to remove the PVT variations within the f_{LC} as it could give false successes in the detection period. As an added feature, this technique produces a reasonable to negligible overhead by operating during the idle state of the CE operations. Furthermore, the intermittent sensor operates at <1%, meaning that the power and performance overheads are reduced significantly. In

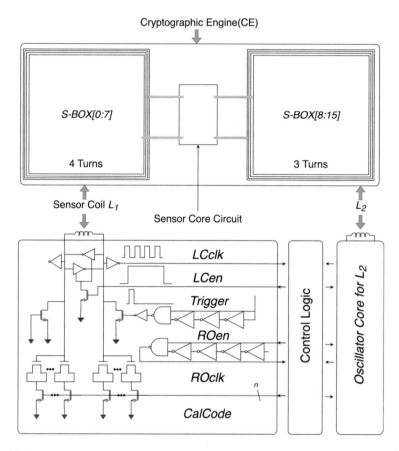

Fig. 15.6 Implementation of a fully digital sensor circuit with a dual-coil sensing

order to be a fully digital process, the calibration system only uses two counters and a LUT to control the output capacitance of the main sensor system.

In terms of innovation, this is the first technique in hardware security that prevents L-EMA by using the attacker's probe as a countermeasure. In addition, the circuit design is novel, one of the first to prioritize a local EMA side-channel attack rather than the broader power analysis attack.

An assumption made by the designers is that the probe would only approach one side of the cryptographic engine. Further, it is assumed that a dual EM probe attack could occur. Still, it is mitigated by having the LC-oscillator coils be of different shapes, making the frequency shifts between both coils inconsistent. Furthermore, even if the attacker knew about the oscillators and had a way to balance the mutual inductance to negate its effects, it would be difficult to circumnavigate around the oscillator coils with two different orthogonal metal layers, as there are different turns hidden beneath strands of interconnects.

The designers claim that using the sensor circuit developed, L-EMA's can be detected on approach and prevented by effectively detecting the EM from the attacker. Furthermore, an L-EMA aims to analyze the device's power consumption and extract the hidden key information. This idea is somewhat significant because it is unique but situational, as later explained. However, given the results, it effectively detects an incoming probe and puts the device in a "locked" mode.

There are several ways that this technique is unique. For one thing, L-EMA's are side-channel attacks that are effective in extracting secret keys from critical areas using the EM radiation of the targeted area. There has been no technology or technique that can counteract this, and the current countermeasures that prevent power analysis are ineffective. This technique is also able to save overhead power consumption during operation. The sensor only needs less than $1\mu s$ to detect an attack making it an efficient technique with few penalties. The sensor system functions intermittently between the idle state of the encryption period and handles attack detection after the cryptographic engine has encrypted data. Furthermore, the interleaving function allows the probe to be detected before the VDD of the chip is turned on. Finally, the technique can be used for side-channel attacks and as a disruptive attack detector. For example, suppose this: the attacker noticed that this technique is being implemented and decided to cut the oscillator coils. In that case, the system will still behave as intended and detect a lower number of oscillations, thereby changing the operation mode of the CE and protecting its secret.

As good as the technique is, it has a few flaws. The method implements CMOS transistors in its oscillator design. With the transistors aging over time, the digital calibration of the sensor will need to consume more power if an attack occurs later in its life cycle. The testing methods that the designers used are limited. During testing, a test chip was fabricated and designed with an AES cryptographic engine mounted on an attack standard evaluation board, the SASEBO R-II. The testing only involved putting an EM probe to only one of the oscillators, L1. As far as the reader is concerned, L2 was never tested, and a simultaneous dual EM probe attack was never proven. Not much detail was provided for the unit under test; the reader is to assume that their test works under a given generic situation. That said, the reader is also free to assume that the technique might not work on a normal device with a CE, and there is no evidence that other functions of the normal device can affect the sensor system. This technique also does not support legacy designs, requiring the hardware to be built around its system. The system also does not provide a way of protecting or detecting tampering prior to the chip's synthesis. The research work also does not provide enough details to replicate the results. The technique mentions that there are two cores, the sensor and the oscillator, but does not give enough detail on the sensor core and its primary functions. After an attack is detected, the chip is supposed to go into a locked state, but the designers fail to provide enough detail on how it becomes locked and if it is possible to unlock the device for further normal operation.

In order to test the technique, the designers fabricated a 0.18 um CMOS transistor with an AES cryptographic engine mounted on a SASEBO R-II evaluation board. An EM probe was manually controlled and monitored using a microscope. The digital calibration performed as intended during its evaluation. The calibration system suppresses f_{LC} within the expected $+/- 1\%$ over the range of temperature from 0 to 60 °C and a range of voltages from 1.6 to 2.0 V. The actual sensor was then evaluated. It performed as expected and changed its operating mode to a locked state when the detector noticed the EM probe approaching the surface of one of the oscillators, L1, with the final measured frequency shift at 5.2%. The technique's power consumption was minimal, at W17 μW in the nominal 1.8 V V_{DD}.

15.3.4 Tamper Detection Using a Temperature-Sensitive Circuit

This tamper detection method can monitor both active and passive intrusion attempts [21]. This method uses a novel mesh built into the printed circuit board. A simple circuit and signal processing chip can be used to monitor for changes in the physical layout of the mesh or for temperature extremes and rapid changes in temperature that may be used to initiate side-channel attacks.

Extreme temperature in integrated circuits can cause Negative Bias Temperature Instability (NBTI). NBTI has a similar effect on transistors as aging can lead to unexpected results and operations coming from the circuit. These can be taken advantage of to steal important data stored in the integrated circuits. Physical intrusion is another method of tampering where the circuit is broken down and probed to reveal valuable data. The mesh method is used to monitor both of these attacks.

This method uses a unique mesh built into the system's printed circuit board. The top layer of the mesh is the signal layer, which has a unique pattern of fine copper traces. A second layer is the ground plane. This mesh has a unique RC constant due to its size and shape. A pulse generator and a signal processing system can then be used to monitor tampering in two different ways. First, if the mesh is somehow altered, such as in the case of a physical intrusion, the RC constant would change dramatically (if it is not shorted or turned into an open circuit). Because of the unique shape and density of the fine copper traces, it would not be easy to work around. The second task is to monitor for temperature changes. The RC constant of the mesh is sensitive to changes in temperature, and temperature extremes or quick temperature changes can generate a tamper detection event. The electronic circuit used in this tamper detection circuit is shown below in Fig. 15.7. The processing circuit monitors tamper detection events, and when it detects one, it sends a signal to the ICs where the sensitive data is stored so the data can be protected or erased.

This design approach assumed that the mesh would work regardless of the shape and size since the shape and size will directly impact the RC constant of the circuit. Another assumption is that the mesh cannot be bypassed and that the mesh modification would not impact temperature detection. However, if the circuit's inputs and outputs could be probed without changes to the mesh and

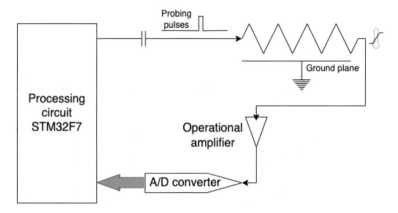

Fig. 15.7 The system described in "Temperature-sensitive active tamper detection circuit"

the target circuit had vulnerabilities outside of temperature variations, then the circuit could still be vulnerable to data leaks and side-channel attacks. The last assumption is that the system's signal processing part cannot be replicated; if the mesh was removed and then simulated with a signal generator (and RC circuit) that generates the expected output, then the device could still be vulnerable to temperature variation attacks. This attack could require substantial engineering to perform without activating the tamper detection signal, however, and if the signal were generated, the valuable data could be deleted.

A benefit to this tamper detection approach is the simplicity of the design. The unique design of the mesh creates a unique capacitance and resistance of the circuit, creating a unique RC circuit constant. In addition, the mesh has fine traces of copper that will vary in size with temperature, allowing the RC constant to be monitored for any changes. Breaking the mesh will also change the RC constant if a short circuit or open circuit is not created, all of which would trigger a tamper detection event.

15.3.5   Advanced Tamper Detection Using a Conductive Mesh

The improved circuit architecture outlined in the previous section was used in this tamper technique. This circuit design provides a solution for safeguarding security circuits such as cryptography modules and other circuits with comparable needs. This approach employs an active tamper detection circuit (ATDC) and a conductive mesh (CM). A conductive mesh is an adjustable or stiff structure composed of many layers with conductive traces separated by dielectric layers. CM completely covers the protected security circuit (SC) and the ATDC. ATDC is a connected electrical component that detects physical invasions and takes necessary action in the event of an intrusion. The schematic diagram of the ATDC, CM, and SC system is depicted

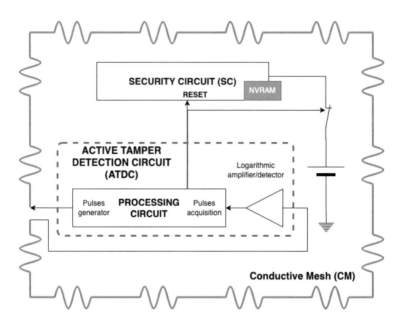

Fig. 15.8 Diagram of the security system, enclosed by the conductive mesh

in Fig. 15.8. The active tamper detection electrical circuit can probe the mesh with signals that an enemy cannot [22].

Pulses are generated by ATDC at specific intervals and sent to CM's input port. The output port (detector and logarithmic amplifier) is provided with the signal being passed through the mesh. This amplifier generates a continuous signal with an amplitude proportional to the input signal level from the pulses. ATDC periodically probes the CM. If deviations are detected from the reference values, it wipes the secret data and resets the SC. This approach is easily scalable and can detect tampering throughout the entire chip. This scalability aids in the prevention of unequal tamper detection throughout a chip and removes the need for sophisticated internal architecture such as switching buses. When a chip detects tampering using this manner, it wipes the secret data, leaving it useless or unusable depending on the application. A separate battery powers the non-volatile RAM, which often stores secret data. Removing this power supply causes irreversible data loss and is the fastest way to wipe data, although it necessitates using a backup power supply. The backup power supply is also critical to power the ATDC when the main supply does not power the equipment (including the SC). ATDC is designed to utilize less power because it must function continuously to secure the secret data.

The CM is created with three layers that surround the target circuit. An innermost layer is a ground plane used for reference and shields the circuit from tamper detection. The middle layer is a single-path sensory grid with a single input and output. The outermost layer directly matches the middle layer, except the grid is short-circuited in key areas to divide it into independent (i.e., not interconnected) zones. Each zone consists of a single loop. A dielectric material separates each layer in this grid, and as a result, the middle and outermost layers are capacitively and inductively coupled with the middle layer, affecting the transfer characteristics.

The ATDC system sends pulses at varying frequencies into the conductive mesh. It then amplifies the output and compares it to a reference value. If any tamper resulted in modifying a trace in the outermost layer, the resultant voltage would differ, indicating a tamper. Any tamper detection causes the ATDC to disconnect the power to the NVRAM, erasing its contents.

The combination of a CM and ATDC successfully detected "tamper" events that broke any path in the outermost layer. This behavior was successfully observed with three different dielectric thicknesses (0.1 mm PES dielectric foil, 0.3 mm PCB, and 0.6 mm PCB). However, these CMs appear to be tested as standalone devices, not integrated into an IC. Each CM was tested with 32 frequencies, starting at 50 MHz with a 5 MHz step. Each frequency pulse lasted 250 ms, and measurements were taken after 20 ms of a pulse. Notably, even though each CM could detect tampering, the CM with the thinnest dielectric was only successful in detecting up to 90 MHz.

This tamper detection design is a low-cost solution that can effectively detect tamper events in a system that a CM surrounds. This approach can be used in many applications and could likely detect macro-scale tamper events (such as on PCBs). Additionally, the presence of the ground plane helps shield the protected system from environmental interference and interference from the tamper detection circuit.

Unfortunately, there are some limitations and drawbacks to this design. First, this design requires an auxiliary power source to keep the ATDC system operational. This results in power overhead, leading to unintentional memory clears if the system is not handled appropriately. Second, integrating this methodology into a chip package will require wafer overhead for the ATDC system (basic microcontroller/FSM, ADC, logarithmic amplifier). It may even need new, highly specialized foundry methods for creating the CM for each chip (resulting in additional cost). Third, if the reference values and sequence for the CM become known by an attacker, the output signal from the CM could be spoofed to bypass the system. Finally, this design is not tested for the effect of temperature on the system—something that would require further testing [9].

15.4 Conclusions

Tamper detection has a wide range of applications in electronics security, varying between devices, environment, and level of invasiveness. In addition, various technologies are being researched as a bulwark against a multitude of hardware security attacks. First, we covered the FLATS architecture, whereby utilizing unused LUTs as IR watermarks lead to the prevention of post-synthesis attacks and tampering. The second method uses the addition of the digital signature to ICs as a countermeasure for recycling used ICs, further sifting through the tampered ICs in the market. Though, an RO-based method is sufficient to identify a recycled or a new IC if the NVM is impossible to tamper. Nevertheless, it is not the case in the real world; hence, this method requires a digital signature to verify NVM contents for tamper detection. Finally, the next method discussed in this chapter uses an LC-oscillator circuit to detect and counteract against L-EMAs to crypto-engines. However, it does not prevent tampering prior to the fabrication of the sensing coils. Next, we discussed a tamper detection method that uses a combination of a conductive mesh and sensing circuit, which effectively increases security by actively detecting tampering across the entirety of the IC. However, the overall power consumption and unintentional memory clearing can be detrimental to this method. Finally, we covered a method that enhanced an approach over the previously discussed method that actively monitors intrusion in the system's physical layout by using a mesh that encompasses the PCB of a system. However, it is situational because it works if the temperature changes are used before tampering. Despite its downsides, each technique effectively detects and further secures tamper-sensitive hardware.

References

1. Ahmed B, Bepary MK, Pundir N, Borza M, Raikhman O, Garg A, Donchin D, Cron A, Abdel-moneum MA, Farahmandi F, et al (2022) Quantifiable assurance: from IPs to platforms. Preprint. arXiv:220407909
2. Alam MM, Tehranipoor M, Forte D (2016) Recycled FPGA detection using exhaustive LUT path delay characterization. In: 2016 IEEE international test conference (ITC). IEEE, Piscataway, pp 1–10
3. Alam M, Chowdhury S, Tehranipoor MM, Guin U (2018) Robust, low-cost, and accurate detection of recycled ICs using digital signatures. In: 2018 IEEE international symposium on hardware oriented security and trust (HOST). IEEE, Piscataway, pp 209–214
4. Becher A, Benenson Z, Dornseif M (2006) Tampering with motes: real-world physical attacks on wireless sensor networks. In: International conference on security in pervasive computing. Springer, Berlin, pp 104–118
5. Bhunia S, Tehranipoor M (2018) Hardware security: a hands-on learning approach. Morgan Kaufmann, Burlington
6. Cruz J, Farahmandi F, Ahmed A, Mishra P (2018) Hardware trojan detection using ATPG and model checking. In: 2018 31st international conference on VLSI design and 2018 17th international conference on embedded systems (VLSID). IEEE, pp 91–96
7. Duncan A, Rahman F, Lukefahr A, Farahmandi F, Tehranipoor M (2019) FPGA bitstream security: a day in the life. In: 2019 IEEE international test conference (ITC). IEEE, Piscataway, pp 1–10
8. Duncan A, Skipper G, Stern A, Nahiyan A, Rahman F, Lukefahr A, Tehranipoor M, Swany M (2019) FLATS: filling logic and testing spatially for FPGA authentication and tamper detection. In: 2019 IEEE international symposium on hardware oriented security and trust (HOST). IEEE, Piscataway, pp 81–90
9. Farahmandi F, Huang Y, Mishra P (2020) Automated test generation for detection of malicious functionality. In: System-on-chip security. Springer, Cham, pp 153–171
10. Guin U, Forte D, DiMase D, Tehranipoor M (2014) Counterfeit IC Detection: Test Method Selection Considering Test Time, Cost, and Tier Level Risks
11. Guin U, Huang K, DiMase D, Carulli JM, Tehranipoor M, Makris Y (2014) Counterfeit integrated circuits: a rising threat in the global semiconductor supply chain. Proc IEEE 102(8):1207–1228

12. Guo Z, Xu X, Tehranipoor MM, Forte D (2019) EOP: an encryption-obfuscation solution for protecting PCBs against tampering and reverse engineering. Preprint. arXiv:190409516
13. Hazari NA, Niamat M (2017) Enhancing FPGA security through trojan resilient IP creation. In: 2017 IEEE national aerospace and electronics conference (NAECON). IEEE, Piscataway, pp 362–365
14. Khaleghi B, Ahari A, Asadi H, Bayat-Sarmadi S (2015) FPGA-based protection scheme against hardware trojan horse insertion using dummy logic. IEEE Embed Syst Lett 7(2):46–50
15. Mazumder Shuvo A, Pundir N, Park J, Farahmandi F, Tehranipoor M (2022) LDTFI: layout-aware timing fault-injection attack assessment against differential fault analysis. In: IEEE computer society annual symposium on VLSI (ISVLSI). IEEE, Piscataway
16. Miura N, Fujimoto D, Tanaka D, Hayashi Yi, Homma N, Aoki T, Nagata M (2014) A local EM-analysis attack resistant cryptographic engine with fully-digital oscillator-based tamper-access sensor. In: 2014 symposium on VLSI circuits digest of technical papers. IEEE, Piscataway, pp 1–2
17. Pundir N, Li H, Lin L, Chang N, Farahmandi F, Tehranipoor M (2022) Security properties driven pre-silicon laser fault injection assessment. In: International symposium on hardware oriented security and trust (HOST)
18. Ravi S, Raghunathan A, Chakradhar S (2004) Tamper resistance mechanisms for secure embedded systems. In: Proceedings of the 17th international conference on VLSI design. IEEE, Piscataway, pp 605–611
19. Shi Q, Vashistha N, Lu H, Shen H, Tehranipoor B, Woodard DL, Asadizanjani N (2019) Golden gates: a new hybrid approach for rapid hardware trojan detection using testing and imaging. In: 2019 IEEE international symposium on hardware oriented security and trust (HOST). IEEE, Piscataway, pp 61–71
20. Vashistha N, Lu H, Shi Q, Woodard DL, Asadizanjani N, Tehranipoor M (2021) Detecting hardware trojans using combined self testing and imaging. IEEE Trans Comput Aided Des Integr Circuits Syst. 41(6):1730–1743
21. Vasile D, Svasta P (2017) Temperature sensitive active tamper detection circuit. In: 2017 IEEE 23rd international symposium for design and technology in electronic packaging (SIITME). IEEE, Piscataway, pp 175–178
22. Vasile DC, Svasta P (2019) Protecting the secrets: advanced technique for active tamper detection systems. In: 2019 IEEE 25th international symposium for design and technology in electronic packaging (SIITME). IEEE, Piscataway, pp 212–215
23. Villasenor J, Tehranipoor M (2013) Chop shop electronics. IEEE Spect 50(10):41–45
24. Villasenor J, Tehranipoor M (2013) The hidden dangers of chop-shop electronics: clever counterfeiters sell old components as new threatening both military and commercial systems. IEEE. Spectrum (cover story). https://scholar.google.com/citations?view_op=view_citation&hl=en&user=n9JsBeAAAAAJ&cstar t=600&pagesize=100&citation_for_view=n9JsBeAAAAAJ:X9ykpCP0fEIC
25. Whitworth D (2011) Gaming industry lose "billions" to chipped consoles. BBC Radio 1
26. Xiao K, Tehranipoor M (2013) BISA: Built-in self-authentication for preventing hardware trojan insertion. In: 2013 IEEE international symposium on hardware-oriented security and trust (HOST). IEEE, Piscataway, pp 45–50

Chapter 16
Counterfeit and Recycled IC Detection

16.1 Introduction

Security has grown increasingly crucial in recent years in many parts of computing. It is especially true regarding hardware, which is usually referred to as the "root of trust" for a device. Counterfeit ICs and PCBs themselves generate billions of dollars every year [5]. An electronic device is labeled as counterfeit as it is recycled, remarked cloned, or overproduced copy of the original device [10]. These counterfeit electronics can jeopardize consumer safety, cost electronic design houses significant revenue, and damage brand recognition [17]. According to a report published to the US Congress by the US Armed Services Committee, 15% of all electronic components that are spare and replacements that are acquired by the Pentagon are counterfeit [6]. In addition, a well-known mobile and computing device manufacturer discovered that a recycling facility is illegitimately reselling over a hundred thousand devices with a market value of twenty-three million [4]. Furthermore, counterfeit items pose serious safety and national security concerns in mission-critical systems such as space and military [20]. An OEM is always concerned about protecting its business revenue and brand value in the supply chain. Other than the original equipment manufacturer, entities such as IC distributors can sell counterfeit ICs. Therefore, any entity other than the OEM can introduce counterfeit devices into the supply chain. This chapter examines numerous ways to identify and prevent the entry of recycled chips into the semiconductor supply chain. A recycled IC is a discarded chip after an electronic device or system has reached its end of life, and after that, it is reintroduced into the market, modified, and marketed as a new chip.

Among all kinds of counterfeit chips, recycled chips account for more than 80% of all counterfeits, which bring billions of dollars worth of counterfeit chips into the supply chain each year [21]. A hardware security problem with this severity level necessitates the development of distinctive and novel solutions. A typical solution is to employ ring oscillators (ROs). The frequency of these ROs decreases over time, implying that the chip has been in use. Inverters are often arranged in a circular chain. The wear and tear on traces and transistors in a circuit causes the change in RO's frequency over time. This technique, however, has certain drawbacks, such as chip-to-chip fluctuation and limited accuracy. Furthermore, because of its isolation from the rest of the chip's functionality, the RO is an easy target for modification and, in some situations, removal.

This chapter further discusses various strategies aimed at improving sensor technology and providing a unique solution to the recycled IC problem. Each of these approaches, such as the ROs, has advantages and disadvantages. All of the solutions recognize that for a sensor to be effective, it must have a small footprint, low power consumption, and be simple to verify the degree of utilization of a chip under test (CUT). The examined sensors may be used as comprehensive solutions to the problem and can potentially be used as IP cores in an existing design.

Finally, by thinking beyond hardware security primitives, this chapter discusses the counterfeit chips detection approach based on a consortium-style blockchain called *eChain* (electronic Chain). *eChain* can map semiconductor supply entities as peers of the blockchain and can be used for verifying the authenticity of ICs.

16.2   Background

Various researches address methods to detect used ICs that have been resold as new. There are two possible ways to combat this problem: first, detecting whether an IC has been used based on its circuitry. Ring oscillators, anti-fuse memory, and fuse memory can be used as security primitives to detect prior usage. Second, providing each IC a unique ID that can be checked from a distributed secure and tamper-proof database to verify whether an IC has been sold more than once. Clock sweeping, path delays, and unique IDs (on-chip markings and RFID tagging) can be used for storing IC information in a distributed database, aka blockchain.

ROs are odd-numbered gates where altering one gate's output causes the next gate's output to change to the opposite value [2, 13, 18]. These are typically shown as a series of inverters, although they may easily be NAND or NOR gates if properly set. Because the frequency of their oscillations lowers with age, ROs are useful for detecting aged semiconductors. This is helpful because a design can have two oscillators, a reference, and a strained oscillator. The frequency difference between their oscillations determines the age of the strained oscillator. The stretched oscillator gets its name from the fact that it is constantly on while the semiconductor is in normal operation. The reference oscillator, on the other hand, is only used when comparing the stressed oscillator or during particular manufacturing steps.

Two variables that cause IC components to deteriorate are hot carrier injection (HCI) and negative bias temperature instability (NBTI). When a carrier, such as an electron or a hole, accumulates enough kinetic energy to break the potential energy barrier that stops it from passing from the conducting channel to the gate dielectric of a MOSFET, HCI occurs. As a result, leakage current rises, and dielectric material deteriorates. Over time, this activity will lower the transistor's threshold voltage, the current, and the circuit's operating frequency. A negative gate voltage traps holes in the MOSFET's dielectric or creates interface states, which are subsequently positively charged, resulting in NBTI. The effects of NBTI can be recovered after the transistor no longer sees the stressing voltage. This is considered in the design of the ring-oscillator circuit, which is discussed further in the studies.

Anti-fuse memory was developed to enable a more accurate age measurement for integrated circuits. It operates by counting and measuring the clock cycles on various nets within the integrated circuit. The circuit will write to the next lowest address of anti-fuse memory after a specified amount of oscillations from either the clock or the nets. Because anti-fuse can only be written once and cannot be deleted, hence it is temper resistant. Other forms of memory are generally either volatile (i.e., they are deleted when the circuit is turned off) or rewritable (i.e., they may be written several times). The precision of this technique is limited by the fact that any time spent in use before the next memory write operation is lost if the IC is turned off. So, if the IC were intended to write to the anti-fuse memory every hour, but it was only switched on for 40 minutes at a period, it would appear that it had never been utilized. Because anti-fuse memory blocks might take up a significant amount of the IC, the frequency of write operations must be matched with the maximum number of planned write operations.

The fuse memory technology has the benefit of being extremely compact, allowing it to be used in small digital, analog, and mixed-signal ICs [23]. The fuse technology is based on the notion of having a single, one-time use fuse that blows when the chip is utilized for the first time in the field. It will be feasible to check the chip's operation without blowing a fuse by setting the voltage of the test pin to 0 during any production testing. Even though the remainder of the chip will operate equally regardless of the voltage provided to the test pin, this technology relies on whoever is using it to follow the designer's intentions and set the voltage of the test pin to V_{D_D} when they use it.

Instead of attempting to establish whether a chip has previously been used, another option is to issue each chip a unique ID that will be maintained in an immutable database. This allows the user to look up the ID of their chip and see if it has been sold before in the database. This approach includes clock sweeps to estimate route delays and serial IDs (markings, ECID, and RFID tagging).

The method of adjusting the clock to different frequencies to identify values that cannot be transmitted via the circuit is called clock sweeping [22]. It can be used to calculate the path's delay. Due to manufacturing process variation, path delays differ from one IC to another IC. The threshold voltage, gate length, and oxide thickness for a particular transistor will differ from one IC to another IC, resulting in different low-to-high and high-to-low timings. If an IC's design has X routes and there are N ICs that need unique IDs, then N IDs of length X-1 will be created. This ID is created by sorting the route delays of an initial chip in ascending or descending order. The pathways for this first chip are utilized for all future chips in the same sequence. When a delay is bigger than the delay before it, a value of 1 is assigned to the chip's ID; otherwise, a value of 0 is assigned. The chip's ID is made up of these

1s and 0s linked together. This approach picks the pathways to employ based on the path delays observed from chip to chip having a specific level of variability.

Radio frequency identification (RFID) tags are miniature radio transponders that broadcast a digital ID when activated by an external electromagnetic pulse. Active RFID tags, which require an internal battery, and passive RFID tags, which are powered by an external electromagnetic pulse, are the two types of RFID tags. Because of their smaller size, lower cost, and longer lifespan, passive RFID tags are frequently preferred over active RFID tags. These tags may be included in a chip package, requiring no changes to the IC's architecture.

Blockchains are decentralized systems in which each peer retains a same copy of a distributed database known as a ledger. After all, once the transaction's peers have approved it, a transactional record on the blockchain is finalized on the ledger. The distributed ledger is immutable, which means that previously created records on the ledger cannot be changed or deleted. This ledger can store an IC serial marking (or any other unique identifier).

16.3   Recycled and Counterfeit IC Detection

In this section, we will cover various methods to identify recycled and other counterfeit chips.

16.3.1   IC Fingerprinting Using Lightweight On-chip Sensor

It would be challenging to differentiate recycled ICs from unused ICs with identical functionality and packaging. On the other hand, a hardware security engineer can use various types of lightweight sensors that are on the chip to determine recycled ICs by circuit usage time measurement, namely aging in ring oscillators and anti-fuse (AF-based) [22]. For sensors based on RO, statistical data analysis may be performed to isolate the impacts of temperature fluctuations and process variations on the sensor due to sensor aging in the ICs. Furthermore, in AF-based sensors, counters and integrated one-time programmable memory are utilized to capture the runtime of ICs by calculating the cycle of the system clock or the switching actions of a certain amount of nets that are in the design. Next, the evaluation of usage time of an IC saved in anti-fused sensors exhibits that recycled chip may be correctly detected. Finally, simulation and silicon findings from 90-nm test circuits employing 90-nm technology illustrate the usefulness of ring-oscillator-based sensors for recognizing recycled ICs.

16.3.1.1   RO-Based Sensor

The RO-based sensor was the first ever primitive developed for recycled IC detection. This on-chip sensor is composed of two ROs, the stressed RO and a reference RO. Recycled integrated chips are those chips that have been used before, cleaned up, and packed to be offered as a new chip. Using the CDR sensors, the two ROs will have the same frequency when the chip is new. However, during the usage of ICs, the stressed RO will start to slow down due to aging. The aging effect in the stressed RO is caused due to HCI and NBTI effects on repeated use of the IC. Only the stressed RO goes aging effect because it is tied to the IC's functionality and is switched on during normal chip operation, whereas the reference RO is only turned during the frequency comparison.

Figure 16.1 shows the reference circuitry for CDR sensor [22]. The sleep transistors, which are situated between inputs and V_{DD} and V_{SS}, regulate the ROs. The mode signal controls these transistors, which selects between manufacturing, normal, and authentication modes. The ROSEL signal determines which oscillator output is transmitted via the MUX during authentication to be monitored. The counter keeps track of how many oscillations occur within the period allotted by the timer. To test the effectiveness of their design, the researchers performed both simulation and silicon tests. HSPICE MOSRA from Synopsys developed the simulation findings to estimate the impacts of aging on the transistors. One thousand chips were made with a Monte Carlo variant and aged for 24 months in 1-month increments. Due to the proximity of ROs in the system, temperature and voltage changes are ignored for the initial batch of data. The initial study compared 21-stage ring oscillators to 51-stage ring oscillators and discovered that they could correctly recognize one month of aging as a completely recovered IC 100% of the time. In addition, it has been discovered that utilizing large-stage oscillators reduces

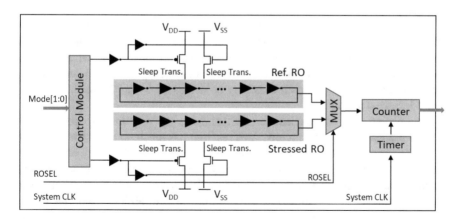

Fig. 16.1 Schematic of the ring-oscillator-based sensor

the difference in frequency between the reference and stressed oscillators, but not to the point where it affects the capacity to detect old transistors.

Next, the simulations are repeated on the 21-stage oscillators with higher temperature and process variation. The success rate for detecting chips after one month fell to 95.2% simply by changing the settings of the process variation factors. When the temperature was added on top of that, the percentage fell even lower to 92.3%. These simulation findings were 100% successful in detecting used chips after six months. The researchers thought their process variation and temperature estimates were prudent because, in actuality, they will build the oscillators to be as close together as possible to reduce both impacts.

The chip includes 96 delay chains that operate in ring-oscillator mode by manipulating various input signals [16]. Six of these ROs are chosen to develop three RO-based sensors.

Three types of 201 chain CDRs were employed in the silicon testing conducted by the researchers. It is worth noting that they have 201 steps with complicated gates (NAND, buff) against the simulation's 21-stage inverter gates:

1. The RO-based1 architecture consists of two identical ROs (R_{RO1} and S_{RO1}), each having 200 SVT buffers and one SVT NAND.
2. The RO-based2 comprises two ROs (R_{RO2} and S_{RO2}), comprising of one HVT NAND gate and 200 HVT BUFs. These two ROs are identical.
3. The RO-based3 consists of two ROs (R_{RO3} and S_{RO3}). Each RO consists of 201 HVT NAND gates.

Fifteen test chips were aged for 80 hours at 135 °C with a 1.8 volt supply and then restored to 25 °C for authentication. These parameters of aging were used to accelerate the aging process. After 80 hours of use, the researchers could recognize the recovered ICs with 100% accuracy. The efficacy of the three RO-based sensors examined was in this order: second design, third design, and first design. The HVT gates outperformed the SVT gates, while the buffers outperformed the NAND gates in terms of aging time. It is also worth noting that RO-based1 and RO-based2 were detected quicker than RO-based3. This faster detection rate could be due to spatial differences between ROs that are not placed close to each other, which caused some ROs to be faster than others. For the most effective RO-based sensors, the detection rate variance between ROs may be decreased by putting both ROs in a single localized module.

This design uses an efficient circuit that takes up very little space in more extensive digital integrated circuits. As a result, it has much potential for detecting completely recycled ICs. However, the RO-based sensor's applicability is restricted since older technology nodes, such as those used by the military and space sectors, do not age as fast as newer technology nodes. This design can be further improved by altering the ring oscillators to experience the effects of NBTI all the time rather than half of the time. The main drawback of this design is that it is incompatible with smaller or analog ICs, and it only detects completely recovered ICs rather than preventing them from accessing the market. It is also ineffective against other counterfeit ICs, such as cloned or overproduced ICs.

16.3.1.2 Anti-fuse (AF)-Based Sensor

The RO-based sensor is constructed by connecting the inverters of reference and stressed ROs. This is done to reduce the impact of intra-die process variation. However, it is challenging to eliminate the inter-die process variation effect on the sensor completely. The age of the recycled IC is detected by measuring the difference in frequency between the stressed RO and the reference RO. As a result, depending on how effectively the frequency of stressed RO degrades with time, the minimum number of days after which a recycled IC can be recognized is constrained. For example, the results in [16] showed that the sensor could not detect recycled ICs if they had been used for less than 30 days.

To eliminate the problem of process variation and determine recycled ICs that are used for a short period of time, for example, one day, two AF-based sensors, i.e., CAF (clock anti-fuse)- and SAF (signal anti-fuse)-based sensors, can be used. The main difference between the two designs is that CAFs use counters to track the high-frequency system to clock the IC under operation (see Fig. 16.2). This method can provide accurate time for how long recycled ICs have been used, with an adjustable measurement scale and total measurement time depending on the IC's application. In contrast, the SAF, which has lower accuracy than CAF, counts the switching activity of specific nets in the design. It then compresses this information to reduce area overhead for storage on the chip's memory block (see Fig. 16.3). Both devices have a second counter that may regulate the time intervals the sensor detects (T_s):

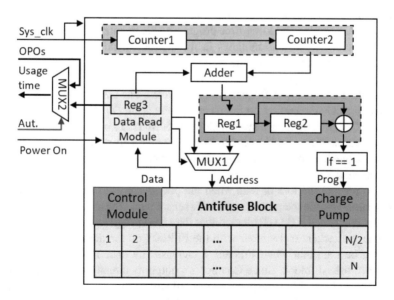

Fig. 16.2 Circuit illustrating the structure of CAF-based sensor

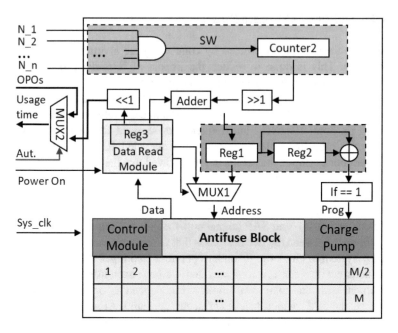

Fig. 16.3 Circuit illustrating the structure of SAF-based sensor

1. **Clock Anti-fused-based Sensor**: The circuit illustrating the configuration of the CAF sensor is shown in Fig. 16.2. This sensor consists of a data read unit, two counters, an adder, and a one-time programmable memory block. The high-speed system clock to various modules in the sensor, including the anti-fuse block, registers, and data read module, is represented by $Sysclk$ in the figure. Counter 1 is used as a frequency divider to generate a lower frequency clock, whereas Counter 2 counts lower frequency signal cycles. The sizes of these two counters are adjustable depending on the measurement scale (T_s) and total measurement time (T_{total}). Assuming a system clock of 50MHz, T_s of one hour, and T_{total} of one year, then one would need a 38-bit Counter 1 to measure usage time from 20ns to one hour and a 14-bit Counter 2 for measuring usage time from one hour to one year.

   To prevent the loss of data when the power supply is turned off, one would need non-volatile memories to store the sensor data. The usage time data is stored in the embedded AF-OTP block rather than field programmable ROM (FPROM) because attackers can tamper with the FPROM on the device. From the figure, it can be seen that when Counter 2 increases by one, 1'b1 is assigned to the prog in the AF block. Since Counter 2's output is directly connected to the AF block's address, the corresponding AF cell will be directly programmed as 1'b1. The usage time of the chip under test (CUT) is based on the largest address whose content is 1, which is determined by Counter 1's measurement scale setup. This setup of two counters helps to reduce the AF block's size. However, the same address signals are shared between program and read operations, which are selected using a multiplexer (MUX1 in Fig. 16.2) controlled using the data read module. The AF block briefly enters read mode when the power supply is turned on. During this time, the data read module generates read addresses that are passed through MUX1, and the traversing binary tree is used to traverse all AF cells.

The important thing to note is that after each power on the counters starts from zero, however, to compute the total usage time, it is essential to know the last usage time. Therefore, once the last usage time is determined, it is stored in Reg3 and sent to the adder to add to the total usage time. Furthermore, the adder's data is sampled using Reg1, Reg2 buffers (or delays) the Reg1 data by a clock cycle, next XOR gate is used for comparing Reg1 and Reg2 data. If the usage time increases, the data from Reg1 and Reg2 will differ, and the data from Reg1 will be routed to the address in the AF block via the MUX1. As a result, a new anti-fuse cell with a larger address will be programmed in conjunction with the value in Counter 2 (the usage time after power-on), and the anti-fuse one-time programmable block will be updated with the new total duration. By designing the sensor in this manner, the likelihood of tampering or altering attacks on the anti-fuse-based sensor is reduced. Also, by using another MUX2 and authentication pin to send out usage time to IC's output pins, one can avoid the need for additional authentication pins on the chip. Therefore, no additional output pins are needed, and the pins from the original design can be reused. The routing of data happens as follows. In normal operation mode, regular primary outputs are routed through MUX2. In read mode (authentication mode), MUX2 routes the usage time, and in manufacturing test mode, pins can be used to apply fault test patterns to the sensor.

2. **Signal Anti-fuse-based Sensor**: For smaller designs, the area overhead of a clock anti-fuse-based sensor with two counters could still be considered large. The area overhead can be reduced by using a sensor that is signal anti-fuse-based that is built on signal switching activity. The block diagram of signal anti-fused-based sensor can be depicted from Fig. 16.3. Compared to the design in Fig. 16.2, the structures of the SAF-based and CAF-based sensors are similar to each other. CAF-based sensors count system clock cycles to measure IC utilization time. In contrast, SAF-based sensors measure the number of positive edges (switching activity) of a defined amount of nets in the design. Simulations are used to select a precise number of nets for the input of an AND gate. According to the nets selection rule, the AND gates' switching activity must fulfill the criteria of the measuring scale. For example, if T_s is 1 hour, one of the options could be four nets with SW(N_1) = 30/60 minute, SW(N_2) = 24/60 minute, SW(N_3) = 25/60 minute, and SW(N_4) = 24/60 minute, respectively. The SW, on the other hand, could be significantly different with different functional inputs. As a result, only signals with consistent SW under varied inputs are considered for developing a signal anti-fused-based sensor. Counter 2 in the sensor counts the positive pulse of the AND gate output (SS signal in Fig. 16.3).

The area overhead can be reduced even further by dividing the value in counter 2 by 2, and the largest address of anti-fuse cells with 1 represents [SW/2]. The switching activity is calculated by [SW/2] 2 using a one-bit left shifter. The recorded SW will represent IC utilization time. As a result, the number of anti-fuse cells in sensors that are signal anti-fuse-based would be lower than that in CAF-based sensors. However, the CAF-based sensor is more accurate than the SAF-based sensor because of the following reasons: (1) SAF-based sensor is based on the switching activity of a specific amount of nets in the netlist, and in contrast, the clock anti-fuse-based sensor calculates the number of cycles of the system clock and (2) the signal anti-fuse-based sensor loses some utilization time details due to the shifters.

For testing the anti-fused-based sensors, a scan-chain-based [14, 15] testing technique can be considered. Because of the counters and the AF OTP block, the area overhead of the two AF-based sensors is more considerable than that of the RO-based sensor. As a result, the manufacturing test price would be higher based on the measurement scale and area overhead. However, in comparison to the millions of gates in current integrated circuits, the area and test overheads are insignificant [7–9]. In contrast to RO-based sensors, the consumption time recorded in anti-fuse-based sensors by AF-based sensors is unaffected by technologies (new technology nodes age faster than older technology nodes), packages, assemblies, or process variances. As a result, the anti-fused-based sensor can precisely predict the utilization time of the chip under test, though different foundries fabricated the design at different times. Furthermore, due to the tiny measurement scale, AF-based sensors are capable of detecting ICs that have been used for less than a day.

Because their memory blocks limit the capacity of the CAF and SAF sensors to identify recycled IC, an overhead study was conducted on a design with 500 gates and 12 kb of system programmable memory. However, this comes at the expense of precision since the RO sensor is only expected to resolve for up to one month. Although the CAF sensor does not appear capable of collecting power-on durations at 1-hour intervals, considering the little overhead at the 1-day resolution, it is plausible to estimate the operational capacity is not much more than an hour. On the other side, the SAF sensor could catch power-on at 1-hour intervals.

The SAF and CAF sensors' fundamental limitation is that they can only determine power-on times above Ts. A study was undertaken to assess the actual usage time vs. the anticipated usage time about a time interval, T_s, to examine the influence of this on the IC usage time accuracy. When the power-on time (T_{pon}) is less than T_s, the projected use time (T_{esm}) is presumed to be zero. The sensor precisely detects all of the system's power-on time. For $T_s, > T_{pon}$, the CAF-based sensor outperforms the SAF-based sensor since the SAF cannot detect times under $2T_s$ owing to the lower overhead caused by the one-bit right shifter. Thanks to this feature, CAF may consistently beat SAF by a factor of two in terms of time resolution. Furthermore, because the SAF sensor is dependent on the chance of detecting the output of the switching from the gate(p), the probability that the switch occurs several times during T_s is 1p. Even if the estimations are off, the sensors can identify recycled ICs at T_s and $2T_s$ for the CAF and SAF, respectively.

Finally, the numerous sensors involved were subjected to an attack analysis. For the RO sensor, as discussed, tampering strategies such as removal and manipulation of the system were judged to be conceivable but not cost-effective. Similarly, it was judged that the burn-in mechanism used by attackers to age the reference RO would be too costly to justify. However, it is worth mentioning that a careless attacker may use the burn-in approach to destabilize a chip without being discovered. A timer can be placed on the reference RO pins as a countermeasure to this attack. For the AF sensors, an adversary can attempt to turn them off to hide the usage time, but because the memory is maintained while the power is on, this was ruled impossible without damaging the device. Furthermore, there had been no successful reverse engineering assaults on the AF OTP block at the time of publishing. Hence, attacking the actual memory block was judged unsuccessful. Disabling the link between the counter/signal connection was the final attack vector, which was judged difficult without previous knowledge of the net connections.

16.3.2   NBTI-Aware CDIR Sensor

Numerous integrated circuit types (digital, analog, and mixed signal) and sizes (big or small) make it exceedingly difficult to create a complete solution for recycled IC prevention and detection. This counterfeit IC detection technique introduces solutions to combat die and IC recycling (CDIR) [11]. Recycled integrated chips are those chips that have been used before, cleaned up, and packed to be offered as a new chip. Lightweight, on-chip sensors based on ring oscillators (RO-CDIR), anti-fuses (AFCDIR), and fuses are among these solutions (F-CDIR). Each sensor is adapted to various components' individual needs and limits, resulting in good recycle part detection. HSPICE simulation results utilizing 90nm technology illustrate the usefulness of this method for identifying ICs utilized for a concise amount of time using NBTI-aware RO-CDIR.

The RO-CDIR aims for the reference RO to age as little as possible, while the stressed RO ages as much as possible during regular chip operation. Unfortunately, the stressed RO did not age as well as it could have under the old design. Because NBTI is one of the critical causes of aging in the RO transistors, the only way to maximize the impacts of aging on the stressed RO is to have these transistors endure this effect all of the time when the chip is in regular operation. Because every other gate's input was grounded in the prior configuration, only half of the transistors were affected at any moment. This should more than double the effect of NBTI on their aging since it is twice the time they are exposed to NBTI and eliminates the NBTI recovery that occurs when the transistors are turned on. Every gate was linked to the ground during normal operation using the modified configuration illustrated in Fig. 16.4. This is performed by connecting all of the inverters with a pass transistor. During normal operation, they ground each input, but during authentication, they

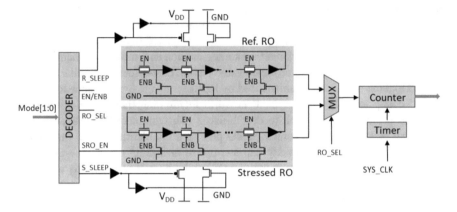

Fig. 16.4 Circuit illustrating the architecture of NBTI-aware RO-CDIR design

enable each inverter's output to be transferred into the input of the next one. For
current medium and large ICs, the space overhead and power consumption for this
circuitry are low, especially if the counter and timer are off-chip and utilized during
authentication.

This design introduces the concept of low-cost fuse technology (F-CDIR) for
tiny and analog ICs. The only extra input required by this design is a Test pin
on an IC. The concept behind the fuse technology is that the first time the chip
is utilized in regular functioning mode by whoever purchased it, a one-time use
fuse will blow. If the resistance between two pins on the IC is low, the fuse has not
yet blown; otherwise, the IC has fully recovered. The structure shown in Fig. 16.5a
is made up of a switch and a fuse. It features three terminals. The two terminals
are connected to V_{DD} or GND pins. The Test pin on the IC controls the control
terminal. The MOSFET serves as the switch in this design. During manufacturing
and burn-in testing, the pin Test will always be "0," indicating that no current
will flow through this structure. The Test pin will be connected to V_{DD} when the
component is installed in the system for usual operation. When the MOS is turned
on, current flows through the fuse, resulting in an open circuit. The device will then
function normally as it was intended. Figure 16.5b depicts the use of this structure
in differential designs. The structure is located between the differential output pins,
O+ and O−. The Test pin is connected to the control pin. Figure 16.5c depicts
our second F-CDIR structure proposal. There is only one semiconductor fuse in the
design. The sensor's terminals are connected to the Test and GND pins.

Counterfeit (in-field) components will be detected by measuring the resistance
between: (i) V_{DD} and GND pins while setting the Test pin to V_{DD} for F-CDIR
I and (ii) Test and GND for F-CDIR II. For a new component, the measured
resistance should be negligible. However, the measured resistance will be high if
the component has previously been used (infinite). Chips can be authenticated as
new and authentic using low-cost devices such as a multimeter.

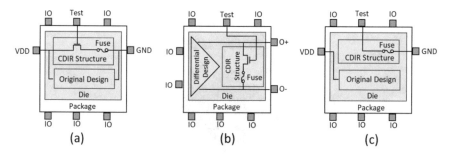

Fig. 16.5 (**a**) AF-CDIR design I (**b**) AF-CDIR design I with differential input (**c**) AF-CDIR design II

The results reveal a significant improvement over the initial design since completely recycled ICs are identifiable after 15 days in virtually all situations, rather than one month, and the 51-stage PV0 simulations had a minimal misprediction rate.

Smaller nodes can use the upgraded NBTI-sensitive RO-CDIR to cover medium and large digital ICs. Finally, the F-CDIR is agnostic of node size and covers all sizes of ICs, both digital and analog. However, the F-CDIR's ability to achieve this is limited since it operates as a binary operator to signal if the chip has been used previously. For it to be useful, the system integrator must utilize it as intended and check its status prior to regular operational usage. While these strategies effectively detect recycled ICs, they do not prevent it from happening. Instead, they increase the likelihood that it will be recognized. Unfortunately, they are still ineffective in fighting other types of counterfeiting.

16.3.3 IC Tracing in Supply Chain

This recycling detection method involves embedding an RFID chip on-chip packaging to trace them from the manufacturer via distributors to the system integrator [21]. The technique involves placing a tiny non-volatile memory (NVM)-equipped radio frequency identification (RFID) tag on the packaging. The chip must also have an electronic chip ID (ECID) and a ring oscillator (RO). The registration data in the tag, which includes the RO frequency and frequency measurement conditions, can be saved in a database. To prevent tampering with the records (RD) and ECID, a digital signature computed on the records (RD) and ECID must be stored in the tag. By analyzing/comparing the measured chip values with the RO frequencies in the tag, recycled ICs can be detected. This signature is then added to a blockchain ledger of previous entries and saved in a manufacturer's database. Each entity's read, verify, and update procedure (see Fig. 16.6) not only check the chain of custody through the database, but it also uses a ring oscillator to identify usage across entities based on a change in frequency connected with chip aging as stated in other studies.

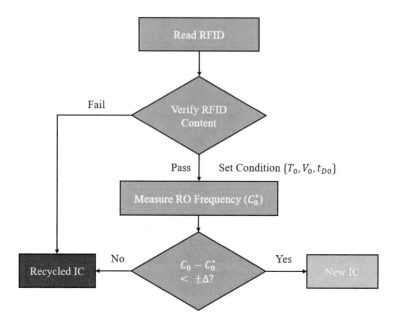

Fig. 16.6 Read, verify, and update process for RFID tag

In terms of implementation, the approach has much potential. However, one may argue that a chain-of-custody verification is pointless when used with a ring oscillator because regardless of where the chip's chain of custody ends if the ring oscillator has aged, the chip has been utilized while it was intended to be new. As a result, while the RFID implementation is new, it is a gimmick. On the other side, this proof of chain of custody may be used to identify and hold responsible untrustworthy parties in the supply chain, a characteristic that is impossible to achieve with simply a use detection mechanism, such as a ring oscillator. Furthermore, this system has a single point of failure that impacts all of the chips: the database. If the database storing the chains of custody was hacked and the cryptographic verification keys were taken, the entire system would be rendered untrustworthy since chains of custody might be faked. Finally, it seems this method is not a proper implementation of blockchains, which uses a distributed ledger that is immutable and can only be updated by a voting process involving all the stakeholders. A more robust and secure approach to end-to-end traceability is covered in the upcoming Sect. 16.3.4. The end user can authenticate the RFID tag content to determine the precise route of a chip. Because digital signatures protect RFID tag data, any modification or tampering with it can be easily detected, presuming the database is safe and accurate. The computational complexity of the hashing technique employed in signature verification also protects against dictionary attacks. This approach neglects to address the relative inefficiency of integrating an

RFID chip onto low-cost chips. The additional expense of incorporating the chip tag may be as much as the cost of the original chip.

16.3.4   Thwarting Counterfeit IC Using Blockchain

A typical semiconductor supply chain involves several steps such as design, manufacturing, PCB assembly, system integration, and end users. Figure 16.7 depicts semiconductor supply chain entities and paths followed by different kinds of counterfeit devices involved with every step. First, OEMs (original equipment manufacturers) design and manufacture fabricate ICs. Two types of OEMs exist, integrated device manufacturers (IDMs), who own the foundry, and fabless semiconductor companies. A fabless OEM shares GDSII design with an offshore foundry for fabrication.

On-chip security primitives are suitable for technologically competent entities in the electronics supply chain. However, these approaches lack practicability of verification by technologically incompetent supply chain entities and cannot help in performing end-to-end traceability of ICs. Furthermore, these approaches rely on hardware security modules such as silicon odometers and physically unclonable functions. Therefore, these verification approaches are not suitable for distributors who are technically incompetent, and only one IC can be verified at a time, which requires unboxing (or un-packaging) of shipment packages. Given the challenges mentioned earlier, a consortium-style blockchain network, i.e., *Chain* (electronic Chain) [12, 19], can be used for verifying the genuineness of ICs using provenance and smart contracts [1]. A blockchain network stores information about assets and transactions between different peers using a distributed database. This distributed ledger ensures that each peer's storage systems contain identical information. If a peer wishes to change a record, it must be authorized through a polling process known as consensus. *eChain's* distributed database stores IC ownership records, and it can be accessed using IC identifiers such as on-chip serial number marking and ECID.

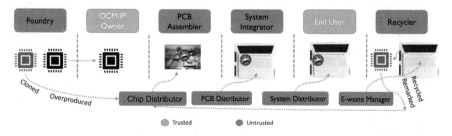

Fig. 16.7 Supply chain entities and various types of involved counterfeits

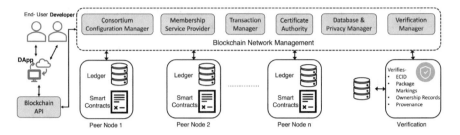

Fig. 16.8 *eChain* architecture

16.3.4.1   eChain Architecture

Various supply chain entities (see Fig. 16.7) are mapped as peers in the *eChain* network. Each peer has a copy of the current distributed database. In addition, any entity can access the blockchain network using a graphical user interface called distributed application (DApps) to post buy–sell transactions and perform device verification. The following components are included in the *eChain* architecture (see Fig. 16.8):

- **Consortium Configuration Manager** specifies the eChain network regulations and guidelines. This component configures and manages enrollment. In addition, it prevents peers (except OEM) from enrolling ICs into the blockchain database.
- **Certificate Authority** creates digital certificates to blockchain peers. Peers need this digital certificate for posting transactions to the blockchain.
- The **Membership Service Provider** verifies, permits, and handles the identity of blockchain peer.
- **Data Privacy Manager** encrypts transactions and database to prevent access of buy–sell to non-relevant peers.
- A **Smart Contract** is an automated agreement that converts the provisions of a peer-to-peer contract into software code. For example, smart contracts could execute supply chain tasks such as the generation, updating, and ownership verification of blockchain records for IC provenance.
- **Verification Manager** is not a real element of the *eChain*. It uses smart contracts to manage IC enrollment, supply chain transactions, and ledger record verification.
- The **Transaction Manager** is crucial to the execution of blockchain transactions. First, it collects simulated transaction results from blockchain peers involved in the transaction. Then, after achieving a consensus, it executes the approved transactions using the majority polling method. Finally, it adds the permitted transactions to the blockchain database as blocks.
- **Blockchain Application Programming Interface (API)** A blockchain developer can use API to manage blockchain peers, create smart contracts, process transactions, traverse nodes to search ledger records, and broadcast events to peers using distributed application [3].

16.3.4.2   IC Supply Chain Transactions

Smart contracts can enlist IC (or semiconductor chips) into the blockchain and execute supply chain transactions between various peers. In the semiconductor supply chain, an IC travels through several entities before reaching an end user. The following are the primary entities and events that happen during the movement of the chips in the semiconductor supply chain:

- OEM ensures that every chip has a unique ID and is enrolled into the *eChain*.
- OEM dispatches chips to distributors for sale in the market.
- IC distributor sells chips to PCB assemblers.
- (Discretionary) A PCB distributor sells PCBs to system integrators.
- System integrators build a system using PCB.
- (Discretionary) A system distributor sells systems (such as computers, tablets, and phones) to end users.
- End users purchase a system directly from the system integrator.

16.3.4.3   IC Verification Using *eChain*

The *eChain* can detect counterfeit ICs embedded in any system using provenance records from the blockchain ledger. This section covers algorithm for detecting recycled, cloned, overproduced, and remarked ICs. Figure 16.9 depicts *eChain's* workflow for various counterfeit detection:

1. **Recycled IC** An IC will be classified as recycled, if an owner has reported an IC at the end of its life cycle. For example, let us assume an adversarial IC distributor purchases used chips from a recycler and sells this as new chips. Later, the PCB assembler purchases the chips and sends an authenticity verification request. The *eChain* network will identify the current phase of the chip as the end of its life cycle, whereas the IC distributor claims ownership; thus, it is classified as a recycled chip.
2. **Remarked IC** During the enrollment step, *eChain* stores the package marking of the authentic ICs in the blockchain with the respective chip ID. The external package marking will differ for a remarked IC, but the ECID cannot be changed. Therefore, an IC can be verified against the grade mentioned on its package against its ECID from the blockchain, whether it is a remarked IC or an authentic grade.
3. **Cloned IC** *eChain* detects cloned ICs by analyzing the provenance and ownership information received from the blockchain. Using provenance, an IC can be traced back to its origin, typically OEM. The IC is classified as cloned if the selling entity does not appear as an owner in the blockchain database.
4. **Overproduced IC** Overproduction refers to the production of more quantities of goods beyond the terms of the manufacturing contract. Overproduction frequently results in significant revenue loss for the OEM. *eChain* can detect overproduced ICs with valid but duplicate IDs by verifying the serial IDs with the

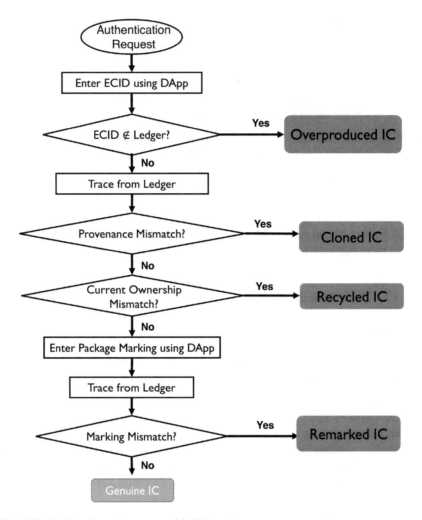

Fig. 16.9 Workflow for various counterfeit IC detection

current owner name from the blockchain ledger. For example, if an adversarial distributor tries to sell overproduced chips with duplicate or invalid IDs, the *eChain* can quickly identify these chips because they are not enrolled in the blockchain.

16.4   Conclusions

We discussed ways to detect recycled integrated circuits using lightweight on-chip sensors. The chip is classified as recycled by computing the difference in frequencies between stressed ROs and reference ROs in the sensor. Furthermore, it can be easily identified for how long the chip has been used by measuring the utilization time saved in the AF memory of the AF-based sensors. These designs can be expanded in the future to: (1) investigate the influence of aging recovery on the performance of RO-based sensors and (2) detection of recycled analog and RF designs using on-chip sensors.

Furthermore, we discussed several structures for identifying recycled and noted ICs of various sorts and sizes, including the NBTI-aware RO-CDIR and F-CDIR. These structures can apply to new technology nodes as well, whereas AF-CDIR is only applicable to older technology nodes. Finally, a structure based on a low-cost fuse is discussed that can apply to both digital and analog circuits. In combination, these structures can detect recycled ICs over a wide range.

Following that, we discussed a dependable and low-cost method for identifying recycled integrated circuits (ICs) retrieved from outdated electronic systems. By building a trust chain between the manufacturer, distributors, and system integrator, this approach provides traceability. It employs a small passive RFID tag that must be inserted into the chip package. Any supply chain organization can validate a chip's authenticity using a hand-held commercial RFID reader. During the verification process, a distributor is not necessary to turn on a chip. The final verification, which includes powering up the chip for RO frequency measurement, must be completed at the system integrator's location.

Finally, a blockchain-based framework for identifying electrical device counterfeiting is explored. The architectural design highlights the framework's utility in counterfeit detection. It comprises recycled, commented, cloned, or overproduced counterfeiting. It can, however, be developed to detect other sorts of counterfeiting, such as falsified documents and malevolent behavior in the supply chain, in order to pinpoint the specific counterfeiting business.

References

1. (2022) What are smart contracts on blockchain? https://www.ibm.com/topics/smart-contracts
2. Amsaad F, Pundir N, Niamat M (2018) A dynamic area-efficient technique to enhance ROPUFs security against modeling attacks. In: Computer and network security essentials. Springer, Berlin, pp 407–425
3. Androulaki E, Barger A, Bortnikov V, Cachin C, Christidis K, De Caro A, Enyeart D, Ferris C, Laventman G, Manevich Y, et al (2018) Hyperledger fabric: a distributed operating system for permissioned blockchains. In: Proceedings of the thirteenth EuroSys conference, pp 1–15
4. Carrique F (2020) Apple sues recycling partner for reselling more than 100,000 iPhones, iPads, and watches it was hired to dismantle. https://www.theverge.com/apple/2020/10/4/21499422/apple-sues-recycling-company-reselling-ipods-ipads-watches
5. Daniel B (2020) Counterfeit electronic parts: A multibillion-dollar black market. https://www.trentonsystems.com/blog/counterfeit-electronic-parts
6. DeBobes RD, Morriss et al DM (2012) The committee's investigation into counterfeit electronic parts in the Department of Defense Supply Chain. https://www.govinfo.gov/content/pkg/CHRG-112shrg72702/html/CHRG-112shrg72702.htm

7. Farahmandi F, Mishra P (2017) Validation of IP security and trust. In: Hardware IP security and trust. Springer, Cham, pp 187–205
8. Farahmandi F, Mishra P, Ray S (2016) Exploiting transaction level models for observability-aware post-silicon test generation. In: Design automation and test in Europe. ACM/IEEE, New York/Piscataway
9. Farahmandi F, Huang Y, Mishra P (2020) Automated test generation for detection of malicious functionality. In: System-on-chip security. Springer, Cham, pp 153–171
10. Guin U, Huang K, DiMase D, Carulli JM, Tehranipoor M, Makris Y (2014) Counterfeit integrated circuits: a rising threat in the global semiconductor supply chain. Proc IEEE 102(8):1207–1228
11. Guin U, Zhang X, Forte D, Tehranipoor M (2014) Low-cost on-chip structures for combating die and IC recycling. In: 2014 51st ACM/EDAC/IEEE design automation conference (DAC). IEEE, Piscataway, pp 1–6
12. Hossain MM, Vashistha N, Allen J, Allen M, Farahmandi F, Rahman F, Tehranipoor M (2022) Thwarting counterfeit electronics by blockchain. https://scholar.google.com/citations? view_op=view_citation&hl=en&user=n-I3JdAAAAAJ&citation_for_view=n-I3JdAAAAAJ: 9ZlFYXVOiuMC
13. Rahman MT, Forte D, Fahrny J, Tehranipoor M (2014) ARO-PUF: an aging-resistant ring oscillator PUF design. In: 2014 design, automation & test in Europe conference & exhibition (DATE). IEEE, Piscataway, pp 1–6
14. Rahman MS, Nahiyan A, Amir S, Rahman F, Farahmandi F, Forte D, Tehranipoor M (2019) Dynamically obfuscated scan chain to resist oracle-guided attacks on logic locked design. Cryptology ePrint Archive
15. Rahman MS, Nahiyan A, Rahman F, Fazzari S, Plaks K, Farahmandi F, Forte D, Tehranipoor M (2021) Security assessment of dynamically obfuscated scan chain against oracle-guided attacks. ACM Trans Design Autom Electron Syst 26(4):1–27
16. Reddy N, Wang S, Winemberg L, Tehranipoor M (2011) Experimental analysis for aging in integrated circuits. In: IEEE North atlantic test workshop
17. Sengupta A, Rathor M (2020) Enhanced security of DSP circuits using multi-key based structural obfuscation and physical-level watermarking for consumer electronics systems. IEEE Trans Consumer Electron 66(2):163–172
18. Suh GE, Devadas S (2007) Physical unclonable functions for device authentication and secret key generation. In: 2007 44th ACM/IEEE design automation conference. IEEE, Piscataway, pp 9–14
19. Vashistha N, Hossain MM, Shahriar MR, Farahmandi F, Rahman F, Tehranipoor M (2021) eChain: a blockchain-enabled ecosystem for electronic device authenticity verification. IEEE Trans Consumer Electron 68:23–37
20. Villasenor J, Tehranipoor M (2013) Chop shop electronics. IEEE Spectrum 50(10):41–45
21. Zhang Y, Guin U (2019) End-to-end traceability of ICs in component supply chain for fighting against recycling. IEEE Trans Inf Forens Security 15:767–775
22. Zhang X, Tehranipoor M (2013) Design of on-chip lightweight sensors for effective detection of recycled ICs. IEEE Trans Very Large Scale Integr Syst 22(5):1016–1029
23. Zhang X, Tuzzio N, Tehranipoor M (2012) Identification of recovered ICs using fingerprints from a light-weight on-chip sensor. In: Proceedings of the 49th annual design automation conference, pp 703–708

Chapter 17
Package-Level Counterfeit Detection and Avoidance

17.1 Introduction

Counterfeit electronic devices have recently emerged as a multi-billion-dollar black market [16]. A counterfeit part is an illegitimate replica, imitation, substitute, or altered component that has been intentionally misquoted as a genuine component from a legitimate manufacturer. A counterfeit part can also be a used electronic component that has been modified and willfully misquoted as new to the intended customer without disclosure of its prior usage [8].

According to the official definition by the US Department of Commerce, a counterfeit component: (a) is an unauthorized copy; (b) does not conform to (original equipment manufacturer) OEM design, model, or performance standards; (c) is not produced by the OEM or their authorized contractors; (d) is an off-specification, defective, or used OEM product sold as "new" or working; or (e) has incorrect or false markings or documentation [9]. Counterfeit products in wide circulation greatly impact the industrial sector, especially the semiconductor industry, because they are less reliable and secure than their original counterparts. Figure 17.1 shows the failure characteristics between new and counterfeit devices [29]. It is clear that counterfeit devices are more likely to fail than new devices.

Semiconductor devices play a crucial role in various critical systems in industries such as medicine, defense, aerospace, automobile, banking, and energy. The repercussions can be devastating when these systems fail due to counterfeit components. Counterfeit products on a larger scale can threaten economies as legitimate electronic companies lose about $100 billion of global revenue yearly due to counterfeits [17]. The use of counterfeit devices can also threaten user safety, sabotage health, and sometimes lead to loss of lives and properties. Though harmful to the economy, counterfeiting is a very lucrative business because it generates easy money. It is very easy to counterfeit components because of the availability of raw materials due to the ever-increasing electronic waste, easy availability to reverse engineering tools, and commercial-off-the-shelf (COTS) ICs [4, 13].

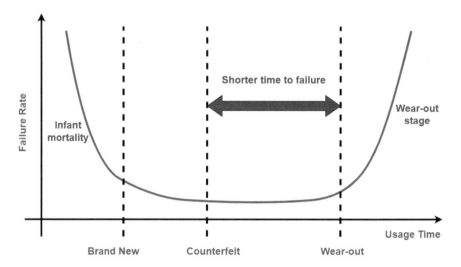

Fig. 17.1 Graph showing the typical device failure characteristics for different types of ICs

Besides generating easy money, counterfeiting can also be done to sabotage competitors. The counterfeiters can damage a company's reputation or eliminate a competitor's market share by flooding the market with counterfeit products. Counterfeiters can also belong to a radical organization trying to cripple a nation's defense system. Therefore, developing an efficient method for counterfeit detection is crucial, and several efforts have been made toward that. However, it is becoming more challenging to recognize the counterfeit products among the authentic ones because the counterfeiters have significantly improved their techniques and technologies in achieving their aim. This problem creates a necessity for improving the technologies and methodologies used to detect and avoid these counterfeits.

This chapter discusses some of these methodologies developed to combat the counterfeiting issue. It constitutes detection methods involving physical inspection of the device, X-ray imaging-based techniques, low-power visual inspection, microblasting analysis, scanning acoustic microscopy, scanning electron microscopy, and DNA markings [1, 17, 19, 27]. It also briefly discusses electrical inspection techniques such as parametric, functional, burn-in, and structural tests [36]. Next, we cover the overview of IC counterfeiting challenges and the limitations of various detection techniques.

17.2 Background

Figure 17.2 shows package-level IC counterfeiting at various stages of the electronic supply chain [5]. Dishonest SoC integrators can clone the design from IP owners and sell them as their IPs or produce more chips than are authorized by IP owners to

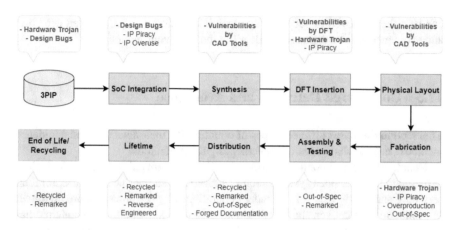

Fig. 17.2 Hardware supply chain with security issues in red bulletin points and trust/counterfeit issues in blue bulletin points

reduce development or licensing costs. Untrusted foundries can fabricate more chips than they are committed to and sell them through the gray market by tampering with the yield data. Overproduced counterfeit ICs threaten OEM/IC design houses because they can lose revenue. From Fig. 17.2, it can be seen that recycled and remarked counterfeits are the most common threats in every step of the supply chain because they are easily accessible to attackers at a low cost.

Examples of a remarked counterfeit include counterfeiters modifying the markings on a lower grade chip to sell it as a higher grade for higher profit. A remarked IC with modified speed/grade can cause reliability problems in the system designed for high performance and higher tolerance to adverse environmental conditions. IC remarking is a prevalent problem because it is easy to remark an IC to such an extent that remarked and original ICs are indistinguishable. In case of recycled ICs threat, used ICs from PCBs or electronic devices are extracted and modified to be sold as new. Since these ICs have been used in the system, they have poor performance and a shorter lifespan due to the aging effects. Similarly, forged documentation is a problem where specification documents of purchase components are falsified such that they do not match those of the OEMs. Defective/out-of-spec components are those parts that fail in structural tests or fail to perform to specifications.

Counterfeit detection standards such as SAE G-19A Test Laboratory Standards Development Committee [5, 17] are developed to guide the detection of counterfeit ICs. The SAE G-19A standard developed an inspection and test matrix plan for different classes of Electrical, Electronic, and Electromechanical (EEE) commodities to detect counterfeit components. The standard evaluates risk levels associated with different EEE commodities, provides guidance on the testing that should be performed, and recommends sampling plans for the tests. However, these standards can only be utilized for recycled and remarked counterfeit detection. Some common solutions to prevent cloned counterfeiting include watermarking, but watermarks can be either removed or successfully cloned. Techniques for counterfeit detection and avoidance will be explained in detail in the following sections.

17.3 Counterfeit Detection and Avoidance

This section discusses different types of counterfeits, their detection and avoidance methods, and the challenges of implementing these methods.

Counterfeiting will continue to grow to fulfill the demand for electronic devices to maintain long-serviced systems such as aircraft and submarines, because, for their servicing and repair, they rely on older generations of electronics that are no longer in production. Moreover, the recent global chip shortage due to the pandemic has created a greater risk of IC counterfeiting [6]. The demand for electronic devices has increased, while regular chip production has decreased. System manufacturing companies are now relying more on IC supplies from different third-party suppliers without proper verification to keep their business running. The shortage of IC supplies gives adversaries more opportunities to produce counterfeit ICs to make profits by filling the demand–supply gap. Unfortunately, the existing attempts to combat counterfeiting are insufficient. Electronic component counterfeiting has emerged as a significant problem in the semiconductor industry. Therefore, hardware security researchers must develop specific counterfeit detection, prevention, and avoidance methods for the industry. The defective, counterfeit parts can be classified as physical and electrical (as shown in Fig. 17.3), and further, they can be sub-classified based on the defect's characteristics [29]. For example, Fig. 17.4 demonstrates that the physical defects can be divided into two sub-categories: exterior and interior. On the other hand, the electrical defects can be of two types: parametric and manufacturing (as illustrated in Fig. 17.5).

17.3.1 Methods and Challenges for Counterfeit Detection

This subsection will discuss several other techniques and methodologies used for counterfeit detection in the supply chain. The low-power visual inspection (LPVI) method employs a microscope, a DSLR camera, or an infrared imaging source to scan any modifications of important information labeled on the package-level

Fig. 17.3 Major categories of defects in counterfeit components

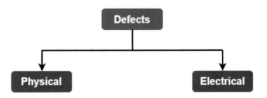

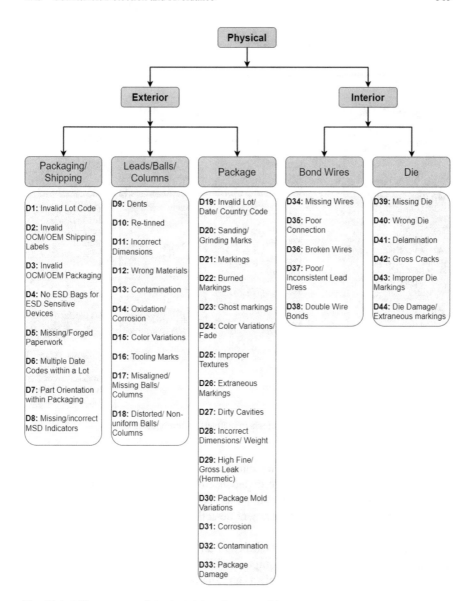

Fig. 17.4 Different types of physical defects in counterfeit components

electronics and discover signs of reused or recycled products such as scratches on the package or solder residuals on the IC leads. In addition, X-ray imaging, a non-invasive testing method, is performed to detect anomalies on internal packages, dies, and bond wires compared to a reference part. Microblast analysis is implemented to discover whether scratches or markings on recycled/counterfeit parts have been removed purposefully through the dry blasting technique. Another technique used

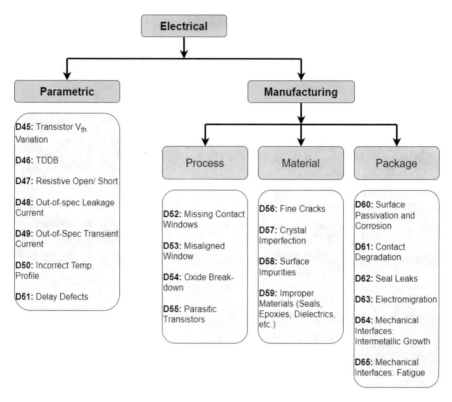

Fig. 17.5 Different types of electrical defects in counterfeit components

to detect die cracks and bond wire anomalies is scanning acoustic microscopy
(SAM), which records the acoustic impedance of a component at different depths
when submerged in iso-propyl alcohol (IPA). This technique can also be applied to
delamination, or die–package attachment detection, as well as ghost marking and
sanding marks detection.

Scanning electron microscopy (SEM) is an invasive imaging technique that
generates electron beams on a sample and then scans secondary emissions of
electrons on target areas to detect defects and anomalies on a counterfeit component.
SEM imaging takes images of die, package, or leads with good resolution after
depackaging and delayering the IC. X-ray fluorescence (XRF) spectroscopy uses
radiation emission through X-ray absorption on material surfaces to generate a
spectrum of elements as unique fingerprints on the package of a component
for authentication purposes. Fourier Transform Infrared (FTIR) Spectroscopy is
an infrared-based spectroscopy method that creates its spectrum on component
material as fingerprints to verify any residual from a chemical process or material
remnants from the sanding process for detection of remarked or recycled products.
Energy Dispersive Spectroscopy (EDS) uses X-ray spectrum excited from the

bombardment of a high-energy beam on a component's surface as fingerprints during packaging. Terahertz Time-Domain Spectroscopy (THz-TDS) technique uses a laser pulse at the THz frequency range (100 GHz–30 THz) to inspect the internal structures of a component under test for counterfeit detection. A counterfeit IC's refractive index is different from an authentic IC when a THz pulsed laser is applied for imaging.

Electrical inspection techniques for counterfeit detection can be parametric, functional, burn-in, and structural tests. Parametric tests include two types of tests, DC and AC [11, 32]. A series of these tests are conducted through either instrumentation or automatic test equipment (ATE) to automatically detect any shifts in electrical parameters of a counterfeit or reused IC. Functional tests are the best options for testing the functionality of a component. These tests require a PCB-based functioning system to test the functionality of components to detect any counterfeit parts that are remarked to a higher grade. Burn-in tests can be used to find infant mortality of either an authentic or a counterfeit device, which tells the reliability of a device [17]. Finally, structural tests require internal scan chains can be used to find structural defects or anomalies on a device.

Two methods related to aging-based statistical fingerprints can be used to detect recycled counterfeit IC [17]. The first method, named Early Failure Rate (EFR), is performed by collecting parametric measurements of brand-new devices during burn-in tests at different time points (for analysis of aging degradation over time) and then training a one-class classifier for anomaly detection on counterfeit ICs. This technique is based on the fact that a counterfeit IC has a different speed of aging degradation compared to a new one. The circuit path-delay analysis method leverages the difference between path-delay distribution of recycled and new ICs to identify recycled ICs. This analysis can be done by choosing the paths based on the aging rate, measuring the paths' delay, and finally authenticating the device using statistical analysis. Generally, the more time an IC is used, the more aging effects can be detected through this method [28, 35].

Counterfeit detection techniques suffer from several challenges. For physical inspection, one challenge is that many techniques are destructive. Sample preparation and decapsulation of ICs are required. Another challenge is that many physical detection tests are incredibly time-consuming and costly. In addition, automation is not available for most tests, and no metrics are available to evaluate against a set of counterfeit types, anomalies, or defects [17]. The test results, in this case, depending on the interpretation of the subject matter experts. This human-centric decision step makes the decision-making process susceptible to errors and imperfections [31].

Electrical testing techniques also have their limitations on performance:

1. Electrical parameters can vary due to the manufacturing process or environmental variations.
2. Functional tests are not viable for obsolete and active devices, as generating test programs for these devices with limited knowledge about the functionality is challenging.

3. Both functional tests and burn-in tests are costly. Burn-in tests also require a long time to perform.
4. Limited access to the scan chain and little knowledge about the design and layout of an IC make structural testing ineffective [24, 25].

17.3.2 Methods and Challenges for Counterfeit Avoidance

This subsection discusses several counterfeit avoidance techniques proposed in the literature. Various methods are devised for counterfeit avoidance, such as physically unclonable functions (PUFs), secure split test, hardware metering, poly fuse-based technology for recording usage time, combating die/IC recovery (CDIR), and electronic chip ID (ECID) [16]. Silicon PUFs take advantage of the physical variations that inherently exist in integrated circuits. Since these variations are random, they are uncontrollable and unpredictable. As a result, PUFs make suitable candidates for IC identification and authentication. There are different versions of PUFs, such as RO (ring oscillator) PUF and SRAM PUF. The cons are that PUFs can have a high bit error rate. The bit error rate (BER) can be used to describe a PUF's reliability or reproducibility. BER is defined as the percentage of flipped (error) bits (also known as measurement noise) in response bits as a result of noise [12]. A PUF should, ideally, always maintain the same challenge–response pairs (CRPs) across varied operating circumstances and times, resulting in a zero-bit error rate [5]. However, since noise, such as environmental changes, circuit instabilities, and aging, is inescapable, there are always unpredictable factors influencing the response. Therefore, IC designers and engineers must choose a suitable PUF design to reduce this error rate. A limitation of PUFs is: that they rely entirely on the process variations within the IC and cannot be reproduced. If an entity or company wanted to, for example, design ICs with specific challenge–response pairs, this would be impossible due to the nature of PUFs.

Hardware metering is a system of security techniques and protocols allowing the design house to maintain some control over the ICs after fabrication. It can be active or passive hardware metering. Active metering works by locking each IC until the IP holder unlocks them. There are several approaches for implementing this locking: (a) ICs are locked when they are powered on [2], (b) XOR gates are scattered throughout the design [26], and (c) a finite-state machine (FSM) is added that is originally locked and can only be unlocked by the correct key input [3]. On the other hand, passive metering works by uniquely identifying each IC and registering them using the challenge–response pairs. For both active and passive metering, the pros are that it allows the IP owner to have more significant control over the post-fabrication of chips. In addition, they both aim to respond to the number of counterfeits that enter the market. However, the different entities along the supply chain put some limitations on this approach. For passive metering, the IP holder must keep track of the challenge–response of ICs. For active metering, the IP owner is susceptible to the foundry lying about the amount of yield or quantity that needs to be activated because once an IC is unlocked, the foundry can still sell the IC through black markets.

Secure split test protects the manufacturing test process, giving IP owners more control over the process and preventing counterfeiting [7]. This method allows IP owners to protect their IPs by metering them. It is done by adding cryptographic hardware modules and blocking an IC's correct functionality until the IP owners activate it. The pros are that this allows the IP owners to have greater control of their chips and the amount of the chips being produced. One disadvantage is that the hardware for cryptography must be added to the design, which would increase production cost and design time.

CDIR helps in preventing counterfeits using old parts. In this technique, the aging of the chip is captured using a small circuit/sensor embedded into the chip [30]. This type of sensor works by utilizing the aging effects of MOSFETs to alter the frequency of a ring oscillator compared to a golden one planted in the chip. The advantage of CDIR is that the wear-out mechanisms of NBTI (Negative Bias Temperature Instability) and HCI (Hot Carrier Injection) cause a change in the sensor's frequency, allowing for a simple read-out of the value [18]. The limitations are the available space for the sensor within the chip and the increased area overhead. Similar to CDIR, poly fuse-based technology can be used to record the aging of the chip. Poly fuse-based technology consists of an embedded anti-fuse (AF) memory block with integrated counters. The counters record IC usage time, and its value is recorded in the AF memory block. Two different types of AF-based sensors, such as CAF (clock anti-fuse)-based sensors and SAF (signal anti-fuse)-based sensors, can be used to measure the IC usage time [34]. A CAF-based sensor records a system's cycle count, whereas a SAF-based sensor employs a circuit that serves as the trigger for a counter. Both types of sensors contribute overhead to the IC, although SAF-based sensors typically have a smaller footprint than CAF-based sensors.

ECID is a unique ID added to each chip that allows each IC to be tracked throughout the device's lifetime. The advantage of ECID is that it can be easily programmed into a non-programmable memory. Furthermore, the disadvantage is that ECID must be added post-fabrication, which takes additional time and resources. The limitations of ECID are that it can be easily cloned and is limited by the space of the non-programmable memory space.

Counterfeit avoidance methods provide better solutions to the problems of long testing time, high cost, and metrics for testing performance evaluation [10, 17]. Combating Die and IC Recycling (CDIR) sensors are utilized to detect recycled and remarked ICs by either finding RO frequency difference (a reflection of aging) between reference and stressed ROs or counting the usage time of ICs, which can be stored in a memory block. In the SST technique, all chips are locked, and a True Random Number (TRN) is generated for each chip to be fabricated. The IP owner then uses these TRNs to create keys for unlocking each chip and send the keys to the foundry for testing. The test results are sent back to the IP owner, who will determine which die passes. The IP owner will then send the electronic chip IDs (ECIDs) of the passing die to the foundry. The passed dies are then sent to assembly for packaging by the foundry. The IP owner can only unlock chips to achieve correct functionality, ensuring an effective countermeasure against overproduced, cloned, and out-of-spec/defective parts. Hardware metering uses digital fingerprints stored in the IC or a PUF to prevent cloned parts passively. It can also be used by the IP owner to actively track each IC by unlocking each IC only by themselves after fabrication and testing to prevent overproduced ICs.

Split manufacturing is a technique to split the IC fabrication process into two parts to reduce the risks or threats during fabrication. Typically, the more expensive process of fabricating the transistors and active layers (with few metal layers) is performed in an untrusted foundry, whereas the less expensive part of fabricating the upper metal layers is performed in a trusted foundry. In IC camouflage, dummy contacts are added to an IC layout to fool attackers unfamiliar with the IC's functionality. It is based on the assumption that an adversary cannot identify a logic cell by etching a single device layer. This technique prevents the reverse engineering of circuits. For hardware watermarking, a unique fingerprint is created for IP authentication. Constraint-based watermarking creates easily seen and unre-producible patterns and adds time constraints to sub-paths of a logic path. Physical Unclonable Function (PUF) uses the unique and unclonable nature of manufacturing process variations to create hardware security measures against overproduced and cloned ICs. Pairs of challenges and corresponding responses are used for testing and authentication purposes. Package ID is the only counterfeit avoidance technique that can target active and obsolete ICs with no design documents accessible. DNA markings and nanorods [17] are two ways of creating package ID [23].

There also exist challenges in counterfeit avoidance techniques [15, 17]. One is that attackers in the foundry can claim low yields but produce more ICs and sell them to gain profits. Another is that a PUF is very sensitive to environmental variations and the aging effect, which sometimes makes it not a reliable counterfeit avoidance measure. In addition, entry of out-of-spec and defective ICs into the supply chain cannot be prevented by the design house [17, 20, 33]. Furthermore, DNA markings have authentication, validation speed, and cost limitations when applied at a large scale.

17.3.3 Counterfeit PCB Detection

PCBs are also highly vulnerable to counterfeiting. An adversary can reverse engineer the design and create counterfeit copies of a PCB by assembling the components. Even a complicated multi-layer PCB can be completely reverse-engineered using low-cost solutions in a reasonably straightforward manner. Assembling components on counterfeit PCBs is also straightforward since most PCBs use active/passive components readily available in the market. So, detecting and avoiding counterfeits in PCBs is a challenging task. In this section, we will cover a counterfeit detection approach for PCBs.

A new counterfeit PCB detection approach based on surface fingerprints exploits PCB inter-layer connection vias that can be highly unpredictable, unclonable, and immeasurable. These features can be vias through the surface or drilled holes of different sizes, shapes, angles, orientation, and misalignment of via and solder mask centers due to PCB manufacturing process [21].

For the following reasons, PCB vias are considered good candidates for PCB surface fingerprints. First, variations in PCB via surface and drilled hole characteristics and via–solder mask misalignment are tough to control and predict in the manufacturing process. Second, PCB vias are usually very small and metallic, making them good options for generating fingerprints of electronic devices that can survive a catastrophe such as a fire accident. Fire alarm counterfeit detection is one of the applications for this PCB via-based counterfeit detection method [21]. Third, fingerprints based on inter-layer vias will not be affected by the placement of components on the PCB. Lastly, vias are clearly visible, making them easily observable and highly measurable for counterfeit detection. The process flow of the PCB via-based authentication approach is illustrated through a flow chart shown in Fig. 17.6 [21]. It is universally applied to all surface fingerprint-based counterfeit detection and device authentication methods. The first step in the process chain is to capture details of minor variations on PCB pads, vias, and traces, followed by high-resolution optical-imaging techniques. For example, X-ray computed tomography (CT) can be a high-resolution imaging tool for electronic devices packaged with plastic covers. These high-resolution optical-imaging techniques make the surface footprint method an ideal approach for counterfeit detection. The following steps ensure visibility of the surface patterns (pads, vias, or wires) by reducing noise through multiple sequences of averaging of the images and minimizing misalignment-induced geometric distortions [21]. The next step is to identify the regions of PCB via fingerprints. Here, the template matching technique is used to segment and find desired surface patterns by matching the patterns captured through imaging with predefined surface pattern image templates registered in the database. Grayscale images are used for reduced data size. In the final counterfeit detection phase, fingerprint templates are created as a reference for anomaly/counterfeit detection. Normalized cross-correlation (NCC) is implemented on a pair of templates and testing image sets. The two-dimensional NCC is computed for each pixel on the template and testing images. The highest NCC value among all pixels is then compared to a threshold set during experiments to determine

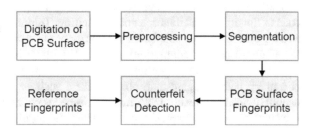

Fig. 17.6 The flow chart illustrating the authentication flow using PCV via-based method

whether the testing PCB surface patterns are counterfeit or not. This counterfeit detection technique can be applied to any electronic system that involves security or money, such as payment systems or mobile phones [21].

In addition to those aforementioned shining points of using PCB vias as surface fingerprints for counterfeit detection and PCB authentication, it has been demonstrated that the template-based segmentation technique is also occlusion and lighting variation resistant [21]. Besides, the NCC computation method used in decision-making for counterfeit detection gives better results than standard feature extraction-based techniques in a way that complicated surface and fingerprint patterns can be matched to detect counterfeits. However, template-based fingerprints have bigger size that makes them more computationally expensive than those based on feature detection. Further, an advanced level reader can refer to the three experiments' results discussed in [21]. The first experiment used 60 via patterns on 60 PCB surfaces, 50 of which had templates already registered in the database. The results [21] clearly show that authentic and counterfeit testing images can be identified and classified correctly. In the second experiment, five images were taken for each 100 via patterns on different days. 100 of those 500 images were then used as templates stored in the database, while the remaining 400 images were used for testing purposes [21]. The error rate, including false positive and true negative, is low at around 10% when a threshold is optimized for the highest prediction accuracy for counterfeit detection. The third experiment employed six via patterns from each of 60 PCB surfaces. Three hundred sixty images were taken at different time frames of multiple days. The result shows some false positive errors when only focused on counterfeit surface detection. So, the low error rates of the three experiments prove that this individual via surface-based counterfeit detection method can be successfully employed for electronic device counterfeit detection. Furthermore, since it is assumed that having many vias on a PCB will not cause problems, the performance of scalability was better than using a single via pattern. Another assumption to support the good performance was that principal component analysis (PCA) was used to reduce image size in the experiment while keeping only essential and principal information.

17.3.4 DNA Marking and Authentication

IC markers or taggants can authenticate the originality of ICs by providing a unique code or fingerprint and thus deter counterfeiting. This technique uses botanical DNA as a taggant to provide supply chain security of electronic parts in the supply chain to safeguard against counterfeiting [19]. The effectiveness of an IC mark as an evidential tool rises in proportion to the amount of its information content. So, DNA is a much better taggant than other markers such as mineral-based marking and nanoparticles. As a result, DNA marking impedes counterfeiting strategies because it cannot be copied, recreated, or simulated. In addition, the probability for false positives is less than one in a trillion when DNA is used to identify uniquely.

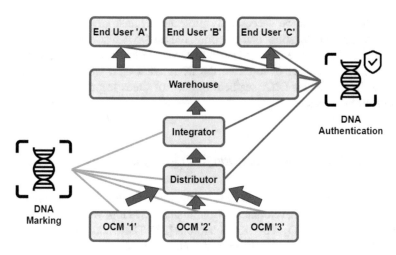

Fig. 17.7 The flow diagram illustrating the DNA authentication

Applied DNA Sciences (APDN) developed this anti-counterfeit technique that uses botanical DNA to mark electronic components to protect against counterfeiting. This technique also protects the DNA mark in the harsh operating conditions of the IC [19]. Forming the identification marking using the DNA is also quite simple. The chosen botanical DNA strand is applied on the chip's surface at low concentrations, forming a structured DNA marking. One added benefit is that the formation of the IC marking does not require significant changes in the chip manufacturing process. In addition, the DNA sequence information is shared with chip owners and securely stored in a server maintained by the APDN for future reference. Figure 17.7 shows the flow of the DNA marking and authentication process. The procurement unit in the microchip manufacturing industry often serves downstream end customers while enlisting the company's aid, Applied DNA Sciences, for the DNA marking procedure. Starting with the original chip manufacturer (OCM), who supervises the chip's marking process, Applied DNA collaborates with all of the trusted supply chain organizations. The authentication criteria are then decided by the confidence levels that have been established at various stages across the supply chain. After that, the different chips are authenticated in the lab. Finally, by proclaiming a clear forensic judgment, DNA lab analysis is utilized to identify genuine and counterfeit components completely. The ultimate consequence of these methods is verifying the flow of genuine components through the microchip supply chain, with counterfeit components wholly segregated. This separation is backed up by forensic evidence if legal action is required. The benefits of utilizing DNA markers include:

- Resistance to reverse engineering or replication.
- The technique is characterized by its low cost and high accuracy.
- The technique is broadly applicable and scalable.

The microchip can be successfully protected against key counterfeiting techniques such as sandblasting, blacktopping, reused DNA blacktopping, and repackaging due to the unique characteristics of DNA marks. In addition, counterfeiters cannot reverse engineer or replicate the DNA mark as it would require them to replicate the particular DNA segment billions of times to pass the detection and identification process.

17.3.5 Advanced Anti-counterfeiting Applications

New hydrothermal methods are now being implemented for anti-counterfeiting security inks with the development of multi-stage (379 nm, 980 nm, and 1550 nm), highly luminescent Y_2O_3:Yb^{3+}/Er^{3+} nanorods [22]. These inks have a strong capacity to produce hypersensitive green and solid red colors. Because semiconductor particles such as cadmium sulfide, cadmium selenide, and cadmium telluride are hazardous, have broad emission bands, and are soluble in dangerous solvents, researchers are working to produce a safer and more cost-effective security ink for future anti-counterfeiting applications [14]. This security ink is a unique method to produce a phosphor with all the necessary characteristics such as low toxicity, sharp emission, good solubility, and a reasonable cost, which was difficult to achieve earlier. Instead, a transparent security ink with multi-stage coding is introduced in a single host lattice based on a nanostructure that is highly luminescent, economical, and beneficial for high-end protection.

As mentioned earlier, the hydrothermal method synthesized the nanorods to create the security ink. The doping optimization of the concentration of rare-earth ions was necessary for using them to dope the nanorods. The process comprises stages such as dissolving a stoichiometric quantity of Y_2O_3 in nitric acid and stirring at 80 °C to generate a clear nitrate solution. After numerous tests, it was discovered that a pH of around 12 favored the reaction, resulting in long, homogeneous nanorods. This technique yielded material with a high degree of uniformity across the sample in more than 90% of the cases. A conventional PVC gold medium was used to disseminate the synthesized dual-mode nanorods before they were used in security printing applications. It was crucial for two reasons: (1) to distribute rare-earth metals without forming clumps in a medium and (2) to obtain better printing. 200 mg of Y_2O_3:Yb^{3+}/Er^{3+} nanorods were combined with 50 mL of PVC gold medium solution in a beaker in a regular experiment. After mixing the product ultrasonically at 45 kHz for 30 minutes, a satisfactory dispersion was produced. The absorption is increased by adding Yb^{3+} ion because the Yb^{3+} ion becomes excited by efficiently absorbing the energy. Therefore, the doping concentration of both the dopant and the co-dopant affects the energy transfer. In addition, the surface morphology was investigated using SEM, and it was observed that a highly dense nanostructure with a consistent size distribution in the produced nanorods. In the magnified version of the SEM image, the dimensions of the nanorods in the nanometer range (50–90 nm diameter and length of roughly 1–3 m under ideal growth conditions) are apparent. The authentic proof of its uses for high-end protection displays the practical scheme based on the collected spectroscopic data. $Y_2O_3:Yb^{3+}/Er^{3+}+$ nanorods exhibit acceptable structural and optical characteristics, according to the results in [22]. They are also very much soluble in PVC gold media. Furthermore, large scale manufacturing of these nanorods for commercial printing applications is simple. These characteristics confirm that nanorods are the best solution for anti-counterfeiting security ink applications with several stages. As a result, $Y_2O_3:Yb^{3+}/Er^{3+}$ has been demonstrated to provide a new viewpoint for the production of low-cost multi-stage security ink for sophisticated anti-counterfeiting applications.

17.4 Conclusions

In this chapter, a variety of methods relating to package-level counterfeit detection are discussed. The Sect. 17.3.3 focuses on dealing with counterfeit detection of electronic devices using PCB surface fingerprints. The PCB surface imperfections caused by the PCB manufacturing process are investigated as potential PCB surface fingerprints for counterfeit detection. The PCBs can be authenticated using surface fingerprints and normalized cross-correlation as a similarity measure. Real PCB surfaces with no assembled components are considered for performance evaluation. For image capture, a digital microscope is used to capture fine details on a surface.

In Sect. 17.3.4, a technique for marking and authenticating ICs developed by Applied DNA Sciences (APDNs) botanical DNA technology is covered. It can be implemented in a microchip supply chain environment to see how effective it is against current known counterfeiting strategies.

Section 17.3.5 discusses the development of a highly luminescent nanorod-based transparent security ink. These luminous nanorods were generated utilizing a large scale hydrothermal process, and they could radiate both ultra-sensitive green and powerful red at different spectral regions in a single host lattice. Furthermore, by varying the sintering temperature, these luminous nanorods displayed significant properties, such as the capacity to modify emission colors from red to green. As a result of integrating luminous nanorods, new possibilities for high-end multi-stage excitable transparent security ink, particularly valuable for counterfeit detection, are opened up.

With the increased use of technology, counterfeiting has become more prominent and raises issues in multiple areas of the semiconductor supply chain and market. Counterfeiting is expected to grow in the coming years, and researchers and organizations must scramble to increase avoidance and detection of counterfeits. Furthermore, counterfeit parts pose dangers to both consumers and IP owners. Various detection, avoidance, and challenges of counterfeits are also discussed here according to recent and relevant research and corresponding literature. In addition, the limitations, proposed methodology, and, if applicable, results of the conducted research activities are also analyzed in depth. Counterfeits will continue to be a threatening issue in the coming years, but researchers are developing ways to mitigate those risks and stop counterfeiters.

References

1. Alam MA, Roy K, Augustine C (2011) Reliability- and process-variation aware design of integrated circuits — a broader perspective. In: 2011 international reliability physics symposium, pp 4A.1.1–4A.1.11. https://doi.org/10.1109/IRPS.2011.5784500
2. Alkabani YM, Koushanfar F (2007) Active hardware metering for intellectual property protection and security. In: 16th USENIX security symposium (USENIX security 07). USENIX Association, Boston
3. Alkabani Y, Koushanfar F, Potkonjak M (2007) Remote activation of ICs for piracy prevention and digital right management. In: 2007 IEEE/ACM international conference on computer-aided design, pp 674–677. https://doi.org/10.1109/ICCAD.2007.4397343
4. Asadizanjani N, Rahman MT, Tehranipoor M (2021) Counterfeit detection and avoidance with physical inspection. In: Physical assurance, Springer International Publishing, Cham, pp 21–47
5. Bhunia S, Tehranipoor M (2018) Hardware security: a hands-on learning approach. Morgan Kaufmann, Burlington
6. Blyler J (2021) Staggering chip shortages have led to counterfeit Technical Can't We Test for Fakes?
7. Contreras GK, Rahman MT, Tehranipoor M (2013) Secure split-test for preventing IC piracy by untrusted foundry and assembly. In: 2013 IEEE international symposium on defect and fault tolerance in VLSI and nanotechnology systems (DFTS). IEEE, Piscataway, pp 196–203
8. Daniel B (2020) Counterfeit electronic parts: A multibillion-dollar black market. https://www.trentonsystems.com/blog/counterfeit-electronic-parts
9. Department of Commerce US (2010) Defense industrial base assessment: Counterfeit electronics
10. Farahmandi F, Mishra P (2017) Validation of IP security and trust. In: Hardware IP security and trust. Springer, Cham, pp 187–205
11. Farahmandi F, Huang Y, Mishra P (2020) Automated test generation for detection of malicious functionality. In: System-on-chip security. Springer, Cham, pp 153–171
12. Gao Y, Ranasinghe DC, Al-Sarawi SF, Kavehei O, Abbott D (2015) mrPUF: a novel memristive device based physical unclonable function. In: Malkin T, Kolesnikov V, Lewko AB, Polychronakis M (eds) Applied cryptography and network security. Springer International Publishing, Cham, pp 595–615
13. Ghosh P, Bhattacharya A, Forte D, Chakraborty RS (2019) Automated defective pin detection for recycled microelectronics identification. J Hardw Syst Secur 3(3):250–260
14. Guin U, Forte D, Tehranipoor M (2013) Anti-counterfeit techniques: From design to resign. In: 2013 14th international workshop on microprocessor test and verification. IEEE, Piscataway, pp 89–94
15. Guin U, DiMase D, Tehranipoor M (2014) A comprehensive framework for counterfeit defect coverage analysis and detection assessment. J Electron Testing 30(1):25–40
16. Guin U, DiMase D, Tehranipoor M (2014) Counterfeit integrated circuits: detection, avoidance, and the challenges ahead. J Electron Testing 30:9–23
17. Guin U, Huang K, DiMase D, Carulli JM, Tehranipoor M, Makris Y (2014) Counterfeit integrated circuits: A rising threat in the global semiconductor supply chain. Proc IEEE 102(8):1207–1228. https://doi.org/10.1109/JPROC.2014.2332291
18. Guin U, Zhang X, Forte D, Tehranipoor M (2014) Low-cost on-chip structures for combating die and IC recycling. In: 2014 51st ACM/EDAC/IEEE design automation conference (DAC). IEEE, Piscataway, pp 1–6
19. Hayward J, Meraglia J (2011) DNA marking and authentication: A unique, secure anti-counterfeiting program for the electronics industry. Int Sympos Microelectron 2011:000107–000112. https://doi.org/10.4071/isom-2011-TA3-Paper5
20. Hossain MM, Vashistha N, Allen J, Allen M, Farahmandi F, Rahman F, Tehranipoor M (2022) Thwarting counterfeit electronics by blockchain. https://scholar.google.com/citations?view_op=view_citation&hl=en&user=n-I3JdAAAAAJ&citation_for_view=n-I3JdAAAAAJ:9ZlFYXVOiuMC

21. Iqbal T, Wolf K (2017) PCB surface fingerprints based counterfeit detection of electronic devices. In: Media watermarking, security, and forensics

22. Kumar P, Dwivedi J, Gupta BK (2014) Highly luminescent dual mode rare-earth nanorod assisted multi-stage excitable security ink for anti-counterfeiting applications. J Mater Chem C 2:10468–10475

23. Miller M, Meraglia J, Hayward J (2012) Traceability in the age of globalization: a proposal for a marking protocol to assure authenticity of electronic parts. In: SAE aerospace electronics and avionics systems conference

24. Rahman MS, Nahiyan A, Amir S, Rahman F, Farahmandi F, Forte D, Tehranipoor M (2019) Dynamically obfuscated scan chain to resist oracle-guided attacks on logic locked design. Cryptology ePrint Archive

25. Rahman MS, Nahiyan A, Rahman F, Fazzari S, Plaks K, Farahmandi F, Forte D, Tehranipoor M (2021) Security assessment of dynamically obfuscated scan chain against oracle-guided attacks. ACM Trans Design Autom Electron Syst 26(4):1–27

26. Roy JA, Koushanfar F, Markov IL (2008) EPIC: Ending piracy of integrated circuits. In: 2008 design, automation and test in Europe, pp 1069–1074. https://doi.org/10.1109/DATE.2008.4484823

27. Shahbazmohamadi S, Forte D, Tehranipoor M (2014) Advanced physical inspection methods for counterfeit IC detection. In: ISTFA 2014: conference proceedings from the 40th international symposium for testing and failure analysis, ASM International, p 55

28. Tehranipoor M, Salmani H, Zhang X (2014) Counterfeit ICs: Path-delay fingerprinting. In: Integrated circuit authentication. Springer, Cham, pp 207–220

29. Tehranipoor M, Salmani H, Zhang X (2014) Counterfeit ICs: Taxonomies, assessment, and challenges. In: Integrated circuit authentication. Springer, Cham, pp 161–178

30. Tehranipoor MM, Guin U, Forte D (2015) Chip ID. In: Counterfeit integrated circuits, Springer, Cham, pp 243–263

31. Tehranipoor MM, Guin U, Forte D (2015) Counterfeit test coverage: An assessment of current counterfeit detection methods. In: Counterfeit integrated circuits. Springer, Cham, pp 109–131

32. Tehranipoor MM, Guin U, Forte D (2015) Electrical tests for counterfeit detection. In: Counterfeit integrated circuits. Springer, Cham, pp 95–107

33. Vashistha N, Hossain MM, Shahriar MR, Farahmandi F, Rahman F, Tehranipoor M (2021) eChain: a blockchain-enabled ecosystem for electronic device authenticity verification. IEEE Trans Consum Electron 68:23–37

34. Zhang X, Tehranipoor M (2013) Design of on-chip lightweight sensors for effective detection of recycled ICs. IEEE Trans Very Large Scale Integr Syst 22(5):1016–1029

35. Zhang X, Xiao K, Tehranipoor M (2012) Path-delay fingerprinting for identification of recovered ICs. 2012 IEEE international symposium on defect and fault tolerance in VLSI and nanotechnology systems (DFT), pp 13–18

36. Zhang F, Hennessy A, Bhunia S (2015) Robust counterfeit PCB detection exploiting intrinsic trace impedance variations. 2015 IEEE 33rd VLSI test symposium (VTS), pp 1–6

Chapter 18
Side-Channel Protection in Cryptographic Hardware

18.1 Introduction

Cryptographic hardware has emerged as a solution to the high computation cost of cryptographic operations on the software. Using hardware to perform cryptography is much faster because hardware accelerators can process computationally expensive code in a reasonable amount of time. However, cryptographic hardware accelerators are often not designed with security in mind. Vulnerabilities in such hardware accelerators result in the leakage of sensitive information, thereby reducing the efficacy of encryption. A notable adversarial invasion specific to cryptographic hardware is side-channel attacks. These attacks take advantage of hardware-specific vulnerabilities, such as power side-channel leakage [20]. This chapter covers various advanced methods for protecting cryptographic hardware, specifically restricting side-channel attacks and glitches. The existing solutions focus on secret sharing schemes, namely masking. A disadvantage to including these security measures is that hardware accelerators cannot boast faster processing speeds; therefore, some solutions include the development of more compact cryptographic hardware. Although the methods discussed in this chapter were groundbreaking at the time of their development, there are still several problems analyzed within these methods that could render the solutions vulnerable in the present and future. A significant problem addressed was the inability to secure cryptographic hardware accelerators effectively from glitches.

18.2 Background

Side-channel attacks are non-invasive attacks that try to leak confidential information in an integrated circuit (IC) chip through careful analysis of side-channel features such as power, timing, and electromagnetic radiation. During a side-channel attack, the perpetrator often tries to discover the encryption algorithm's secret key. During the implementation of encryption algorithms, particular vulnerabilities could be introduced, which leave an IC more at risk of being subjected to these attacks.

Differential power analysis (DPA) attacks are well-known side-channel attacks. This SCA attack can be launched easily, as it does not require prior knowledge of the hardware to succeed. For example, DPA has been shown to break AES in mere minutes successfully. To perform this attack, the adversary must first collect data from a large number of power traces using a variety of input patterns. Then the attacker uses statistical methods to determine the key [18]. Another type of attack emerges in the form of glitches. Glitches can inject faults that cause confidential information to manifest in a side channel, known as a fault injection attack. The attacker can perform this by changing the operating voltage or using a laser [13]. Changing the operating voltage could prevent some transistors from turning on because they cannot reach the threshold or could cause transistors to overheat and draw more current. Similarly, the laser can induce transient photocurrent at the reverse-biased PN junctions of the transistor, causing them to flip [13, 19]. After the change has been made, an adversary can observe how the circuit reacts to the injected fault and analyze it to discover secret information. Power is the most common source of side-channel leakage because CMOS circuits' power consumption can depend on their operation [1, 6, 12, 14].

Solutions to these attacks primarily focus on secret sharing. Secret sharing schemes divide sensitive or deterministic data related to the secret keys utilized in cryptographic algorithms into partitions such that all shares must be present to obtain the data being split into shares. These shares are assumed to be uniformly distributed. The "threshold" is the number of shares required to garner valuable information from the data [17]. Masking is a typical approach for secret sharing in which shares are masked with randomized values to make the intermediate outcomes of the cryptographic algorithm independent of the secret key. Masking may be used at both the algorithm and gate levels. These schemes can secure standard algorithms, such as DES and AES encryption. Most previously developed methods for side-channel attacks consider first-order masking schemes. However, attackers are becoming more advanced and are gaining the capabilities to perform higher-order attacks, requiring higher-order masking schemes as a countermeasure. Researchers in 2016 presented the first provably secure higher-order masking scheme for AES with reasonable overhead for software implementation [17]. This method was significant at its invention because it provided the first solution for d-th order masking schemes where d is greater than or equal to three. For example, a d-th order masking scheme, where the threshold is $d + 1$, can protect against a d-th order side-channel attack, where the adversary has d pieces of sensitive data. Masking can be used in several security measures. Aside from masking, threshold implementation (TI), a recently developed approach, is a strategy that uses Boolean masking and is based on a form of multi-party communication [2]. Some important properties of TI are that it has provable security against first-order SCA, does not need many assumptions of hardware leakage, and gives the ability to make realistically sized circuits without much intervention or iterations. However, TI cannot protect against uni-variate mutual information analysis or uni-variate higher-order attacks [2].

It is essential to analyze the AES algorithm because the hardware implementations analyzed are meant to accommodate this algorithm. AES is a symmetric cipher block, and AES-128 handles 128 bit of data blocks. In the first step, the input of the algorithm is replicated into the state array arranged in a square matrix. The state array is 16 bytes. This array finally is copied to the output. In encryption mode, the first key is added to the input, called the initial round. Then, with a modified final round, nine repeats of a typical round are completed. Each regular round has numerous operations, which are listed in the following order: SubBytes, ShiftRows, MixColumns, and AddRoundKey. The final round does not compute the MixColumns stage. For the key used in each round, Key Expansion iteratively derives 10 round keys. The S-box substitutes byte-wise to derive round keys, divided into four words that are shifted to the left cyclically by one byte.

Another more efficient threshold implementation method can be used to protect against side-channel attacks even when glitches are present [11]. This method was made more efficient by using more random values at the start-up and then using the same values for re-masking. This method also protects against higher-order attacks based on mean power consumption. In addition, researchers have developed a compact AES-128 hardware implementation, which improves the area overhead of the previous version by 23% [10]. Furthermore, the threshold countermeasure can be applied to the S-box of the AES in order to protect from first-order side-channel attacks [11]. Further researchers demonstrated that their AES randomization technique is secure in a formal model [3]. This technique used random masks to secure against first-order attacks where the attacker has access to a single intermediate result in arbitrary algorithms that meet some requirements. Finally, the researchers fabricated an ASIC that contained masked and unmasked AES-128 encryption engines [9]. Their first-order DPA attacks on the ASIC showed that masking is insufficient to obfuscate the data when glitches occur in the circuit.

The countermeasures mentioned earlier against the side-channel attacks are covered in detail in the upcoming sections.

18.3 Threat Model

The main focus of this chapter is side-channel resistant cryptographic hardware because hardware accelerators are vulnerable to side-channel attacks. Before delving deeper into these attacks, addressing general information about hardware attacks is essential. One of the main factors to analyze is the adversary, who could be an individual or an organization. Either way, they must have the proper tools to carry out such attacks. It is safe to assume that the adversaries have sophisticated expertise in multilevel analysis at the software, gate, and circuit levels. It is also safe to assume they know the current exploitation using glitches in masking schemes. As for resources, they must have access to high-level analysis equipment to collect the data, a plenty of processing power to run the analysis, and automated analysis software to analyze the data to find the cryptographic key. It would also benefit the adversary to know current anti-side-channel attack techniques to circumvent them and related methods such as FPGAs to simulate attacks. All side-channel attacks aim to extract secret information from cryptographic processes, i.e., the encryption key. As mentioned earlier, this can be achieved by measuring various signals or variables that may unintentionally leak information. For instance, the technology presented at this time is susceptible to leaking information due to glitches. Most anti-side-channel attack methods are implemented at the software/algorithm or gate level.

Gate-level masking schemes have proven unreliable as they can be broken. Even in software implementations, masking techniques have been proven unreliable as they are not secure against higher-order attacks. These shortcomings are the main focus of hardware-level attacks. The main challenge in addressing this threat is that the capabilities of attackers are growing and changing every day. Due to this reason, hardware must be provably secure against higher-order attacks for any AES implementation.

18.4   A Compact Threshold Implementation of AES

This AES-128 hardware architecture was designed with the intention of a small area as possible [10]. When compared to prior designs, such as [7, 8], it is possible to arrange the time–area and power–area trade-offs differently, favoring a more compact hardware realization. Furthermore, a holistic approach can optimize the overall design rather than each component individually to achieve the goal of a smaller area. This implementation needs just 2400 GE and 226 clock cycles, that is 23% smaller than any previous implementations [7, 8]. Likewise, a threshold-based side-channel countermeasure was applied to the implementation of the lightweight AES S-box, which improves resistance from first-order SCA (side-channel attacks), the primary focus of an adversary. This design is a continuation of previous design work that looked into implementing the S-box of the AES in hardware with negligible area overhead [4]. The problem with this previous work was the presence of first-order leakages; however, it is improved by the current design approach by targeting a compact implementation area of [10]. Furthermore, numerous masking strategies have been developed to produce a AES S-box that is masked by using either multiplication- or addition-based approaches. Because of the inversion's linearity in $GF(2^2)$, a typical technique for an additive masking strategy is to employ the representation of the tower field. Furthermore, when employing "Canright's representation, the only non-linear function is multiplication in $GF(2^2)$" [4]. Previous research motivates this approach to include a countermeasure in the unprotected AES implementation. A comprehensive side-channel evaluation was done for this design utilizing real-world power traces collected from SASEBO. Also, the achieved level of resistance was investigated for this implementation against first-order DPA attacks using various power analysis attacks, even if an attacker can measure 100 million power traces. Finally, the threshold-based countermeasure is applied to the AES S-box to share the non-linear function of multiplication with a minimal number of shares. For example, n masks can prevent up to n-th order differential power analysis attacks [5]. This design further demonstrates that if non-linear functions were implemented with this masking, provable security could be achieved against side-channel attacks. More specifically, the researchers were focused on correlation power analysis (CPA) since it can reveal the secrets of comparing measurements to estimations from the theoretical power model. During the development of this countermeasure, the goal is to avoid a dependent relation between the power consumption and the encryption technique. However, it is assumed that perfect protection does not exist currently against DPA attacks. The AES hardware consumes most area when storing the intermediate state, which requires flip-flops. In order to design a compact AES implementation, a designer can use scan flip-flops, a lower number of MUXes and the reorder of input and output bytes. Reordering is not a big problem, but it helps in compact implementation [15, 16]. This implementation's unprotected version is called *Profile 1*.

The data path and the key schedule will be shared for the acquiring threshold version (*Profile 2*). In this case, the implementation requires four randomly generated masks where the state of these masks needs to be maintained, and a shared S-box module replaces the S-box with stages accommodating the unmasking and maintenance of the scheme. The main security point is that the pipeline must be refreshed and empty during every encryption round. The threshold implementation constitutes the second profile of the experiment. Both profile architectures are displayed in Fig. 18.1. For this profile, the dashed lines represent the additional hardware requirements. For this profile, the four randomized masks that are generated $(m_{d1}, m_{d2}, m_{k1}, m_{k2})$ are XORed with the chunks of data and key. The phase of unmasking is as easy as XORing all the three shares, resulting in output (data out). All the mask's state must be preserved, requiring two further instantiations of both the state and the key modules ($maskm_{d1}, maskm_{d2}, maskm_{k1}, andmaskm_{k2}$). The S-box has also been replaced with a S-box module that is shared with five pipelining stages. As a result, round key computation is delayed, and the pipeline must be emptied after each encryption round. Profile 2 uses a tiny FSM to create the control signal and needs 25 clock cycles to complete one loop (77 GE).

The use of scan flip-flops for the compact architecture is present in implementations of known ciphers such as PRESENT. Moreover, the architectural choices are sound in that operations of the AES algorithm are represented through the hardware implementations. For example, the choice of parallelism in the MixColumns approach is related to the architectural choice of one additional MUX per row instead of one of every cell of the state array. The problems with this approach come from the division of goals. To elaborate, the first goal of this design was to make a compact AES implementation, but a compact implementation limits many countermeasures against side-channel attacks because these countermeasures incur more considerable power and area overheads. These overheads are meant

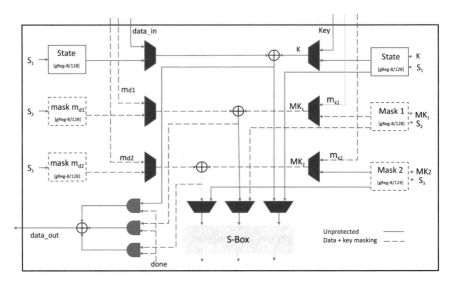

Fig. 18.1 Hardware architectures of serialize AES-128 encryption core

to be avoided by the approach. The results of the experiment are as such: Profile 1, which aimed to reduce area cost, has an area footprint primarily taken up by storing the key and data state, which is not much space if compared with earlier implementations; Profile 2 increases this area more than four times the amount of Profile 1 due to the shared S-box, which includes pipelining stages. As for the main security experiment, the entire AES encryption design was executed on an FPGA-based platform to measure the resistance of the implementation to first-order DPA attacks by analyzing power consumption traces [10]. The measurement platform SASEBO (Side-channel Attack Standard Evaluation Board) was used to measure the implementation's resistance to these attacks. Mask generators were shut off, and masks were at zero for this portion of the experiment to correctly judge the resistance. CPA attacks broadly differ based on the model, according to the results. For example, HW model CPA attacks could not recover the secrets in the secret sharing scheme, regardless of whether the model predicted the input or output of the S-box. CPA utilizing an HD model, on the other hand, was predicted to be a successful attack because it anticipated bit flips on the state register when outputs of the S-box overwrite each other.

Further results show that when RNGs and masks are off, it took 30,000 traces for a successful CPA attack, whereas, with the protections, implementation resisted with even 5 million traces. Using other attacks, it was found that correlation collision attacks do reveal the secret of around 3.5 million traces. Correlation collision attacks recover first-order leakages in combinational circuits, demonstrating that the shared S-box in the AES implementation contains first-order leakage. It is also discovered that the intermediate value registers of the S-box that is shared are not consistently distributed. This uniformity issue may be addressed by re-masking, which involves randomly placing new masks into each pipeline stage as needed. The experiment was re-simulated with 100 million traces, and CPA and mutual information analysis (MIA) attacks could not recover a secret, not even a third-order CPA attack; however, an MIA attack using an HD model on an S-box input did manage to recover the secret after 80 million traces. The experimental results were promising: a compact AES implementation can be made smaller than previous implementations, and several DPA-based attacks can be thwarted by the suitable secret sharing scheme implemented on hardware.

18.5 Efficient AES Threshold Implementation

This threshold implementation of AES-128 claims to be faster and more compact than the implementation discussed in Sect. 18.4. In addition, this implementation addressed the uniformity problem of threshold implementations and the need for re-masking more thoroughly [2]. This implementation covers a scenario when every function does not always satisfy the uniformity property, and fresh randomness is required for re-masking. The researchers concluded that a TI utilizing re-masking does not need sharings that are uniform to be safe from first-order attacks. It is a significant finding since removing the criterion of homogeneity allows the total shares and size of the TI to be decreased.

The primary motivation for this new TI-supported AES-128 scheme is to reduce the space and unpredictability overhead incurred by secret sharing. Regarding the general data flow, the new TI uses the serial implementation method proposed earlier [10]. This serial implementation only requires instance of one S-box and loads the plaintext and key in the byte-wise and row-wise order. The new TI also changes the number of shares used throughout the operations in the block cipher. It uses two shares, the minimum number possible, for the affine operations, and then the number of shares used is either increased or decreased according to the requirement for the non-linear layer. With this change, there is already a significant decrease in the area for the new TI.

Regarding the TI of the AES S-box, it is instantiated once and utilized by the key schedule, and it is also utilized by the state update. The S-box implementation is similar to that discussed in Sect. 18.4, in which the design follows the tower field approach, but the current design uses $GF(2^4)$ instead of utilizing the $GF(2^2)$. By taking this approach, the $GF(2^4)$ inverter can be seen as a four-bit permutation. Also, the $GF(2^4)$ multiplier can be seen as a four-bit multiplier. As a result, for these non-linear blocks, the uniform TIs may be obtained individually, meaning that when these uniformly implemented components are combined, less fresh random bits are utilized. On top of using fewer fresh random bits, this approach also makes the S-box calculation faster as it takes fewer clock cycles to complete.

While the $GF(2^4)$ multiplier could not be optimized and had a set required area, the $GF(2^4)$ inverter had two possible representations. It might be expressed with four shares and the function divided into three uniform sub-functions, or

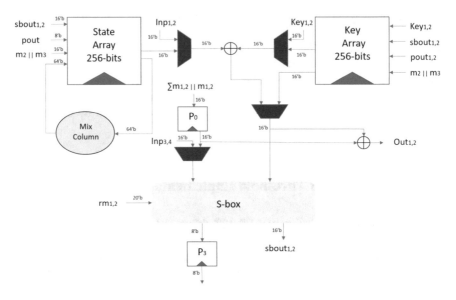

Fig. 18.2 Architecture of serialized TI of AES-128

it could be represented using five shares without decomposition of the function. After comparing the two representations, this implementation used five shares representation and determined that while both had similar area requirements, the five shares version took fewer clock cycles. Then, for sharing this module, this implementation uses a method that varies from direct sharing, and further, it is implemented with fewer hardware logic gates. When combining the sub-blocks in the new TI, the designers faced two issues: first, maintaining uniformity in the pipeline registers when combining the sub-blocks, and second, they needed to keep the uniformity as the number of shares changed. The problem concerning uniformity in the pipeline registers was addressed and solved earlier [10] with re-masking, so the designers also chose to go with re-masking to avoid facing this problem. By adding new masks before the S-box operation and at the conclusion of the second phase, the problem of maintaining uniformity as the number of shares changes may be overcome. All in all, through the construction of this new threshold implementation, shown in Fig. 18.2, and it is optimized a lot of the aspects of the previous TI [10].

The results of this new threshold implementation are shown in Table 18.1, and it can be observed that the new threshold implementation required 8% fewer random bits, it is 18% smaller, and 7.5% faster when compared to the previous design [10]. On top of being much more optimized, it was also concluded that this implementation is resistant to first-order attacks, as it took attackers to collect 10 million traces in the worst conditions. Furthermore, regarding higher-order attacks, this new TI was also proven to be somewhat resistant in more realistic and less favorable conditions with much more noise and an unstable clock since it took

Table 18.1 Synthesis results for different versions of AES TI

Design	State array	Key array	S-box	MixCol	Ctr	Key Xor	MUX	Other	Total	Cycles	Rand bits
[10]	2529	2526	4244	1120	166	64	376	89	11,114	266	48
[10] Compile_Ultra	2529	2526	4244	1120	166	64	376	89	11,031	266	48
[2]	1698	1890	3708	770	221	48	746	21	9102	246	44
[2] Compile_Ultra	1698	1890	3003	544	221	48	746	21	8171	246	44

600,000 traces to break under conditions conducive to adversaries. Overall, this implementation can be a faster and more compact threshold implementation of AES-128 with provable security compared to the implementation presented in Sect. 18.4 [10].

18.6   Masking Based on Secret Sharing, Threshold Cryptography, and Multi-party Computation Protocols

This threshold implementation technique protects against SCA and glitches by masking the data with two or more random values rather than simply one. This modification necessitates additional random values at the start but does not necessitate new random values after each non-linear transformation. There is no need for extra random values to mask the intermediate outcomes of the calculation with this enhanced masking approach since each intermediate result is uncorrelated to the input parameters. Secret sharing schemes (n,n) are used; hence, this method is related to secret sharing, where all n shares are required to decide an arbitrary variable x uniquely. These secret sharing (n,n) techniques are combined with a (1,n,n) ramp scheme, where there is one malicious party that can obtain information about the secret and where the conditional probability distribution is uniform and represented by the following equation [11]:

$$\bigvee \overline{X} : Pr(\overline{x} = \overline{X}) = c\,Pr(x = \oplus_{i=1}^{n} X_i). \tag{18.1}$$

This method is also related to threshold cryptography because the secure implementation of the linear transformation requires $z = L(x)$ over GF($2m$) and requires n shares to be processed independently. Through this implementation, no leaked information could be used in a SCA, despite glitches being present. A common aspect of this technique is that each output share, Z_i, is dependent on just one input share of each variable $(X_i, Y_i, \ldots)$. Furthermore, this approach is also related to threshold cryptography in implementing non-linear transformation, $z = N(x,y\ldots)$. These non-linear transformations also have a similar property to the linear transformations, where if Z_i does not depend on an input share, it cannot be correlated to that share, and the output computation will not leak information about the input share value. By imposing additional constraints, the designers ensured no correlation to the output z. They also show that if two properties of non-linear transformations, non-completeness, and correctness, are fulfilled, it can result in a secure circuit. To meet these requirements, which set a lower restriction on the number of shares n, a minimal number of shares necessary to implement a product of s variables are $n \geq 1 + s$. However, since there might be other solutions that have even fewer shares, the largest number of shares required to execute a function N of u variables over GF(2^m) is $1 + 2^{mu}$ [11]. Also, as the number of shares increases, so will the number of gates, and to combat this large number of gates problem, pipelining is often used to speed up hardware implementations. This approach is also connected to multi-party computational protocols since all intermediate outcomes are independent of the inputs and outputs, as is the mean power consumption of a variable. This novel masking strategy is especially significant since it can guard against SCA even when glitches are present.

Additionally, this method does not result in the information leakage provided that the attacker is restricted to looking at one function at a time. It can be explained through multi-party computation, where an adversary can simultaneously corrupt a function but only at most one function. Also, if the input shares satisfy the non-completeness property, where the mean power consumption is independent of the input shares, it holds, even in the presence of glitches. There are many benefits to this method of implementation that counteract side-channel attacks. One of those benefits is that there is no longer a need for fresh random values to be generated after every non-linear transformation. Another advantage is that even when glitches are present, verifiable security against first-order attacks was established. Lastly, specific higher-order attacks are resisted by this scheme. Some downsides of this strategy include an increase in data storage needs owing to the increased number of shares required, as well as an increase in computing complexity. Another problem with this method, it does not prevent template attacks but makes it more difficult because computing the n shares in parallel improves the signal-to-noise ratio. In terms of limitations, this design did not deal with extremely complicated circuits, which would lead to an even more significant increase in circuit complexity; however, other approaches from proactive secret sharing systems might be applied to decrease this greater circuit complexity, although this would be restricted by the requirement for new random values.

$$Pr(\overline{x} = \overline{X}, \overline{y} = \overline{Y} \ldots) = c\,Pr(x = \oplus_i X_i, y = \oplus_i Y_i, \ldots). \tag{18.2}$$

This implementation demonstrated that this method is secure and resistant to side-channel and fault injection attacks. Furthermore, this method further demonstrates its security against first-order attacks by proving the theorem based on the condition in Eq. 18.2. If the non-completeness and correctness properties are fulfilled, then all the intermediate outcomes are independent of the inputs and output, and the power consumption is also independent of those variables. Also, this method is resistant to higher-order attacks, specifically leveraging statistical data from power consumption traces through another theorem's proof, which made the circuit's mean power consumption always the same.

18.7 Provably Secure AES Randomization Technique

This algorithmic countermeasure is designed for AES to focus on the timing and power attacks [3]. This approach can work with hardware countermeasures to provide more security. The goal of using algorithmic countermeasures is to provide a mathematically provable solution to side-channel attacks. This approach is somewhat novel because it refers to two other custom masking methods but points out the failings of these methods to meet the standards presented in this approach, and then it is mathematically proven that this algorithm protects against first-order side-channel attacks. A couple of assumptions are used in the algorithm to define the security notion of a perfectly masked algorithm. First, a portion of the computation runs in a protected environment. Second, an adversary has limited access to intermediate results. Lastly, any deviations in the distribution of an intermediate result that depended on the secret key or the plaintext could completely reveal the system to the attacker. According to the researcher, an *algorithm* can be defined as "provably secure" against a *d-th* order attack if the joint distribution of any d intermediate results is secret key independent and the plaintext. Therefore, the algorithm needs to guarantee that any side-channel information is random to provide no information to the attacker. This algorithm is a first-order provably perfectly masked algorithm for AES that can be generalized to higher-order attacks using more randomness. The fact that this algorithm can be proved mathematically to be secure and efficient in both area and timing is what the researchers' claim makes this approach novel. Two algorithms can be used for the multiplicative inverses in 256, the primary step in the SubBytes-transformation, to make the process securely masked [3]. The method used to find the inverse of x was to raise x to the 254th power, which was calculated using the square-and-multiply algorithm. The perfectly masked squaring (PMS) and perfectly masked multiplication (PMM) algorithms are created to mask the square-and-multiply algorithm completely. Both of these algorithms can keep the value masked at each intermediate step and keep the value and the mask separable where the mask is only added to the value, not multiplied.

The algorithm squares the intermediate output value, masks it, and then converts it to become the mask for the current step raised to the first power. An advanced researcher can refer to [3] for in-depth details of the algorithm. Both algorithms use a new mask for each step, but only three masks are needed to protect against a first-order attack because of the independence of the masks. These algorithms can be applied to any implementation that uses additions, multiplications, and squares in a field or ring. For example, an inversion operation over $GF(((2^2)^2)^2)$ can be used on an ASIC for efficient implementation. A NOT gate can be used as the normalizing factor for area and delay for the other implementations. The 2-input AND gate is assumed to have twice the area, and the 2-input XOR gate has three times the area of the NOT gate. The delays for all three gates are assumed to be equivalent.

This method has an increase in the area over the previous implementations but only a 20% increase in delay compared to a standard inverter, making it the second best masking method in terms of area and delay. However, the increase in these factors is outweighed by the benefits of having a fully masked implementation of AES. One con of this masking technique is that while the area increase is small, it is still significant. The other con is that each algorithm step requires new random mask values. The limitation of the given algorithm is that it was only proved for first-order side-channel attacks. It would need more randomization to defend against higher-order attacks. The benefit of this approach is its guarantee of security against first-order attacks, which previous techniques could not guarantee.

18.8 Conclusions

When hardware is made without security in mind, vulnerabilities manifest in the architecture, and cryptographic algorithms cannot work on vulnerable hardware because sensitive data will be leaked if an attacker can bypass encryption. Exploiters of these vulnerabilities obtain data using mainly side-channel attacks with some instances of glitches. This chapter analyzed research works that have made notable contributions to side-channel attack resistant cryptographic hardware. Although the solutions provided were novel for publishing at the time of their publication, certain limitations were observed, showing that further research is required to find more effective and efficient solutions to side-channel attacks on cryptographic hardware. An obvious limitation to the proposed solutions is that the security methods cannot be tested with an adversary observing an infinite amount of traces. There is no practical way to test this, but an adversary failing to exploit the device until many traces is not a definitive sign that the proposed security measures will always work. Again most of the side-channel assessments are done in post-silicon by collecting power traces from the hardware, which is very late in the design stages and can incur overhead on time-to-market. There is a need to develop CAD frameworks in the pre-silicon design stages to verify designs for side-channel leakages [1, 12]. Similarly, there is a growing need for post-quantum cryptographic algorithms. Most of the research done applies to traditional cryptographic algorithms. There is a need to perform similar assessments on the upcoming post-quantum cryptography algorithms.

References

1. Ahmed B, Bepary MK, Pundir N, Borza M, Raikhman O, Garg A, Donchin D, Cron A, Abdel-moneum MA, Farahmandi F, et al (2022) Quantifiable assurance: From IPs to platforms. Preprint. arXiv:220407909
2. Bilgin B, Gierlichs B, Nikova S, Nikov V, Rijmen V (2014) A more efficient AES threshold implementation. In: Pointcheval D, Vergnaud D (eds) Progress in cryptology – AFRICACRYPT 2014. Springer International Publishing, Cham, pp 267–284
3. Blömer J, Guajardo J, Krummel V (2005) Provably secure masking of AES. In: Handschuh H, Hasan MA (eds) Selected areas in cryptography. Springer, Berlin, Heidelberg, pp 69–83
4. Canright D, Batina L (2008) A very compact "Perfectly Masked" S-box for AES. In: Bellovin SM, Gennaro R, Keromytis A, Yung M (eds) Applied cryptography and network security. Springer, Berlin, Heidelberg, pp 446–459
5. Chari S, Jutla CS, Rao JR, Rohatgi P (1999) Towards sound approaches to counteract power-analysis attacks. In: Wiener M (ed) Advances in cryptology — CRYPTO' 99. Springer, Berlin, Heidelberg, pp 398–412

6. Dey S, Park J, Pundir N, Saha D, Shuvo AM, Mehta D, Asadi N, Rahman F, Farahmandi F, Tehranipoor M (2022) Secure physical design. Cryptology ePrint Archive
7. Feldhofer M, Wolkerstorfer J, Rijmen V (2005) AES implementation on a grain of sand. IEE Proc Inf Secur 152(1):13–20
8. Hamalainen P, Alho T, Hannikainen M, Hamalainen TD (2006) Design and implementation of low-area and low-power AES encryption hardware core. In: 9th EUROMICRO conference on digital system design (DSD'06). IEEE, pp 577–583
9. Mangard S, Pramstaller N, Oswald E (2005) Successfully attacking masked AES hardware implementations. In: Rao JR, Sunar B (eds) Cryptographic hardware and embedded systems – CHES 2005. Springer, Berlin, Heidelberg, pp 157–171
10. Moradi A, Poschmann A, Ling S, Paar C, Wang H (2011) Pushing the limits: A very compact and a threshold implementation of AES. In: Paterson KG (ed) Advances in cryptology – EUROCRYPT 2011. Springer, Berlin, Heidelberg, pp 69–88
11. Nikova S, Rechberger C, Rijmen V (2006) Threshold implementations against side-channel attacks and glitches. In: Ning P, Qing S, Li N (eds) Information and communications security. Springer, Berlin, Heidelberg, pp 529–545
12. Park J, Anandakumar NN, Saha D, Mehta D, Pundir N, Rahman F, Farahmandi F, Tehranipoor MM (2022) PQC-SEP: Power side-channel evaluation platform for post-quantum cryptography algorithms. IACR Cryptol ePrint Arch 2022:527
13. Pundir N, Li H, Lin L, Chang N, Farahmandi F, Tehranipoor M (2022) Security properties driven pre-silicon laser fault injection assessment. In: International symposium on hardware oriented security and trust (HOST)
14. Pundir N, Park J, Farahmandi F, Tehranipoor M (2022) Power side-channel leakage assessment framework at register-transfer level. IEEE Trans Very Large Scale Integr (VLSI) Syst
15. Rahman MS, Nahiyan A, Amir S, Rahman F, Farahmandi F, Forte D, Tehranipoor M (2019) Dynamically obfuscated scan chain to resist oracle-guided attacks on logic locked design. Cryptology ePrint Archive
16. Rahman MS, Nahiyan A, Rahman F, Fazzari S, Plaks K, Farahmandi F, Forte D, Tehranipoor M (2021) Security assessment of dynamically obfuscated scan chain against oracle-guided attacks. ACM Trans Des Autom Electron Syst (TODAES) 26(4):1–27
17. Rivain M, Prouff E (2010) Provably secure higher-order masking of AES. In: Mangard S, Standaert FX (eds) Cryptographic hardware and embedded systems, CHES 2010. Springer, Berlin, Heidelberg, pp 413–427
18. Schramm K, Paar C (2006) Higher order masking of the AES. In: Pointcheval D (ed) Topics in cryptology – CT-RSA 2006. Springer, Berlin, Heidelberg, pp 208–225
19. Vashistha N, Rahman MT, Paradis OP, Asadizanjani N (2019) Is backside the new backdoor in modern SoCs? In: 2019 IEEE international test conference (ITC). IEEE, pp 1–10
20. Zhang T, Park J, Tehranipoor M, Farahmandi F (2021) PSC-TG: RTL power side-channel leakage assessment with test pattern generation. In: 2021 58th ACM/IEEE design automation conference (DAC). IEEE, pp 709–714

Chapter 19
Fault Injection Resistant Cryptographic Hardware

19.1 Introduction

Fault injection attacks are a growing threat in today's integrated circuits [4, 20, 25]. Cryptographic algorithms are being employed on more and more devices, such as smart credit cards, cell phones, and other devices people interact with on a daily basis. As more and more devices incorporate cryptography in their designs, it becomes harder and harder to mount a successful attack against them with traditional methods. For example, brute force attacks are not feasible, as there are no known methodologies to attack a modern cryptographic algorithm. In addition, side-channel attacks are becoming increasingly harder to successfully implement as chip designers are aware of the problem and incorporate countermeasures in their design [10, 16, 28].

With this increased difficulty in attacking a modern cryptographic device, attackers have turned to a new category of attacks that are known as fault injection attacks. The principle underlying a fault injection attack is to cause a device to act erroneously—i.e., inject a fault—and measure the perturbations in the device's output caused by the fault [30]. With careful applications of faults, the perturbations can leak valuable information to an attacker, up to and including the recovery of the private encryption key [8]. Fault injection attacks have been proven highly effective and can even be combined with other attacks, such as analyzing power consumption after a fault is injected [13]. In order to protect from such attacks, numerous countermeasures have been suggested. This chapter provides a survey of fault injection attacks and their proposed countermeasures.

There are many countermeasures to fault injection attacks. The majority of techniques uses some form of redundancy to detect faults. For example, one solution is to duplicate the entire cryptographic circuit and compare the output of the two at the end of the computation [4]. This solution has the advantage that it is challenging to inject the same fault into two circuits simultaneously, making detection of the fault very likely. However, the area and power overheads are double that of a single circuit. Another common form of redundancy is temporal redundancy—performing a cryptographic computation twice on the same circuit and comparing the results [4]. Again, this has the advantage that it is challenging to inject the same fault simultaneously. The disadvantage is that performance is halved—it takes twice as long to perform one computation.

Another approach is using error detection or parity bits in the computation [4]. This is a less robust approach because it is easier to inject a fault without it being detected by the error detection circuitry. However, the advantage is that the overhead of this technique is significantly reduced from a truly redundant solution. Double data rate (DDR) techniques have also been employed to exploit temporal redundancy without adversely affecting throughput [18]. The idea is to run the cryptographic hardware faster than the rest of the chip, when feasible, and perform the cryptocomputation two times without adversely influencing the performance of the rest of the device.

The need for randomness in fault injection countermeasures is also examined in this chapter. The proper application of randomness can result in a fault injection attack giving garbage data at the chip's output, rendering an attack infeasible [17]. This chapter provides an overview of fault injection attacks and currently effective countermeasures.

19.2 Background

AES encryption is a 128-bit symmetric block cipher, which means it uses a key to encrypt 128 bits of the message into 128 bits of ciphertext. The key can be either 128 bit, 192 bit, or 256 bit with increasing levels of security. The encryption involves substitution to cause confusion, permutation to cause diffusion, and XOR operations with the key to encrypt the plaintext. All of this is done in one round of encryption. There can be 10, 12, or 14 rounds, and the total sum of rounds increases as the number of bits in the key does.

Each round is executed on a matrix known as the state. The state is a 4x4 matrix of bytes since there are 16 bytes in 128 bits. The state is modified in each round until the message has been fully encrypted. The encryption starts by XORing the message with the key. Each round employs a unique key known as a round key. These round keys are generated by utilizing a key scheduler that expands the original key. Then we substitute the bytes, shift the rows, and mix the columns. Finally, the round key is added to the state through an XOR operation. A substitution box is used, which is simply a 256x8 lookup table to substitute bytes. Every possible value for a byte in the state is mapped to a different byte using a non-linear function.

Shifting the rows means the bytes are rotated in a row to the left by the number of the rows they are contained in, i.e., row 0 is not rotated at all, and row 1 bytes are rotated left once, and so on. The mix columns step involves matrix multiplication between a column from the state and another matrix that linearly transforms the bytes. The matrix multiplication in this step uses modulo two multiple and add operations. In the last round, the mix columns step is omitted.

AES encryption is too complex to break with simple brute force, so researchers have developed a class of techniques called fault injection attacks to break the encryption. A fault injection attack involves maliciously altering the functionality of a system to leak information. These attacks can range in difficulty from requiring no technical expertise to requiring the highest education technical knowledge available. Furthermore, they can be classified as either low cost or high cost. Low cost being anything under $3000. Fault injection can be achieved by altering the power supply voltage, clock glitching, and exposing the device to radiation, electromagnetic waves, intense light, or extreme temperatures.

A fault injection attack against AES encryption will attempt to cause a fault in the byte's state by injection of either a single-bit or single-byte fault. Fault can be injected at many different locations and times in the algorithm. This fault injection ranges from the first key addition to the last round [1, 6, 9, 14, 15, 19, 21, 26, 27]. A commonly attacked location is the mix column round step, but the substitution box can also be exploited. The inserted fault impacts the final ciphertext. Then cryptanalysis is used to analyze the faulty and correct ciphertexts to leak the secret key. Research is also being done on how to combat these fault injection attacks. In the case of AES, this means altering the AES architecture to allow for error detection or duplicated execution.

19.3   Fault Injection Countermeasures

A review of current research into fault injection attacks by [4] describes low- and high-cost fault injection attacks, as well as countermeasures that use redundancy (both spatial and temporal) and error detection using parity bits.

Low-cost fault injection techniques are those that require less than $3000 in equipment. The main advantage of these methods is that no expensive equipment or advanced technical skills are required. These techniques include under-powering a device or introducing temporary power supply glitches. A single instruction in the execution of a software program, for example, may be interrupted by a well-timed power supply drop. This attack necessitates very little specialized equipment. Similarly, tampering with the device's clock signal can introduce errors at specific points in the computation, which can be used to glean useful information about the secret key if timed correctly. Altering the device's environmental conditions is another low-cost fault injection technique, for example, operating the device at temperatures above its rated temperature or in the presence of a strong electromagnetic disturbance. However, these techniques have the disadvantage of being less precise in the timing of the fault application, and they may cause device damage. The final low-cost method described is the use of UV light. If the attacker is able to decapsulate the chip, he or she can execute the fault injection by illuminating

Table 19.1 Summary of different fault injection techniques and trade-offs

Technique	Location accuracy	Timing accuracy	Technical skill	Cost	Implementation knowledge required?	Damage to the device
Underfeeding	High	None	Low	Less	No	No
Clock glitch	Less	High	Average	Less	Yes	No
EM	Less	Average	Average	Less	No	Possibly
Heat	Less	None	Low	Less	Yes	Possibly
Voltage glitch	Less	Moderate	Average	Less	Partial	No
Light radiation	Less	Less	Average	Less	No	Yes
Light pulse	Average	Average	Average	Average	Yes	Possibly
Laser beam	High	High	Skilled	More	Yes	Yes
Focused ion beam	Absolute	Complete	Expert	Expensive	Yes	Yes

the die with UV light or even a simple camera flash. Again, this has low precision compared to some more expensive techniques.

The high-cost (above $3000) techniques improve the fault injection precision. Decapsulating the chip and using a tightly focused laser to inject precise faults is one method of attack in this category [29]. The limitation of this method is that gates built using today's processes are much smaller than a wavelength of light, limiting precision. A focused ion beam (FIB) is used to inject faults to overcome this. This is the most precise fault injection technique, but it is also the most expensive technique investigated and requires advanced technical skills. Table 19.1 provides a summary of different fault injection attack techniques and their trade-offs.

There are several countermeasures to the outlined fault injection techniques, each with its own set of trade-offs. However, the overall goal of all defenses is the same: prevent the injected fault from allowing an attacker to obtain sensitive information. This is accomplished in two ways: detecting the injected fault or making device faults result in garbage output. The first countermeasure is to make the device's implementation physically inaccessible, with the device being destroyed if tampering is detected. However, this can become very costly and is thus reserved for the most secure implementations with massive budgets.

Another method for countering fault injection attacks is to duplicate and compare the computation of the encryption or decryption process. The results are discarded if they do not match. Duplication can be spatial (two separate circuits) or temporal (have the same circuit perform the computation twice). The advantage of this approach is that it is incredibly robust because it is challenging to inject the same fault twice. The disadvantage is the added overhead. If an exact duplicate computation is required, either the power and area or the computational time will be doubled, making it quite expensive.

Table 19.2 Comparison of fault injection countermeasures

Countermeasure	Area and power penalty	Performance penalty	Cost	Effectiveness
Tamper proofing	None	None	Very high	High
Spatial redundancy	Very high	None	Low	High
Temporal redundancy	None	Very high	Low	High
Error detecting codes	Low to moderate	Low to moderate	Low	Variable
Inverse operation	None	Low	None	High

A more cost-effective fault injection countermeasure is using error detecting codes (EDCs). Check bits are generated at different times during the encryption process; typically, this is a parity bit. Periodically, the check bits are checked against the actual computed check bits, and a fault is declared if the results do not match. The more often check bits are applied, the more likely a fault will be detected. Increasing the amount of check bits, however, raises the cost.

If a chip is capable of decrypting as well as encrypting, the calculated ciphertext can be decrypted and compared to the original plaintext. If the results match, then a fault has not occurred. This inversion can be done at the operation, round, or cipher levels. This method is very effective because it takes advantage of hardware already on the device while imposing a minimum performance penalty. Table 19.2 provides a comparison of the different fault injection countermeasures examined.

In summary, [4] presented several common fault injection attacks and effective countermeasures. The most effective countermeasures are usually the most expensive, but if the hardware to perform the inverse operation exists on the device, using this hardware to check that the ciphertext was correctly encoded provides very effective fault detection at a low additional cost.

19.4 Algebraic Fault Analysis

Priority in [31] was to see how Algebraic Fault Analysis (AFA) performed on lightweight block ciphers. This was a continuation of previous studies that focused on the work of differential fault analysis, abbreviated as DFA. Another goal was to see how algebraic fault analysis performs on other block ciphers and to standardize how AFA is performed. The specific application's work was done primarily on LBlock, but it was also tested on DES, PRESENT, and TWO FISH. It is critical to continue working on fault injection analysis with a methodical and controlled approach in the future.

The approach breaks down AFA into three levels: target, adversary, and evaluator (as shown in Fig. 19.1). The target level includes different cipher algorithms, the converted code, and the target device. For example, LBlock, DES, and Twofish all have a Fiestel structure. They use a round structure that encrypts half of the data and

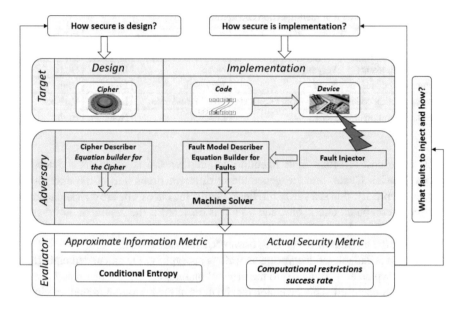

Fig. 19.1 Block diagram illustrating the Algebraic Fault Analysis flow

XORs it with the other half of the data. Similarly, another is an SPN structure that uses substitution in rounds to create encryption, typically used in AES.

The AFA spends most of its time at the adversary level, which houses the describer, fault injector, and machine solver. At this level, describers are used to construct arithmetic equations that take in data from the device before and after the fault and use machine solvers to solve the equation sets from the describers. It also examines how AFA's position and location influence the time and ability to successfully break a key. The position is the state when the fault exists, and the location is the location of the fault within that state [31]. The final level is the evaluator level, which evaluates actual and approximate security levels. At the actual security level, they measure real-world restrictions such as time, space, and data complexity and create a success rate from the data, whereas, in the approximate security, they check the conditional entropy [31].

The testing of AFA on ciphers is done at three levels, i.e., bit, nibble, and byte level. The tests were performed on LBlock during key scheduling and encryption processes. The best findings demonstrated that even a single fault is sufficient to retrieve the LBlock's master key. Please refer to [31] for detailed results at different levels.

At the bit level, the theories were tested for various fault positions and unknown/known fault locations. When the location was unknown, it was observed that slightly moving the fault location and the number of faults improved the solving time significantly. However, when the location is known, there is a hard cut-off at position 26 and 1 fault where the time to solve increases from 15 s to more than

33 min. In nibble level fault models, it was observed that the entropy could be significantly reduced as the location round is decreased. Similar results were found in byte-level fault models.

Finally, faults were injected to modify the round counter. There were two cases in this, faults to change the maximum number of rounds and faults to change the current round counter. There are two subcases in the first situation. When rc' is greater than rc_max, an additional round of encryption is created. It does not, however, return any helpful information. On the other hand, when rc' is less than rc_max, the encryption does not finish completely, reducing the number of rounds the solver must endure cracking the encryption. When modifying the round counter itself, if the rc' is less than the rc and rc_max, then rounds are repeated, and the key can be solved in under 2 min with a 100% success rate [31]. If rc' is greater than rc, then rounds are skipped in computation, and the key can be solved in under a minute with only one injected fault [31].

Researchers also concluded that AFA is also successful on DES and PRESENT. However, it struggled against stronger encryption such as Twofish. It was also observed that using AFA to inject faults and leak keys and other information was more successful than using DFA. However, this method has not been tested on AES, and it may yield useful information about where side-channel attacks against the most common and strongest block cipher exist [2].

19.5 Double Data Rate AES Architecture

A double data rate (DDR) AES architecture was proposed to combat natural and intentional fault injections [18]. Every byte of encrypted data is calculated in two blocks. One being the linear Cell blocks and the other being the non-linear S-Boxes. Each round is calculated in 6 cycles. There already exists an technique for detecting error for this original architecture based on parity code. Unfortunately, the parity code's error detection rate is only acceptable for odd-order faults in the permutation layer.

The AES state registers and computational logic are divided into two independent classes in the DDR design, which are driven by two separate clocks. The one is completely out of sync with the other. The registers driven by these two clocks can be combined to form a single DDR register.

The DDR registers (shown in Fig. 19.2) allow a round to be calculated in only three cycles, down from six. The three unused cycles can be used to generate a second copy of data that can be cross-checked with the initial data. If the two sets of data do not match, an error is assumed to have happened. Naturally, the proposed design will incur additional area overhead and power consumption.

The DDR architecture is a much more robust error detection method than the existing parity code. For a minimal increase in area and power, the DDR was able to vastly increase the number of errors detected. Furthermore, the DDR uses much less area and power when compared to the Hi-Speed architecture, which utilizes

Fig. 19.2 Block diagram
illustrating the structure of
DDR register

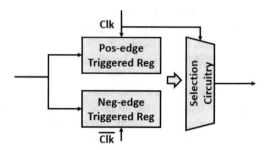

the same amount of clock cycles per round. However, when considering the system clock, the drawbacks of the DDR design become apparent. The DDR architecture cannot support frequencies above 200 MHz without requiring a significant increase in area. Having two synchronized clocks also results in complex clock routing throughout the architecture.

The DDR architecture was able to detect 100% of single- and double-byte faults injected in the linear layer. Faults injected in the non-linear layer were almost perfectly detected as well. The only problem is that the DDR architecture would falsely classify an output as having an error for every correctly identified fault. In addition, it was stated that many false positives could be identified through means outside the scope of DDR. Lastly, one hot encoding was used to assist in the error detection in the control unit to great success in all areas except the inner signals.

19.6   Randomness in Fault Attack Countermeasures

Countermeasures based on randomness are proposed in [17] to thwart two main types of physical attacks on AES, namely side-channel analysis (SCA) and fault attacks (FA). It is shown that then-existing "state-of-the-art" countermeasures that claimed to prevent such physical attacks are prone to a different set of vulnerabilities against attacks of the same class, sometimes applied in combination, and proposed a new set of countermeasures against them.

SCAs are performed by utilizing various types of leakages from a cryptographic operation. Differential side-channel analysis (DSCA) primarily applies to block ciphers [12, 22, 23]. DSCAs are of particular interest because they are susceptible to noise. An adversary's goal, in this case, is to apply SCA to a subset of the input for the same key, develop a hypothesis, run statistical analysis and prediction to correlate results, and then extract the correct key.

Fault attacks extract the key by introducing errors into the cryptographic operation. The adversarial hypothesis is based on the logical effect of errors on intermediate values when the size (bits) is taken into account. The authors introduce the FA bi-product as an invariant. The attack models based on fixed fault diffusion patterns, fixed fault logical effects, and their combination are explained. The fixed

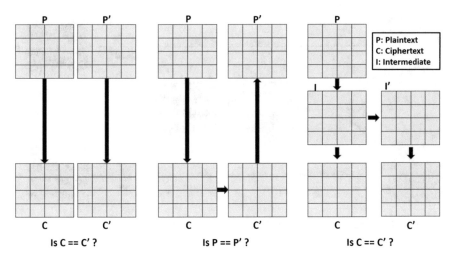

Fig. 19.3 State registers illustrating the three classical detection countermeasures. (Left) Full duplication, (Middle) encrypt then decrypt, and (Right) partial duplication

fault diffusion pattern attack takes advantage of the diffusion property of a random fault introduced into the cryptographic operation's intermediate values. With the help of the localization effect of the fault and the precise selection of faulty intermediate bit sizes, the brute force search space for different versions of AES can stay within the range of 28 to 232, and the correct/faulty ciphertext pair can range from one to a maximum of four pairs to recover the correct key [17]. FA based on a fixed fault logical effect assumes that the logical effect of the fault is fixed on the target intermediate variable, such as the safe error attack, because the adversary does not require erroneous ciphertext and instead relies on the correctness of encryption.

Combined attacks employ both SCA and FA methods. Because fixed fault diffusion patterns are difficult to observe through side channels, combined attacks primarily exploit the fixed fault logical effect of an FA mounted on a fault-resistant design and then perform SCA on the faulty ciphertext to extract the key [3, 11]. An opposing set of attacks can be used in tandem to suppress the masking effect and achieve the same adversarial goal.

When it comes to countermeasures against SCA, classical approaches rely heavily on masking schemes of a different order. FA detection countermeasures work based on fault detection. Three types of detection schemes are available to detect execution/runtime errors: full duplication, encrypt/decrypt, and partial duplication, as shown in Fig. 19.3.

FA infection countermeasures focus on thwarting the exploitation of faulty ciphertexts. The resulting diffusion pattern modifies the logical effect of fault and therefore nullifies the fault invariant. Figure 19.4 shows the flow of generic infection countermeasure.

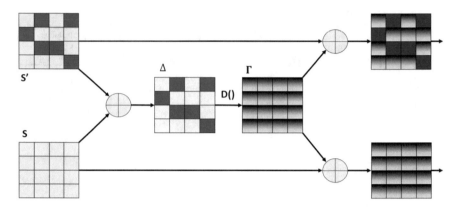

Fig. 19.4 Block diagrams illustrating the classical infection countermeasure

A combination of DSCA and DFA countermeasure schemes may be applied; however, [17] argue that this scheme is flawed. With the various attack scenarios where their classical countermeasures failed to ensure SCA and FA protection and robustness, the authors introduced their arguments on the need for randomness in the existing countermeasures.

In detection-based FA countermeasures, the comparison "if (a=b)..." is carried out on masked ciphertexts. This comparison is performed by either XOR operation or subtraction operation. The leakage will be tied to the combination of a and b, and there is at least 35% overlap in the input space between the leakages from the XOR and subtraction operations, allowing the attack to continue. So, [17] argued that randomness must be introduced to secure the comparison to suppress the leakage-to-input space correlation.

Since FA infection countermeasures aim to kill fault invariant, research has shown that the traditional approach is ineffective against both previously classified FA attacks [17]. In the case of a diffusion-based attack, in the worst-case scenario where a 1-byte fault is injected before the last SubBytes operation, it is possible to precompute 28-1 differences at SubByte input, with a 16-byte discrepancy between C and C'. Assuming I and I' are the correct and faulty inputs before the last SubByte operation, it is possible to establish a hypothetical correlation between pairs using the bijection principle (C to C' and I to I'), which can eventually bypass the countermeasure. In the case of logical effect-based attacks, if the diffusion function is fixed, localization is unaffected, and the attacker can repeat the attack, undermining the countermeasure's masking scheme (DFA). Therefore, [17] proposed modified detection-based countermeasures with added randomness.

Countermeasures are based on two separate unpredictability levels for the comparison of middle round intermediate values—Encrypt/Partial Decrypt and Encrypt/partial Encrypt/Partial Decrypt. This is shown in Fig. 19.5. Although this improves the comparison scheme, it remains the weak point of FA countermeasure.

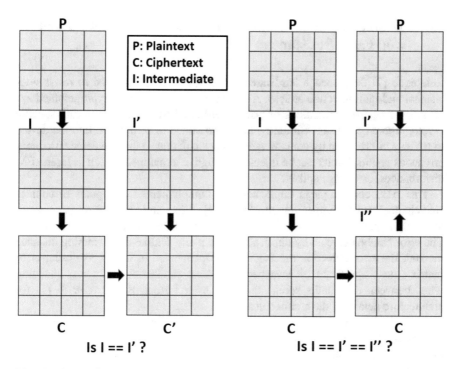

Fig. 19.5 Introducing unpredictability. (Left) Encrypt/partial decrypt. (Right) Encrypt/partial encrypt/partial decrypt

To bolster FA countermeasures based on the infection concept, which cannot suppress the DFA effect, [17] introduced an algorithm such as the detection case, which uses random multiplicative masking. Reference [17] also introduced a concept of manipulating intermediate values further to avoid key predictability with a small number of hypotheses.

Although [17] demonstrated major flaws in classical countermeasures that can be exploited by various attack models and argued that randomness could overcome many of these flaws, they did not consider the attack model based on double fault insertion that can inject the same fault in concurrent encryptions. Further analysis using a more powerful attack model can evaluate the proposed countermeasure. The argument presented is based on the algorithmic implementation of AES-128, but it would be interesting to see how the arguments hold up during practical attacks on the hardware implementation of AES.

19.7   Duplicated and Complemented Paths for Side-Channel and Fault Resistance

Reference [7] developed a hardware AES-128 IC, which claimed to resist side-channel and fault injection attacks. An attacker's goal is to leak cryptographic keys by measuring EM emanations or power consumed by the device during specific cryptographic operations. Similarly, faults are aimed to be injected into the device to retrieve information by contaminating the calculation so that the device produces erroneous results. Faults can be injected through external means such as lasers, EM disturbances, or voltage spikes.

This AES chip employs duplicated and complemented data paths to counter side-channel and fault injection attacks [5, 24]. When an induced error is detected, the design spreads the error, protecting against differential fault analysis (DFA). The error can be detected by duplicating data paths. Unlike the Hamming distance method, the error detection mechanism is implemented in the reverse order error matrix (most significant bit swapped with least significant bit). Cross-changing wires between dual paths spread the error even further. Please refer to [7] for architecture details of the architecture.

Several analyses of the hardware AES module for various attack scenarios were performed. Correlation power analysis was carried out, no data dependency was discovered from each round of calculation, and no distinct peak can be seen. AES mode of operation can be identified using EM radiation measurement; however, no correlation between calculation and emitted EM waves can be established.

The error spreading mechanism is built into the design to combat the fault injection attack. A green laser source is used to introduce a fault into the circuit; however, because duplicated and complemented data paths were used, the targeted error spreads to a larger number of bits, and corrupted ciphertext cannot be used for differential fault analysis.

TR-AES (tamper-resistant AES) chips have demonstrated promising resistance to SCA and FA. However, the design resulted in a 67% overhead from the non-secured AES implementation in the same technology. Furthermore, it would be interesting to see the effect of various environmental conditions such as temperature, voltage, and current on the resiliency of the chip against SCA and FA.

19.8   Conclusions

To summarize, it is clear that there is still much work to be done in developing an IP resistant to side-channel and fault attacks. According to research, if an attacker can inject faults in later rounds, it is possible to break the encryption in under 2 min with 100% accuracy [31]. Furthermore, when attempting to protect the ciphers, it was discovered that there were numerous trade-offs between power and overhead. Area and power increased dramatically as more layers of protection were added. For example, adding DDR protection increases area by 36% and power by 50%, and adding high-speed detection causes more than 20% area overhead [7, 18]. When the fault error is dispersed across the entire design, there is another significant increase in the area overhead.

References

1. Bae K, Moon S, Choi D, Choi Y, Choi Ds, Ha J (2011) Differential fault analysis on AES by round reduction. In: 2011 6th international conference on computer sciences and convergence information technology (ICCIT), pp 607–612
2. Bai Y, Stern A, Park J, Tehranipoor M, Forte D (2021) RASCv2: Enabling remote access to side-channels for mission critical and IoT systems. ACM Trans Des Autom Electron Syst (TODAES) 27(6):1–25
3. Bai Y, Park J, Tehranipoor M, Forte D (2022) Real-time instruction-level verification of remote IoT/CPS devices via side channels. Discover Internet Things 2(1):1–19
4. Barenghi A, Breveglieri L, Koren I, Naccache D (2012) Fault injection attacks on cryptographic devices: Theory, practice, and countermeasures. Proc IEEE 100(11):3056–3076
5. Bhunia S, Tehranipoor M (2019) Side-channel attacks. Hardware security, pp 193–218
6. Courbon F, Fournier JJA, Loubet-Moundi P, Tria A (2015) Combining image processing and laser fault injections for characterizing a hardware AES. IEEE Trans Comput Aided Des Integr Circuits Syst 34(6):928–936. https://doi.org/10.1109/TCAD.2015.2391773
7. Doulcier-Verdier M, Dutertre JM, Fournier J, Rigaud JB, Robisson B, Tria A (2011) A side-channel and fault-attack resistant AES circuit working on duplicated complemented values. In: 2011 IEEE international solid-state circuits conference. IEEE, pp 274–276
8. Dusart P, Letourneux G, Vivolo O (2003) Differential fault analysis on AES. In: International conference on applied cryptography and network security. Springer, pp 293–306
9. Giraud C (2005) DFA on AES. In: Dobbertin H, Rijmen V, Sowa A (eds) Advanced encryption standard – AES. Springer, Berlin, Heidelberg, pp 27–41
10. Güneysu T, Moradi A (2011) Generic side-channel countermeasures for reconfigurable devices. In: International workshop on cryptographic hardware and embedded systems. Springer, pp 33–48
11. He J, Guo X, Tehranipoor M, Vassilev A, Jin Y (2021) EM side channels in hardware security: Attacks and defenses. IEEE Design & Test
12. He M, Park J, Nahiyan A, Vassilev A, Jin Y, Tehranipoor M (2019) RTL-PSC: Automated power side-channel leakage assessment at register-transfer level. In: 2019 IEEE 37th VLSI test symposium (VTS). IEEE, pp 1–6
13. Hutter M, Schmidt JM, Plos T (2009) Contact-based fault injections and power analysis on RFID tags. In: 2009 European conference on circuit theory and design. IEEE, pp 409–412
14. Kim CH, Quisquater JJ (2008) New differential fault analysis on AES key schedule: Two faults are enough. In: Grimaud G, Standaert FX (eds) Smart card research and advanced applications. Springer, Berlin, Heidelberg, pp 48–60
15. Lashermes R, Reymond G, Dutertre JM, Fournier J, Robisson B, Tria A (2012) A DFA on AES based on the entropy of error distributions. In: 2012 workshop on fault diagnosis and tolerance in cryptography, pp 34–43. https://doi.org/10.1109/FDTC.2012.18
16. Lee J, Tehranipoor M, Patel C, Plusquellic J (2007) Securing designs against scan-based side-channel attacks. IEEE Trans Depend Secure Comput 4(4):325–336
17. Lomné V, Roche T, Thillard A (2012) On the need of randomness in fault attack countermeasures-application to AES. In: 2012 workshop on fault diagnosis and tolerance in cryptography. IEEE, pp 85–94

18. Maistri P, Vanhauwaert P, Leveugle R (2007) A novel double-data-rate AES architecture resistant against fault injection. In: Workshop on fault diagnosis and tolerance in cryptography (FDTC 2007). IEEE, pp 54–61

19. Mirbaha AP, Dutertre JM, Tria A (2013) Differential analysis of round-reduced AES faulty ciphertexts. In: 2013 IEEE international symposium on defect and fault tolerance in VLSI and nanotechnology systems (DFTS), pp 204–211. https://doi.org/10.1109/DFT.2013.6653607

20. Murdock K, Oswald D, Garcia FD, Van Bulck J, Gruss D, Piessens F (2020) Plundervolt: software-based fault injection attacks against Intel SGX. In: 2020 IEEE symposium on security and privacy (SP). IEEE, pp 1466–1482

21. Park J, Moon S, Choi D, Kang Y, Ha J (2010) Fault attack for the iterative operation of AES S-box. In: 5th international conference on computer sciences and convergence information technology, pp 550–555. https://doi.org/10.1109/ICCIT.2010.5711116

22. Park J, Xu X, Jin Y, Forte D, Tehranipoor M (2018) Power-based side-channel instruction-level disassembler. In: 2018 55th ACM/ESDA/IEEE design automation conference (DAC). IEEE, pp 1–6

23. Park J, Nahiyan A, Vassilev A, Jin Y, Tehranipoor M, et al (2019) RTL-PSC: automated power side-channel leakage assessment at register-transfer level. Preprint. arXiv:190105909

24. Park J, Anandakumar NN, Saha D, Mehta D, Pundir N, Rahman F, Farahmandi F, Tehranipoor MM (2022) PQC-SEP: Power side-channel evaluation platform for post-quantum cryptography algorithms. IACR Cryptol ePrint Arch 2022:527

25. Pundir N, Li H, Lin L, Chang N, Farahmandi F, Tehranipoor M (2022) Security properties driven pre-silicon laser fault injection assessment. In: International symposium on hardware oriented security and trust (HOST)

26. Roscian C, Dutertre JM, Tria A (2013) Frontside laser fault injection on cryptosystems - application to the AES' last round -. In: 2013 IEEE international symposium on hardware-oriented security and trust (HOST), pp 119–124. https://doi.org/10.1109/HST.2013.6581576

27. Takahashi J, Fukunaga T, Yamakoshi K (2007) DFA mechanism on the AES key schedule. In: Workshop on fault diagnosis and tolerance in cryptography (FDTC 2007), pp 62–74. https://doi.org/10.1109/FDTC.2007.13

28. Tehranipoor M, Lee J (2012) Protecting IPs against scan-based side-channel attacks. In: Introduction to hardware security and trust. Springer, New York, NY, pp 411–427

29. Vashistha N, Rahman MT, Paradis OP, Asadizanjani N (2019) Is backside the new backdoor in modern SoCs? In: 2019 IEEE international test conference (ITC). IEEE, pp 1–10

30. Wang H, Li H, Rahman F, Tehranipoor MM, Farahmandi F (2021) SoFI: Security property-driven vulnerability assessments of ICs against fault-injection attacks. IEEE Trans Comput Aided Des Integr Circuits Syst

31. Zhang F, Guo S, Zhao X, Wang T, Yang J, Standaert FX, Gu D (2016) A framework for the analysis and evaluation of algebraic fault attacks on lightweight block ciphers. IEEE Trans Inf Foren Secur 11(5):1039–1054

Index

© The Author(s), under exclusive license to Springer Nature Switzerland AG 2023
M. Tehranipoor et al., *Hardware Security Primitives*,
https://doi.org/10.1007/978-3-031-19185-5

Printed in the United States
by Baker & Taylor Publisher Services